Applied Electro-Optics

Louis Desmarais

To join a Prentice Hall PTR Internet mailing list, point to:
http://www.prenhall.com/mail_lists/

ISBN 0-13-802711-0

90000

9 780138 027117

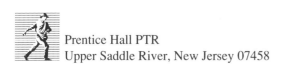

Prentice Hall PTR
Upper Saddle River, New Jersey 07458

Acquisitions Editor: Bernard M. Goodwin
Editorial Assistant: Diane Spina
Editorial/Production Supervision: James D. Gwyn
Cover Designer: John Christiana
Cover Design Director: Jerry Votta
Manufacturing Manager: Alexis R. Heydt
Marketing Manager: Betsy Carey
Compositor/Production Services: Pine Tree Composition, Inc.

Prentice Hall books are widely used by corporations and government agencies for training, marketing, and resale.
The publisher offers discounts on this book when ordered in bulk quantities. For more information contact: Corporate Sales
Department. Phone: 800–382–3419; FAX: 201–236–7141; E-mail: corpsales@prenhall.com; or write: Prentice Hall PTR,
Corp. Sales Dept., One Lake Street, Upper Saddle River, NJ 07458.

Printed in the United States of America
10 9 8 7 6 5 4 3 2 1

ISBN 0-13-802711-0

Prentice-Hall International (UK) Limited, *London*
Prentice-Hall of Australia Pty. Limited, *Sydney*
Prentice-Hall Canada Inc., *Toronto*
Prentice-Hall Hispanoamericana, S.A., *Mexico*
Prentice-Hall of India Private Limited, *New Delhi*
Prentice-Hall of Japan, Inc., *Tokyo*
Simon & Schuster Asia Pte. Ltd., *Singapore*
Editora Prentice-Hall do Brasil, Ltda., *Rio de Janeiro*

To

My wife, Roberta

and

In loving memory of my mother, Anne, and father, Ludger

Contents

Preface

The purpose of this book is to present the rapidly growing field of electro-optics in an applications-oriented manner. This presentation has been kept at an introductory level. To do this effectively, two major areas of scientific study must be considered. These areas are optics and electronics. The book deals with the fundamental principles in optics, semiconductor electronics, and electromagnetics. Optoelectronic devices such as LEDs, diode lasers, and photodiodes are studied in detail. The integration of these devices into useful electronic circuits is also covered in detail.

This book is intended as a text for people with diverse backgrounds. It should be useful to freshman and sophomore college students for a course in electro-optics and to practicing engineers, scientists, or managers who have little or no knowledge of electro-optics. Anyone working in the field of electronics may find this book very useful since most electronic devices now use optoelectronic components. Areas where electro-optical systems are used include the biomedical field, communications, remote sensing, imaging, test and measurement, and surveillance. It is assumed that the reader has a basic knowledge of electronics that includes operational amplifiers. A short tutorial on the basics of operational amplifiers is given in Appendix B for those who need help in this area. Part I of this book provides the necessary background in optics.

Numerous examples with full solutions are given. Many of these examples are taken from practical situations. In later chapters, practical circuit examples are given using manufacturers' data sheets for the optoelectronic components specified. In the last chapter, the electro-optical portion of the compact disc (CD) player is considered in detail. The presentation here relies upon many of the optical and electrical examples presented previously.

Even though electro-optical systems tend to be very complex, we can understand their operation by considering some basic things that they hold in common. For example, the vast majority of receivers used in electro-optical systems rely upon only a few circuit techniques to convert the input optical signal into a useable electrical signal. This electrical signal then undergoes signal conditioning with the help of conventional electronics. This book discusses, in great detail, the most common circuit techniques used to convert the optical signal into an electrical signal. In this way, the reader can use one of these techniques in a particular application. Unfortunately, it would be impossible to consider all of the electronic signal conditioning circuits. But, many common electronic amplifier techniques are discussed in detail.

Acknowledgments

It is a pleasure to acknowledge the many people who have contributed to the production of this book. During the initial writing stages, I received some very positive and helpful comments about the manuscript that encouraged me to go on to its completion. I wish to express my gratitude to Jeffrey L. Cooper and David E. Shipman, who provided these critiques. I would also like to thank Betty Lise Anderson (Ohio State University) for a very thorough technical review of the completed manuscript. Special thanks are extended to Russ Hall, senior editor, who assisted me greatly during all stages of writing this book. He provided many suggestions to enhance its content and organization. This book would not have been possible without the contributions from many companies specializing in optoelectronic components. These contributions, in the form of data sheets, tables, illustrations, and electrical circuit schematics, serve as basic source material. I am very grateful to the following companies, listed below, for providing this critical information.

Burr-Brown Corporation
Cal Sensors
Centronic Inc.
EG&G Judson
EG&G Vactec
Hewlett-Packard Company
Linear Technology Corporation

Melles Griot
Mitsubishi Electric Corporation
Motorola, Semiconductor Products Sector
Nippon Electric Glass Co., Ltd.
Siemens Components, Inc. Optoelectronic Division
Texas Instruments Incorporated

I would also like to thank my wife, Roberta, my daughter, Joanne, and my sons, Robert and John, for their patience and understanding during the writing process. We give thanks to God, who created the universe and the natural laws by which it endures. These laws provide the framework of physics and allow this book to be understood. The existence of these orderly laws gives evidence of divine design. For it is written: "Your faithfulness endures to all generations; You established the earth, and it abides. They continue this day according to Your ordinances, For all are Your servants." (Psalm 119:90, 91)

1

Introduction and Overview

By many accounts, the field of electro-optics is progressing at a faster pace than the field of computer electronics. For example, when we hear of breakthroughs in storage capacities and playback speeds, we know that optical media provide the best performance. With this in mind, anyone new to this field can easily become discouraged when trying to determine, for example, how the compact disc player stores and plays back this information. As with any other area of science, learning the basics will provide a good start to understanding a seemingly complex application. All optoelectronic devices such as diode lasers, LEDs, and photodiodes operate on the physical principle of the interaction of radiation and matter. This basic quantum process is responsible for the production of electromagnetic radiation.

So, why has the use of optoelectronic devices skyrocketed in recent years? One answer that quickly comes to mind is speed. As anyone who works with a personal computer knows, a machine built just a few years ago will have trouble running the latest software applications. The computer must have the ability to handle the increased information throughput that allows for the efficient transfer of electrical signals. Unfortunately, limits exist to how fast these signals can be transmitted in copper wire. As the frequency of the electronic signal increases, its ability to travel through a conductor decreases. This limitation is known as the skin effect. The consequence of the skin effect requires signals with oscillation frequencies in excess of about 100 Megahertz to be transmitted in coaxial cable. Electron flow through a conductor is not the only way to transmit information.

Optical signals, on the other hand, travel through transparent media that do not make good electrical conductors. These signals have no restriction such as the skin effect since photons propagate easily within these media. For example, near infrared optical energy oscillates at a frequency of about 300 Terahertz. This frequency is thousands of times greater than frequencies allowed to propagate electronically in a copper wire. To take advantage of this vast amount of bandwidth, the electronics must be carefully integrated to the optics. This is why we see optoelectronic components such as LEDs, diode lasers, and photodiodes used in electronic circuits in ever increasing numbers. Common everyday electronic devices depend upon these components for their operation. These devices include CD players, wireless remotes, range finders, and barcode scanners.

You will find this book to be different from most on this subject. This book uses a "back to basics" approach to introduce the reader to the science of electro-optics. One of the book's objectives is to provide the reader with information on how to integrate optoelectronic components into useful electrical circuits. This information should not be presented in a cookbook fashion. The variations in the types of optoelectronic components available now are so diverse that it is impossible to cover every application. Instead, if the reader understands the physical concepts behind the operation of these components, this knowledge can then be applied to a particular situation. Both optical and electrical design issues will be discussed. Many circuit applications that use optoelectronic components will be presented and explained in detail.

It's easy to become intimidated by data sheets, especially when they present data with polar plots using Greek symbols, and use terms such as D*. This book will guide you through the various parameters presented by most data sheets. Many of the optoelectronic components introduced in this book will use a manufacturer's data sheet. Detailed explanations for each parameter will be given, and in some cases, actual circuit examples will be presented using the optoelectronic component. This will help the reader to understand data sheets, and to use them as effective design tools.

Data books and application manuals usually present schematic diagrams for the optoelectronic components that they specify. Unfortunately, this provides the circuit designer with only part of the information required to make a functioning circuit. In most cases, component placement, printed circuit board layout, and EMI shielding comprise the completed working circuit. This book will present these often overlooked details of circuit design when using optoelectronic components.

Since electro-optics involves the convergence of optics and electronics, it can best be understood when using both of these disciplines. An electro-optic system can be reduced to the basic sections as shown in Figure 1.1. These sections will be introduced separately in Part II of the book. The transmitter section contains an optoelectronic emitter such as an LED or diode laser. Optical elements such as lenses, apertures, and filters may also be used in the transmitter section with the main purpose being to affect light output characteristics. The emitted optical radiation then propagates through a medium to reach the receiver section. At the receiver, the detector and its associated electronics convert the input optical signal into an electrical signal for further processing.

In the first part of this book, we present the required theoretical background in optical physics. To start with, we consider the macroscopic view of light to explain phenom-

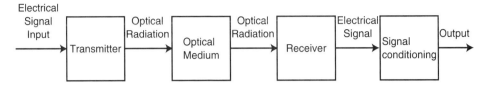

Figure 1.1   Block diagram showing the basic parts of a general electro-optical system.

ena such as reflection, refraction, diffraction, and interference using many examples. Maxwell's equations provide us with a description for the propagation of light as electromagnetic radiation. Understanding the basic properties of electromagnetic radiation will help the circuit designer to optimize, for example, an optical receiver. The more he knows about what he is trying to sense, the easier his job will be. Next, the theory behind the production of light is studied by considering the interaction of radiation and matter. We must shift gears here to consider the microscopic view of light. In this realm, the photon becomes the basic quantum unit involved in this energy exchange. Electrons play a key role in photon production. This provides a natural place to begin the second part of the book.

The second part of the text builds upon the theory and practical examples previously studied to show how to integrate optoelectronic components into useful circuits. The electrical schematic is just one of the many steps involved in this design process. Most importantly, we must consider both the optical and electrical design issues. Electrical design issues given in this book will be specific to optoelectronic devices. The basic circuit theory contained in most electrical design manuals concentrates on devices such as thermistors, thermocouples, and other resistance-varying components. Optoelectronic components, such as photodiodes, vary their current production with input optical energy. Also, a photodiode's internal capacitance varies with reverse-bias voltage. Thus, circuit techniques such as biasing will be different in both cases. Other important considerations include component selection and placement, RFI/EMI shielding, transient protection, environmental factors, lenses, and frequency response.

After we investigate the fundamental operation of optoelectronic components, the next step is to consider important areas such as modulation techniques, circuit board layout issues, and light detection. In the last chapter, we will combine what we have learned from all of these areas to discuss some common electro-optical system designs. Below, we give a brief discussion of the material covered in each chapter of this book.

PART I

Historical Development (Chapter 2)

This chapter provides a brief history pertaining to the development of modern electro-optics. The classical view according to Isaac Newton will be discussed. This classical view required modification when Max Planck made some observations that could not be explained by classical physics. This was the first step necessary for the acceptance of modern quantum theory.

Light and the Electromagnetic Spectrum (Chapter 3)

The nature of light will be introduced here. Maxwell's equations will be used to explain light's wave-like nature as it propagates through space. The electromagnetic spectrum will also be discussed. In this spectrum of electromagnetic energy, radio waves have the longest wavelengths. Cosmic rays possess the shortest wavelengths.

Reflection and Refraction (Chapter 4)

The physical laws governing reflection and refraction will be discussed using simple geometrical diagrams with plane waves. As we know, mirrors and lenses display these important optical phenomena. We will show many examples using Snell's law. After this discussion, fiber optic waveguide will be studied. The physical principles involved with lightwave transmission will be introduced.

Interference (Chapter 5)

Interference phenomena display light's wave-like nature. A practical device, known as the Fabry-Perot interferometer, will be detailed. Since diode lasers also use this basic structure in their operation, many examples will be considered describing how this interferometer works. Other applications considered in this chapter include thin films, interference filters, and non-reflecting glass. These optical components can be found in many electro-optical systems.

Diffraction (Chapter 6)

This chapter continues our study of wave optics. Since light is a wave phenomenon, it will bend around an obstacle such as the edge of a slit. A similar effect can be seen with water waves as they bend around an obstacle in their path. Diffraction effects must be controlled when designing optical systems. These effects can be both beneficial and counterproductive. We will see how diffraction limits the performance of any optical system. Many important applications of diffraction will be discussed in detail. Diffraction gratings take advantage of light's wave-like nature to separate polychromatic light into its spectral components. A CD player uses a diffraction grating to help keep the read head on a particular track without skipping. In Chapter 14, we will consider the application of the diffraction grating in the CD tracking control system. In Chapter 9, a diffraction grating will be used in the construction of a low cost spectroscope.

Polarization of Light (Chapter 7)

According to Maxwell's equations, light is composed of electric and magnetic fields arranged orthogonally to each other. Oscillations occur perpendicular to the direction of propagation, thus making light a transverse wave phenomenon. Optical components such as polarizers and quarter-wave plates prove light's transverse wave nature. Laser light can be modulated in a communications system by controlling its polarization state. In the operation of an electro-optic modulator, the polarization of laser light is controlled after it leaves the laser. We will use this modulation example to serve as an introduction to the modulation process in diode lasers.

Light and Thermal Radiation (Chapter 8)

We find, after studying the last few chapters, that the wave theory of light is not sufficient to describe all observations. It also has a particle-like nature. All objects at a temperature above absolute zero emit radiation. When matter and energy are in equilibrium, the spectral distribution of this emitted radiation follows a blackbody curve. The prominent feature of this curve is a peak intensity at a specific wavelength determined by the temperature of the object. This blackbody curve, first discovered by Max Planck, contributed greatly to our understanding of electromagnetic radiation. The shape of this curve can only be explained when considering the particle nature of light. Light is composed of discrete energy packets known as photons.

Quanta and Optical Spectra (Chapter 9)

After introducing light's particle-like nature in Chapter 8, the time has come to investigate the energy exchange process. The atomic structure, as presented by Neils Bohr, will be used for this purpose. Energy can be absorbed or emitted by an atom in discrete amounts only. These discrete energy packets, known as photons, are fundamental to our study of electro-optics. The construction of a simple spectroscope using a diffraction grating will be presented for the purpose of viewing these photons, collectively. When optical energy is emitted from a source, the discrete energy will show up as emission lines in the spectroscope. We use the spectroscope to determine the wavelength, and thus the energy, of these photons.

PART II

Semiconductor Light Sources (Chapter 10)

In this chapter, we will begin our study of the basic electro-optic system shown in Figure 1.1. The transmitter contains a light source that may produce photons by the interaction of radiation and matter. There are two processes to be considered here: spontaneous and stimulated emission. Light sources convert electrical current (electrons) into photons. Light emitting diodes (LEDs) and diode lasers use semiconductor materials in this photon production process. We use the particle nature of light in describing this process. The operation of the LED can be characterized by the process of spontaneous emission. The diode laser's unique geometry allows lasing to occur by the process of stimulated emission, another way that radiation interacts with matter. We will use the concepts developed in the previous chapters to more fully understand the process involved with the amplification of light. This process can be more fully understood when we consider the wave and particle natures of light. The particle nature helps to explain the process of stimulated emission within the laser structure, while the wave nature helps to explain how a tuned cavity (Fabry-Perot interferometer) is used to amplify light. These semiconductor light sources have the ability to be switched on and off very quickly. This property allows them to modulate the light they emit. The optical and electrical properties of LEDs and diode lasers will be considered here.

Optical Transmitters (Chapter 11)

An optical transmitter uses a semiconductor light source with electronic circuitry to produce a desired optical output. The electronic circuit controls the electron-to-photon conversion process. The subsequent light output must usually be modulated in some fashion for effective information transfer. We will investigate ways to modulate these light sources using electronic components. The electro-optic modulator introduced in Chapter 7 will serve as a guide to understanding modulation using electronic circuits. Many circuit examples will be discussed in detail, including diode laser modulation. These optoelectronic devices require relatively large amounts of current to produce the light levels required for reliable operation. Switching large amounts of current at fast rates produces emission known as electromagnetic interference or EMI. When using these devices, careful attention must be paid to circuit location, grounding, and shielding. Established practices on how to minimize this unwanted noise will be discussed.

Photodetectors (Chapter 12)

The block diagram of the receiver in Figure 1.1 contains a photodetector that converts incident photons into electrons. The process involved here is very much the reverse of the spontaneous emission process discussed in Chapter 10. Optoelectronic components such as photodiodes, phototransistors, optocouplers, and photoconductors will be considered. The basic structure and operation of the photodiode will be studied by considering parameters such as quantum efficiency, spectral response, dark currents, capacitance, noise sources, and response time. The circuit designer must know the various trade-offs available when integrating these devices into useful circuits. For example, to achieve the fastest response time, the photodiode must be reverse-biased by a specified voltage. This practice produces a dark current within the device that shows up as excess noise, reducing the signal-to-noise ratio. Thus, a trade-off between sensitivity and speed of response exists in this situation. Many common trade-off situations will be studied so that they can be applied to the circuit design examples in the next chapter.

Optical Receivers (Chapter 13)

Showing the reader how to integrate a photodetector into a useful receiver circuit is one of the main objectives of this chapter. The electronic circuitry in the receiver provides the current-to-voltage conversion required before signal processing begins. This signal voltage must then be amplified to a useful level. We will start with the basic photodiode circuit discussed in Chapter 12 by looking at its voltage-current transfer curve. This transfer curve tells us how the optical input affects the electrical output. Simple operational amplifier and hybrid circuits will be presented in a way that allows the reader to adapt to his own application. Other topics covered include optimization techniques, bandwidth requirements, circuit board layout precautions, and EMI protection against internal and external noise. Many typical optical considerations are also discussed.

Electro-Optic Systems (Chapter 14)

This chapter considers the completed system as outlined by Figure 1.1. Four interesting examples are presented in this chapter. The operation of the CD player

involves many of the electrical and optical concepts discussed in this book. The optical path in a typical CD player uses photodiodes for many important functions.
We will see that the diode laser makes an excellent optical source for this application. For the next example, we will consider the design of an optical data link between, for example, a personal computer and a printer. A recent development in infrared data transmission has resulted in a standard protocol known as Infrared
Digital Association (IrDA). A typical IrDA transceiver design will be considered by
studying both the electrical and optical design issues. These design issues are common to most electro-optical systems. Chapter 11 provides the background on the
IrDA signaling format. The other systems presented in this chapter include an integrated light sensor and a fiber optic communications system.

PART 1

2

Historical Development

The nature of light has interested man since the earliest times. It was thought since ancient times that light was composed of a stream of particles or corpuscles. This was known as the Corpuscular Theory because it helped to explain the rectilinear propagation of light and why shadows were produced from opaque objects. But this theory could not explain correctly the phenomena of refraction, interference, and diffraction that were noted in the seventeenth century. In 1678, a Dutch physicist, Christiaan Huygens, put forth a theory of light that offered an explanation.

He proposed that light was composed of waves and wavelets. Huygens' theory assumed that light started out as a wavefront in the form of a plane wave in free space. As this wavefront propagated, it produced secondary spherical wavelets along a newly positioned wavefront. With geometrical constructions, he was able to explain the phenomena of reflection and refraction. In the early eighteenth century, Augustin Jean Fresnel used Huygens' principle to explain the phenomena of diffraction.

Isaac Newton, famous for his work on the law of gravitation and mechanics, also did studies on the properties of light. His papers on optics were communicated to the Royal Society between 1672 and 1676. A collection of his papers can be found in his book *Optiks* that was published in 1704. He subscribed to the Corpuscular Theory and rejected Huygens' theory. From his observations, he concluded that Huygens' theory was not compatible with the rectilinear propagation of light. As we know from historic accounts, Newton was appointed Lucasian Professor of Mathematics at Cambridge University. This is the same position that Stephen Hawking, known for his theoretical work on black holes, was appointed to in 1979. Newton's authority was very great at that time due to his many scientific works. In fact, it was so great that for a century, little attention was paid to Huygens' theory.

Newton's work with light initiated the study of spectroscopy. He passed a sliver of sunlight through a glass prism to produce a band of colors that he named the spectrum. He wrote in his book while doing this experiment, "This image or spectrum PT was coloured, being red at its least refracted end T, and violet at its most refracted end P, and yellow, green, and blue in the intermediate spaces." This account was the first to show visible light being composed of many colors or wavelengths. By the late eighteenth century, the astronomer William Herschel found that there was heat beyond the far red end of this spectrum. At about the same period in time, J.W. Ritter detected invisible ultraviolet radiation beyond the violet end. Thus, the spectrum contained energy on both sides of the visible region. The contemporary view of light had to be changed to include these invisible regions of the spectrum.

It turns out that both Newton and Huygens were partly right. Light behaves as a wave and a particle. But for this very unusual combination, no analogy exists in our human experience to help us understand this situation. The wave-particle duality of light is one of the major themes that will be used in this book. It will be discussed in detail in later chapters.

The wave theory of light was established experimentally by Thomas Young in 1802. He demonstrated the interference effects of two light waves. From the interference patterns, he was able to determine the wavelength of light making up the two waves. The only way to explain how this pattern could be produced was by using the concept of wave propagation.

By the mid-1800s, a Scottish physicist, James Clerk Maxwell, put it all together. He concluded that light, composed of electromagnetic radiation, forms a wave as it propagates through space. Put more precisely, this electromagnetic wave has two fields, electric and magnetic, which move together through space at the speed of light. Visible light makes up only a very small portion of the range of radiation called the "electromagnetic" spectrum. The electromagnetic spectrum extends from radio waves to cosmic rays. Maxwell developed four equations that are still used today to describe macroscopically how light propagates through certain solids and free space.

Alexander Graham Bell, noted for the invention of the telephone, made a device that used the electromagnetic properties of light to transmit voice through free space. In 1880, he demonstrated his photophone. The photophone transmitted voices by using a thin, flexible mirror to reflect sunlight to a remote receiver. The flexible mirror modulated the light beam when sound waves disturbed it. This modulated wave was then picked up at a distance by a photocell that converted the light into an electrical current. The photophone did not become a commercial product like the telephone because it was subject to loss of communications whenever it was cloudy or objects interfered with its path.

Up until this point in history, all phenomena involving light could be explained using classical physics. But this changed in the late 1800s when Max Planck found that this classical view could not fully explain some observed phenomena. Classical physics closely approximates reality but is not sufficient to explain how, for example, stimulated emission occurs in a laser. To understand the operation of the laser, we must consider the microscopic origin of light. This involves the realm of atomic physics or quantum theory.

The observation that completely puzzled Planck involved measuring radiation sources such as the sun. He measured the intensity of the radiation as its frequency increased. According to the classical view, when measuring the intensity of light that has many components or frequencies, the intensity should increase with frequency. He did not find this to occur, and tried in vain to find an explanation that did not contradict the classical view.

Finally, after studying this problem more closely, he came up with a mathematical expression that explained his observations. To explain how this could happen, Planck proposed that the energy in the radiation becomes quantized. That is, the energy is not distributed evenly but comes in packets called quanta. This contradicts Newtonian physics that says that energy must be continuously variable. The name given to the quantum of electromagnetic radiation is the photon. This photon is not the tiny little corpuscle that Newton imagined but rather a pure chunk of electromagnetic energy traveling at the speed of light.

This quantum theory took about 20 years of hard work by many scientists before it was widely accepted. It describes the interaction of radiation and matter. To understand this more fully, reactions at the atomic level had to be investigated. In 1905, Albert Einstein demonstrated that shining light of sufficient energy on a metal surface causes the ejection of electrons. This photoelectric effect reinforced the quantum theory even further. Einstein found that as the intensity of the light was increased, the amount of photons emitted increased but the energy level of each photon remained the same. His work on the photoelectric effect won him the Nobel prize for physics in 1921.

The quantum theory gave scientists a new way of thinking about matter and energy. It was already known that matter consists of atoms and molecules. The model of the atom was that of a heavy positively charged nucleus surrounded by electrons. One problem with this model was that it could not explain why radiating electrons do not eventually fall into the nucleus when they lose energy.

A Danish physicist, Neils Bohr, wanted to find the answer to this puzzle, and thus studied the atomic structure of hydrogen, the simplest element. He was convinced from the results of his studies that the origin of the photon or energy packet must be found in the properties of the electron system. Bohr figured out that these electrons don't just orbit the nucleus in a random fashion. They can move around the nucleus in only a finite number of orbits or energy levels. These discrete or acceptable levels that the electron can occupy are stable. The electron does not lose energy gradually when it occupies a stable orbit. It can change orbits or levels only by emitting or absorbing energy in the form of a photon. This atomic model, as proposed by Bohr, contradicted classical physics. According to classical physics, when a charged particle moves through an electrostatic field, energy should be radiated in the process. This radiated energy should eventually result in the electron spiraling into the nucleus.

The Bohr theory of the atom helped to explain why excited hydrogen gas displays well-defined emission lines when studied with a spectroscope. The spectroscope uses a prism to disperse or spread out visible light into the same spectrum that Newton observed.

The first person to detail these emission lines was Balmer in 1885. An emission line is a slit of bright light at a particular location in the spectrum corresponding to the energy of the emitted photon as an electron changes to a lower energy level. This process of the emission of photons is known as spontaneous emission.

Albert Einstein also did some theoretical studies on stimulated emission. He proposed that there are three possible transitions that can occur in the electron system. The first two possibilities involve the emission or absorption of a photon as the electron goes to a lower or higher orbit. Another interesting possible transition can occur if a photon of proper energy strikes an excited atom that has an electron in a higher orbit. In this case, a photon with the same frequency and phase as the original photon will be emitted. The result will be two identical photons. He called this process stimulated emission. Stimulated emission was verified experimentally in 1924. Scientific research on this phenomenon eventually led to the invention of the laser.

One way to demonstrate particle nature is to strike an object with another particle. A particle will obey the laws of conservation of momentum. Arthur Compton did this experiment in 1923 by bouncing high energy quanta off electrons. His results agreed with his predictions by using conservation of momentum. He concluded that light does have a momentum and can thus be treated as a particle. It seemed back then that the theory of light had gone back to the Newtonian view of being composed of particles. But this one experiment did not disprove the wave theory of light. Instead, scientists were forced to think of light as having the properties of both a wave and a particle. This was confusing at first, but it represented hard evidence that could not be ignored. The invention of the laser could not have occurred without this understanding.

The natural tendency of matter, to be in thermodynamic equilibrium, occurs because the atoms and molecules try to be in the lowest possible energy states. This natural energy distribution means that stimulated emission is very rare and needs man's help to occur in appreciable quantities. It wasn't until the 1950s that someone made a device that demonstrated stimulated emission. Charles H. Townes figured out what conditions would be needed to produce stimulated emission of microwaves. He did this by exciting gaseous ammonia molecules with just the right amount of energy. This device was called the maser, short for microwave amplification by stimulated emission of radiation. Emission occurred in the microwave portion of the spectrum. These waves are very short electromagnetic waves, the same kind used in a microwave oven. Yet the waves are still not short enough to occupy the visible portion of the spectrum. With the invention of the maser, the stage was set for the construction of the optical maser or laser.

The first successful laser was constructed in 1960 by Theodore H. Maiman at Hughes Research Laboratory in Malibu, California. Instead of using a gas medium to produce the stimulated emission, he used a solid ruby rod. Most physicists at the time believed that the first lasers would use a gas rather than a solid medium to produce stimulated emission. This was due to the large amount of research work being carried out at the time with masers. Maiman's laser used a helical flashlamp to excite the atoms in the rod-shaped synthetic ruby crystal. One end of the ruby rod was fully silvered and the other end was partially silvered to allow for light to exit the rod. Upon reflecting many times within the rod, the stimulated emission from the medium was amplified. The resultant beam lasted only 300 millionths of a second, but produced 10,000 watts of optical power in the far end of the visible spectrum. A few months later, a helium-neon gas laser was developed at Bell Telephone Laboratories.

The first semiconductor lasers were fabricated in 1962. These semiconductor lasers are made from the same materials as light emitting diodes (LEDs). The early semiconductor lasers had to be cooled to cryogenic temperatures to operate correctly due to the high operating currents required for lasing to occur. By 1970, a much more efficient method of constructing these lasers was developed. This double-heterojunction structure allowed lasers to be operated much more efficiently. Eventually, these lasers operated at room temperatures due to smaller current consumption. Today, you will find a diode laser in every CD and DVD (digital versatile disk) player.

Another important development in the field of electro-optics was optical fiber. The first fiberoptic strands were developed in the 1950s. A year later in London, research was performed to determine the practicality of transmitting images through these fiber strands. These strands were fashioned into bundles of flexible cable and found an immediate medical application for viewing into the human body. At the time when the first lasers were developed, the best optical fibers had an attenuation of about 1 dB/meter. These fibers were adequate for light transmission through a few meters but could not be used for long haul communications due to material losses and scattering within the fiber.

The next improvement to optical fiber was glass-coated glass fibers. These fibers had a core around which a cladding layer was wrapped to help prevent light loss. Fibers made in this way took advantage of total internal reflection to keep optical radiation within the fiber's core. This improvement also helped make possible the development of single mode fibers that greatly increased the amount of information transfer.

The attenuation problem caused by material losses and scattering of light within the fiber still remained a problem. If optical fiber was to be used for long haul communications, this attenuation problem had to be solved. In the 1960's, Charles K. Kao did some work to reduce these losses by using glass with low amounts of impurities. The result of this study showed that glass fibers could be used in long haul communications if these impurities were removed. The first low loss glass fibers became available in 1970. By the mid 1970s, multimode and single mode low loss fibers were used commercially for the purposes of communications.

In the 1970s, semiconductor GaAlAs diode lasers had also become commercially available. These devices were developed specifically for the purpose of lightwave communications because of their ability to be modulated at high data rates. These high rates were required for optical fiber to be competitive with copper wire as a communications medium. The small size and emission wavelength at 0.8 to 0.9 microns made these diode lasers a good match for coupling light into glass fiber. Optical energy losses within these glass fibers had been improved to a few dB/kilometer by the early 1970s. Later in the decade, emitters and receivers were developed that worked efficiently in the 1.1 and 1.4 micron region (near infrared) of the spectrum. This development was important because light displays its lowest attenuation characteristic in optical glass at these wavelengths. Attenuation within the fiber as low as 1 dB/kilometer was achieved using this combination of devices.

SUMMARY

We learned from this chapter that the first successful laser was constructed in 1960. Then about two years later, the first semiconductor lasers were constructed. In fact, when these devices were first developed, they were regarded as an invention looking for an application. But today, we see the diode laser as a very useful optoelectronic component. It is used in every CD and DVD player as a very reliable component. Bar code scanners and laser printers also use diode lasers. Other optoelectronic devices such as LEDs and photodiodes are also used in many different types of electro-optical systems.

Before we learn about these optoelectronic components and how they are used, we must first become familiar with the type of energy exchanged by these devices. In the next chapter, we will begin this process by studying the nature of light. As we progress through this book, we will see why light is so useful and how to control it.

3

Light and the Electromagnetic Spectrum

3.1 THE NATURE OF LIGHT

As we all know, our most important source of visible light is the sun. Those of us who have stayed out in the sunlight on a bright summer's day a little bit too long also know that solar radiation contains damaging ultraviolet radiation. This energy somehow travels across the vacuum of space all the way to earth. But how does it get here? For a long time, many scientists thought that this energy was transmitted through a medium called the ether in a similar way that sound energy transmits through air. It was also proposed that this ether accounted for the finite velocity of light. This idea of an ether was needed to help explain how light energy could travel through the vacuum of space.

In the 1880s, two physicists, A. A. Michelson and E. W. Morley, performed the now-famous experiment bearing their names to try to detect this ether. They used the earth itself as the frame of reference to detect its movement through the ether. On the earth, light was reflected back and forth between two mirrors for a total round trip of 22 meters. This design was first done parallel and then perpendicular to the direction of the earth's motion around the sun. To accomplish this, they performed these measurements at different times of the year, when the earth was moving in different directions with respect to the fixed stars. If this ether existed, there should be a relative difference in the speed of light for the two perpendicular directions at different times of the year.

After completing the experiments to detect the ether, Michelson came to the conclusion that it did not exist. No effect of the motion of the earth could be detected, thus the theory of an ether as a medium for the propagation of light had to be abandoned. This result troubled the scientists at the time because it seemed logical to them that some kind of physical medium was required to allow light to travel in the vacuum of space.

If the energy cannot be found in the proposed ether, then where is this energy stored during the transmission of light through a vacuum? In 1864, James Clerk Maxwell made the discovery that light energy is stored in the wave itself in a similar fashion that energy is stored in a water wave. He found that light energy has two components, an electric field and a magnetic field, that propagate together through space. This electromagnetic wave stores its energy in the electric and magnetic fields as described by Maxwell's equations. Light energy does not need a physical medium such as the ether to propagate through space. Today we know that visible light is only one form of this electromagnetic energy.

Maxwell wanted to write down what was known about electricity and magnetism at that time. This prompted him to combine these concepts into four equations known as Maxwell's equations. These equations describe the behavior of electric and magnetic fields in space and time. The first equation was obtained from Faraday's law of induction. It was Michael Faraday who discovered that a changing magnetic field could produce an electric current. Faraday's experiment consisted of two separate wire windings around a piece of iron. When he placed battery contacts across one of the coils, he noticed a momentary current surge in the other coil. What Faraday witnessed was a changing magnetic field in one coil producing a current in the other coil of wire. Maxwell summarized this result in his first equation:

$$\nabla \times E = -\mu_0 \, \partial H / \partial t$$

In this equation, E is the electric field strength, and H is the magnetic field strength which changes with respect to time. The mathematical operations described in the above equation are the curl of the E field, and the partial derivative of the H field with respect to time. The curl operation can be thought of as a circulation per area. When no rotation is present, no curl exists. In the language of physics, as the angular velocity increases, so does the value of the curl. Since there are three physical dimensions in free space, the curl of E must be taken for all three. A derivative can be thought of as the rate of change of a function with respect to its variable. An example of a derivative is the time rate of change of speed, a description for acceleration. When we specify a partial derivative, as in the above equation, the derivative of the function can have several variables with respect to just one of them. The other variables are treated as constants. In the case of Maxwell's first equation, the partial derivative is taken with respect to time, t. The quantity μ_0 in the above equation is the vacuum permeability. Putting these all together, the above equation basically says that a time varying magnetic field will produce an electric field, E. That's exactly what Faraday witnessed.

The second equation was obtained from Ampere's law. It was well known at the time that when a current flow occurred in a wire, magnetic effects were produced near it. This was demonstrated by placing a compass near a wire having a current flow. In this case, it was noticed that one point of the compass was attracted to the wire while the other compass point was repelled. Maxwell summarized this result by using the concept of fields:

$$\nabla \times H = \varepsilon_0\, \partial E/\partial t$$

In this equation, H is the magnetic field strength, E is the electric field strength. The mathematical operations are the same as described in Maxwell's first equation. You will notice that in this second equation, the curl operation is performed on H, and the partial derivative of E with respect to time is called for. The quantity ε_0 is the vacuum permittivity. This equation basically says that a time varying electric field will produce a magnetic field, H.

The third and fourth equations were obtained by using Gauss' laws for electricity and magnetism:

$$\nabla \bullet D = \rho_v$$

$$\nabla \bullet B = 0$$

The third equation shows how the electric flux density is related to the electric charge density, ρ_v. This flux density will vary with distance from the electric charge or by the number of electrons in a given space.

The fourth equation basically says that there is no such thing as a magnetic charge in nature. If you cut a magnet in half and in half again several times, you will get a smaller and smaller magnet with north and south poles. Magnetic flux lines always form closed loops. Unlike electric field lines, they do not diverge from a point.

Maxwell's first and second equations show a symmetry in nature between the electric and magnetic fields. A changing magnetic field will produce a changing electric field and visa versa. These fields are not independent of each other in the dynamic case. This fact impressed Maxwell, who believed that nature was truly beautiful and elegant. Thus, an important thing to realize from the first two equations is the symmetry involved with the production of the E and H fields of the electromagnetic wave. When all of the mathematical operations are performed as specified in the equations, time varying E and H fields can be described as illustrated in Figure 3.1.

These equations can be used to describe a transverse wave of energy with electric and magnetic field components that are orthogonal to each other as the wave propagates through space at the speed of light. Figure 3.1 gives a description of this wave. A changing electric field, E, traveling in the z direction induces a changing magnetic field, H. Both E and H are vector quantities describing the electric and magnetic fields. The length of each vector gives the magnitude of the field at each moment during its propagation in the z direction. The vector k is the wave vector of magnitude $k = 2\pi/\lambda$, where λ is the wavelength of the electromagnetic energy. This wave vector also defines the direction of propagation of the wave. Thus, all three vectors form a mutually orthogonal triad as shown in the figure. When Maxwell calculated the velocity of various electromagnetic waves, he came up with the same number each time. This velocity turned out to be that of light. Thus, he also considered visible light to be an electromagnetic wave.

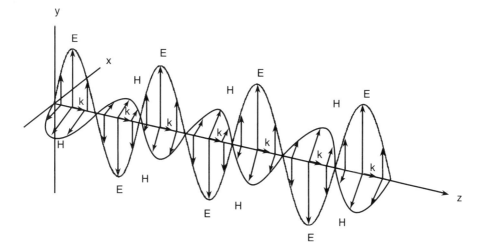

Figure 3.1   The electric and magnetic field distributions for an electromagnetic wave.

Maxwell's equations also describe the unification of electricity and magnetism into one fundamental force called the electromagnetic force. This force becomes important when considering the production of photons. In this process, outer electrons of an atom play a very important role. The electromagnetic force is one of the four fundamental forces in nature. The other three fundamental forces are gravity, the weak force, and the nuclear force. We are all familiar with gravity, the force of attraction that exists between two or more masses. The weak force occurs on the scale of subatomic distances. This force plays an important role in the processes that make the sun shine. The nuclear force holds the positive protons together in the nucleus of an atom. In this book, we will be primarily interested in the electromagnetic force. We will next consider some of the basic properties of wave motion that can be used to describe an electromagnetic wave.

3.2   WAVE MOTION

For simplicity, let's consider one of the waves describing either the electric or magnetic field displayed in Figure 3.1. This wave can be considered as a component of the more complex electromagnetic wave. Figure 3.2a shows this wave separately using x and z coordinates. We will now use this to illustrate some of the mathematical parameters of a wave. The x and z displacement of this wave can be described by the trigonometric sine function. A general equation describing this wave can be written as shown below:

$$x(z) = A \sin kz$$

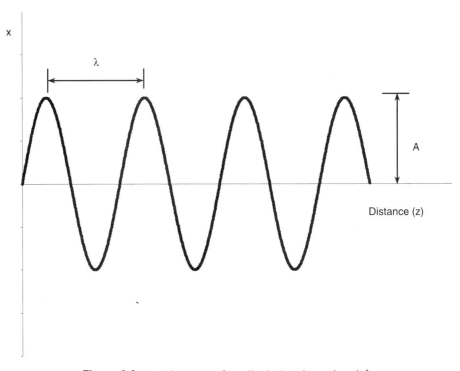

Figure 3.2a   A sine wave of amplitude A and wavelength λ.

In this equation, the amplitude or height, A, defines the maximum vertical or x displacement. The wavelength, λ, is determined by the distance between adjacent crests of the wave.

Now we come to the part of the equation that may be particularly confusing, the quantity kz. Instead of using degree measure, we will use radian measure to define the argument of a trigonometric function. When considering degree measure, the range of 0° to 360° completes one cycle or complete circle. In the case of radian measure, the range of 0 to 2π radians completes the same cycle. As you may remember, a radian is a unit of angular measure equivalent to the angle subtended at the center of an arc equal in length to the radius. In degrees, this angle equals about 57.3°. The complete circle or 360° equates to 2π radians. Thus, radian measure is a ratio of two lengths. For the sine wave in Figure 3.2a, we know that a maximum occurs at 90°. This corresponds to a radian measure of π/2 radians. The first complete cycle to the right makes 2π radians or 360°. At this point on the curve, kz = 2π. This brings us to our next parameter to consider, k. The parameter k or angular wave number defines how fast the wave oscillates with respect to z. Thus k has the units of radians per meter in the MKS system. As the wave number increases, the wavelength decreases. The simple equation below describes this mathematical relationship:

$$k = 2\pi/\lambda$$

In our study of electromagnetic waves, we will also be interested in the time variation of the wave. Going back to our sine function involving displacement, we can easily change this expression to yield the time variation of the wave with amplitude A. The amplitude A now varies with time according to the following expression:

$$A(t) = A \sin \omega t$$

Since the function varies with respect to time, the argument of the sine function must also be changed accordingly. We now use the term ωt in place of kz for the argument. The constant ω, angular frequency, gives an indication of how fast the wave amplitude oscillates as t (time) increases. Since ωt is measured in radians, ω has the units of radians per second. One complete oscillation or period, T, equates to $2\pi/\omega$:

$$T = 2\pi/\omega$$

In this expression, T is the period of oscillation of the wave or the time it takes to complete one cycle. The inverse relationship of the period is known as the frequency of oscillation, ν, which can be expressed in cycles per second or Hertz:

$$\nu = 1/T$$

Figure 3.2b shows the relationship between period, T, and frequency, ν. You can see that as the period decreases in time, the frequency will increase.

Combining the last two equations we get a useful expression for ω:

$$\omega = 2\pi\nu$$

Thus far, we have only considered a stationary wave. To consider the wave's movement, we must also take into account its velocity. Figure 3.3 shows the same sine wave as in Figure 3.2(a). For this wave (lighter shaded line), we assume that the displacement x is zero at z = 0 and t = 0. This wave now travels to the right with a constant velocity known as the phase velocity of the wave. The phase velocity can also be expressed as shown below:

$$v = \omega/k$$

The distance this wave travels in time, t, can be calculated by multiplying the phase velocity by the time as shown in the figure. When the wave moves to the right in Figure 3.3, we must use z − vt for this distance. If we substitute this into our previous expression for displacement of the sine wave, we get a more useful expression. The steps are detailed below:

$$x(z) = A \sin kz$$

Substituting (z − vt) for z,

$$x(z) = A \sin k(z - vt) = A \sin (kz - kvt)$$

Since v = ω/k,

$$x(z,t) = A \sin (kz - \omega t)$$

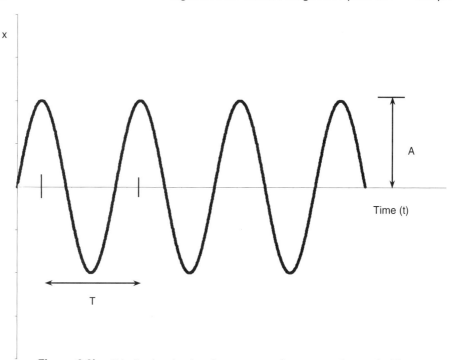

x

Time (t)

A

T

Figure 3.2b   T is the time it takes the wave to make one complete cycle. The frequency of oscillation is $1/T = \nu$.

This expression describes a wave traveling to the right as described in Figure 3.3. The more general form of this equation involves a phase constant φ that gives the amount of phase shift in the z direction (if any):

$$x(z,t) = A \sin (kz - \omega t - \varphi)$$

We will be concerned with this phase constant starting with Chapter 5. For simplicity, we will consider this constant to be zero in Example 3.1 below:

Example 3.1

A traveling wave similar to that described in Figure 3.3 has an amplitude A of .05 meter, angular wave number k of 5.7 rad/m and angular frequency ω of 1.8 rad/sec.

(a) Write down the mathematical expression for this wave:

$$x(z,t) = .05 \sin (5.7z - 1.8t)$$

(b) Find the wavelength and period of this wave:
To find the wavelength, we use the fact that k = 5.7 rad/m, thus:

$$\lambda = 2\pi/k = (2)(3.14)/5.7 \text{ m} = 1.10 \text{ meter}$$

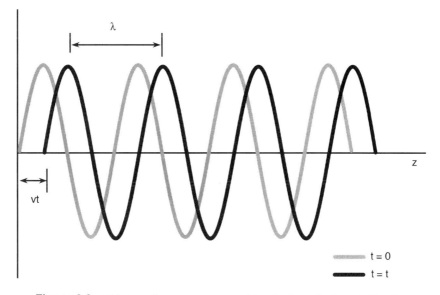

Figure 3.3 This traveling wave moves with velocity v for time t. The distance traveled is vt.

To find the period, we use the fact that $\omega = 1.8$ rad/sec, thus:

$$T = 2\pi/\omega = (2)(3.14)/1.8 \text{ sec} = 3.49 \text{ seconds}$$

(c) Find the frequency of the wave:

$$\nu = 1/T = 1/3.49 = .287 \text{ Hertz}$$

(d) What is the speed of the wave?

$$v \text{ (velocity)} = \omega/k = 1.8/5.7 = .32 \text{ meter/second}$$

(d) What is the x displacement at $z = 5$ meters and at $t = 15$ seconds?

To find this result, we must go back to the equation and enter the parameters specified above:

$$x = (.05) \sin [(5.7)(5) - (1.8)(15)] = .05 \text{ meter}$$

Now we can use what we have learned about wave motion to develop an expression for the E field in Figure 3.1. The result then becomes:

$$E(z,t) = E_o \sin (kz - \omega t)$$

The electric and magnetic field vectors are always perpendicular to the direction in which the wave travels and to each other. The E field vector lies in the yz plane and the magnetic field vector lies in the xz plane. The value of E depends upon z and t. As you can see from Figure 3.1, the E and H fields are in phase with each other at any particular point through which the wave moves. They reach maximum and minimum values at the same instant in time. The spatial pattern that these waves make when they travel through space can be described by Maxwell's equations.

A very important result obtained from Maxwell's equations is the speed of light itself. By using these equations, the following relationship can be worked out:

$$c = (\mu_0 \varepsilon_0)^{-1/2}$$

In this equation, μ_0 and ε_0 are the same constants used in Maxwell's first two equations. Notice that the above relationship does not depend upon the wavelength of the electromagnetic wave but only on the two constants. This shows that the speed of light, c, relates to the electric and magnetic field components of the electromagnetic wave. These quantities, μ_0 and ε_0, did not exist in Maxwell's day but he did however see a relationship between the speed of light and these two fields.

Up to now in our discussion, we have described light as a wave phenomenon. We used Maxwell's equations to help describe the propagation of light. This description treats light as a wave phenomenon. But this description turns out to be incomplete. By the beginning of the twentieth century, evidence began to appear that suggested light also has a particle-like nature. When the interaction of light and matter was considered by scientists at that time, this particle-like property of light could not be explained using classical wave theory.

Eventually, the quantum theory of light was first put forth by Max Planck, Albert Einstein, and Neils Bohr during the first two decades of the twentieth century. The acceptance of this theory occurred as a result of experiments with the interaction of light and matter. Neither a separate wave theory nor a particle theory of light could explain the observations. According to the quantum theory of light, the electromagnetic energy is quantized. This means that energy can be added or taken from the electromagnetic field by discrete amounts called photons during this interaction. Each photon has a specified energy and momentum. When Maxwell's theory and the quantum theory were combined, the study of quantum electrodynamics began.

The modern view of light describes it as having a dual nature. Interference and diffraction effects support the wave-like nature, while the photoelectric effect supports the particle-like nature. There can be no simple description of light because no macroscopic model exists that can be used to explain this wave-particle duality. This wave-particle duality of light will be discussed later in this book.

3.3   SPEED OF LIGHT

Many scientists over the past few centuries have tried to measure the speed of light. In 1676 a Danish astronomer, Ole Roemer, used observations of the moons of Jupiter to determine this speed to be 2×10^8 meters per second. He was puzzled by the variations in the time of eclipse for one of Jupiter's moons. He calculated the speed of light by using the differences in apparent time it took this moon to orbit the planet at two different Earth-to-Jupiter distances.

In 1972, a measurement of the speed of light was performed by K. M. Evanson by using a laser. This remains the most precise determination of c to date. His result is given below:

$$c = 299,792,456.2 \pm 1.1 \text{ meters per second}$$

For our purposes, we will use $c = 3 \times 10^8$ meters per second in this book for all calculations involving the speed of light in a vacuum. Table 3.1 gives the results of past measurements of the speed of light.

An electromagnetic wave propagating in a vacuum will have a wavelength λ given by:

$$\lambda = c/\nu$$

where c is the speed of light and ν is the frequency of oscillation. The visible light spectrum has the approximate wavelength limits of 430 to 690 nanometers. These wavelengths correspond to blue and red light respectively. When using the above formula to calculate the frequencies, we get 6.98×10^{14} Hertz for blue light and 4.35×10^{14} Hertz for red light. Frequency has the units of cycles per second or Hertz (Hz).

Another thing to keep in mind when working with light is that its speed is less in optical media than it is in a vacuum. For example, the speed of light in water is 2.26×10^8 meters per second. A dimensionless constant of the optical medium called the index of refraction, n, is defined as the ratio of the speed of light in a vacuum to its speed in the medium. In the equation below, c is the speed of light in a vacuum, and v is the speed of light in the optical medium:

$$n = c/v$$

Table 3.2 gives some typical substances and their corresponding indices of refraction designated by the letter n. The index of refraction must always be taken into account when doing calculations involving light propagation in a material medium. In fact, the index of refraction of a substance varies with the wavelength of the light used. This property called dispersion accounts for why a beam of white light spreads out into its component wave-

Table 3.1   Measurements of the Speed of Light

Date	Investigator	Method	Result (10^8 m/sec)
1676	Reomer	Moons of Jupiter	2.14
1729	Bradley	Aberration of light	3.08
1849	Fizeau	Toothed wheel	3.14
1879	Michelson	Rotating mirror	2.99910 ± 75 km/sec
1928	Mittelstaedt	Kerr cell shutter	2.99778 ± 2 km/sec
1950	Essan	Microwave cavity	2.997925 ± 1 km/sec
1972	Evanson	Laser method	2.997924562 ± 1.1 m/sec

Table 3.2   Indices of Refraction for Some Common Substances

Medium	Index of Refraction*
Air	1.0003
Water	1.33
Glass, crown	1.52
Glass, flint	1.66
Ethyl Alcohol	1.36
Polyethylene	1.50 – 1.54

*Measurements of these indices of refraction were taken using yellow light. ($\lambda = 589$ nm)

lengths when passed through a prism. Sunlight is polychromatic, meaning it is composed of several colors or wavelengths. Light from a laser can be considered to be monochromatic or of one color or wavelength. Table 3.3 lists values of the index of refraction in crown glass for various wavelengths or visible colors.

The following example gives you a sense of the size of one wavelength of visible light.

Example 3.2

A typical sheet of paper is 0.003 in. thick. How many wavelengths of 589 nm light (yellow) does this distance represent?

First, we must convert the thickness of paper from a familiar English unit of length (inches) to a metric unit of length (nanometers). Thus:

$$\text{Paper thickness} = (0.003 \text{ in.}) \, (2.54 \times 10^7 \text{ nm/in})$$
$$= 7.62 \times 10^4 \text{ nm}$$
$$= 76.2 \, \mu\text{m}$$

Solving for the actual thickness in terms of wavelengths:

$$\text{Thickness of paper}/\lambda = 7.62 \times 10^4 \text{ nm}/589 \text{ nm} = 129 \text{ wavelengths}$$

Table 3.3   Variation of the Index of Refraction for Crown Glass

Color	Index of Refraction
Red	1.515
Yellow	1.517
Blue	1.523
Violet	1.533

3.4 THE ELECTROMAGNETIC SPECTRUM

Visible light occupies a very small slice of the entire electromagnetic spectrum. Its limits in wavelength are approximately 400 to 750 nanometers. To sense the radiation that we cannot see, we need special detectors sensitive to each particular wavelength to supplement what we see in the visible spectrum. Figure 3.4a shows the components of the electromagnetic spectrum starting from long wavelengths such as power and radio waves to the shortest wavelengths known as gamma and cosmic rays. The wavelength is given in meters. Another useful measurement is the photon energy in electron volts, abbreviated as eV. One electron volt represents the energy acquired by a particle having one electric charge moved through a potential difference of 1 volt. All of these wave phenomena travel at the same speed c in free space. The various regions of the spectrum overlap somewhat with no abrupt borders. For example, we can produce radiation having a wavelength on the order of 10^{-3} meter by either infrared or microwave techniques. We will now consider each area of the electromagnetic spectrum, starting with the most energetic type of radiation as shown in Figure 3.4a.

Cosmic rays are extremely high energy protons and atomic nuclei that move through space. They are observed indirectly on earth when they undergo collisions with the air molecules of our atmosphere. When cosmic rays collide with oxygen and nitrogen nuclei, the interaction results in a secondary shower of particles that can be detected on earth. This interaction produces a radioactive isotope, known as carbon 14, when the high energy cosmic ray hits an atom of nitrogen and, in the process, a neutron is absorbed by the nitrogen nucleus. The reaction is shown below:

$$^{14}\text{N} + \text{n} \rightarrow {}^{14}\text{C} + {}^{1}\text{H}$$

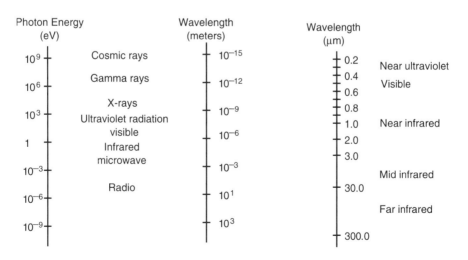

Figure 3.4a The electromagnetic spectrum.

Figure 3.4b The optical portion of the electromagnetic spectrum.

Carbon 14 formed high in the earth's atmosphere becomes oxidized to form molecules such as $^{14}CO_2$. This new product is used in the atmosphere in the same manner as CO_2. Living organisms eventually absorb both of these forms of CO_2. Radiocarbon dating can be performed by knowing the half-life of ^{14}C and its relative abundance. The detectors used for cosmic rays are very similar to those used in particle physics experiments involving high energy particles. At these energies, the particle nature of electromagnetic energy dominates. Techniques to measure this radiation involve particle interactions and the conservation of momentum.

Gamma rays are photons possessing great penetrating power. They can originate from outer space or from nuclear reactions here on earth. Gamma rays also interact with the molecules of our upper atmosphere. They can be detected with particle detectors or by the production of electron-positron pairs in the upper atmosphere.

X-rays result when bombarding a metal plate with high speed electrons. The benefits of x-rays in medicine are well known. But precautions must still be observed when using this form of electromagnetic radiation. The wavelength is comparable to the size of an atom.

Ultraviolet radiation emitted from the sun gets largely absorbed by the ozone layer of our atmosphere. The portion of UV radiation not absorbed is responsible for sun tans and sunburns. UV radiation also finds its use in medicine as a germicidal agent.

The optical spectrum extends from the far ultraviolet to the far infrared as shown in Figure 3.4b, a detail of the UV, Visible, and IR regions. This band is approximately .03 to 300 microns in wavelength. Radiation in this band can be focused and directed by mirrors and lenses. Electro-optical systems operate primarily in this region. The visible spectrum is also included here. Visible wavelengths ranging from 400 to 750nm correspond to the region of sensitivity for the human eye. Photographic film is also sensitive to these wavelengths. The photons of this wavelength are not harmful to humans because their energy is not sufficient to damage living tissue provided that the intensity level remains low. The radiation in this portion of the spectrum displays both particle and wavelike properties. The near infrared region is important in optical communications. For these wavelengths, optical glass fiber provides the lowest attenuation. This property allows light of certain infrared wavelengths to propagate through optical glass fibers nearly unimpeded.

Infrared radiation is often separated into three bands. The first band, near IR, has limits from about 750 nm to 3.0 μm. The next band, mid IR, has limits from about 3 to 30 μm. The far IR extends from 30 to about 300 μm. Near infrared wavelengths can be studied with the same types of instruments used for visible light except that the sensing element is usually made from a different material. We all know mid and far infrared radiation basically as heat. This region of the spectrum finds very important applications in surveillance work. Since all humans emit strongly in this region, they can easily be detected by infrared sensitive detectors.

Radio and millimeter waves are used for radio and television communications. Radiation in this area of the spectrum displays the properties of a wave since the wavelength is relatively long when compared with visible light. This means that diffractive and interference effects characterize this area of the spectrum. Microwaves are also used in communications.

SUMMARY

Light is an electromagnetic wave phenomenon. This transverse electromagnetic wave has two component fields, electric and magnetic, which are orthogonal to each other as the wave propagates through space. James Clerk Maxwell developed four equations that describe the relationship between electricity and magnetism. These equations are listed below:

$$\nabla \times E = -\mu_0 \, \partial H / \partial \tau$$
$$\nabla \times H = \varepsilon_0 \, \partial E / \partial \tau$$
$$\nabla \bullet \Delta = \rho_\varpi$$
$$\nabla \bullet B = 0$$

The first two equations can be used to describe the propagation of light. These two equations show that the electric and magnetic fields are not independent. The last two equations are obtained from Gauss' laws for electricity and magnetism.

An expression for the magnitude of the electric field component of electromagnetic wave as it propagates is given below:

$$E(z,t) = E_o \sin (kz - \omega t)$$

The magnitude of the electric field vector, E, depends upon z and t. This assumes that the wave propagates in the z direction with t being time. In this expression, k is the wave vector of magnitude $k = 2\pi/\lambda$. The constant ω is known as the angular frequency. The angular frequency indicates how fast the wave amplitude oscillates with respect to time.

The speed of light, c, in a vacuum has been accurately measured. This result is given below:

$$c = 299{,}792{,}456.2 \pm 1.1 \text{ meters per second}$$

In this book, we will use the value of 3×10^8 meters per second for the speed of light instead of the more accurate result given above. Since light is a transverse wave with a wavelength λ, and frequency ν, the speed of light in a vacuum can be determined by using the following simple relationship:

$$c = \lambda \nu$$

When light propagates in an optical medium such a glass or water, its speed will be less than it was in a vacuum. A dimensionless constant of an optical medium known as the index of refraction can be used to determine the speed of light in that optical medium. The index of refraction, n, is the ratio of the speed of light in a vacuum, c, to its speed, v, in the optical medium. This simple expression is given below:

$$n = c/v$$

It is important to note that the index of refraction of an optical medium will vary with wavelength. This fact accounts for speading out of white light as it passes through a prism.

The electromagnetic spectrum comprises electromagnetic energy which spans in wavelength from radio waves to cosmic rays. Visible light makes up only a very small slice of this spectrum. The limits of visible light range from about 400 to 750 nanometers in wavelength. In this book, we will be primarily concerned with the optical portion of the electromagnetic spectrum. This band has wavelength limits from about .03 to 300 microns.

4

Reflection and Refraction

4.1 GENERAL REMARKS

In this chapter, we will consider what happens to light as it passes from one homogeneous optical medium to another. This will provide insight into how the materials used to construct optoelectronic components modify the properties of optical radiation. In this chapter, we will be concerned with three ways in which light interacts with matter on a macroscopic scale. Specific examples are given below.

1. Light can be reflected from a mirror.
2. Light can be transmitted through matter such as glass.
3. Light can be absorbed by matter such as sand.

Light never gets reflected 100% from a surface. When reflection of light occurs from a mirror, most is reflected and a small amount is absorbed. The specific amounts depend upon the reflectivity of the mirror's surface. We will find later in this book that even partially reflective mirrors are useful. An example can be found with the mirrors used in the construction of the laser. The laser's resonant optical cavity has two partially reflective mirrors placed opposite each other. During the light amplification process within this cavity, the laser beam will exit one or both of the mirrors. The beam can only do this after obtaining enough optical power to go through the partially reflective mirrors. The ratio of the light reflected from a surface to the total incident amount is known as reflectance. This ratio can be expressed as a percentage. A very good mirror may reflect about 98% of the light striking its surface while the other 2% becomes absorbed. A typical mirror in a diode laser may have a reflectance of about 30%.

4.2 LAWS OF REFLECTION AND REFRACTION

To assist us in the understanding of how light interacts with an object such as a mirror or a lens, we will use the concept of wavefronts. As you remember from Chapter 3, Maxwell's equations describe light as an electromagnetic wave. When light leaves an optical source, it can be represented by wavefronts moving out from the source as shown in Figure 4.1. The rays or vectors show the direction in which the wavefronts travel. The wavefronts are shown by lines passing perpendicularly through these rays. They correspond to either the maximum or a minimum of the wave. Thus, the distance between the wavefront lines equals one wavelength. For our initial discussions on reflection and refraction, we will use the special case where the wavefronts are parallel to each other as shown to the left in Figure 4.1. The parallel rays represent light that originates effectively at infinity. This special case of plane waves makes it much easier to understand the laws of reflection and refraction, so we will use it here. The general case of spherical wavefronts will be considered later in this chapter to help us understand cases when the light rays are not parallel. This generally occurs for lenses and mirrors at close distances.

When we present discussions in the next few sections involving lenses and mirrors, no obstacles or apertures will be used. The dimensions of the lenses and mirrors will be considered to be very large compared to the wavelength of the light used. When these requirements are met, we can consider this to be the study of geometrical optics where light

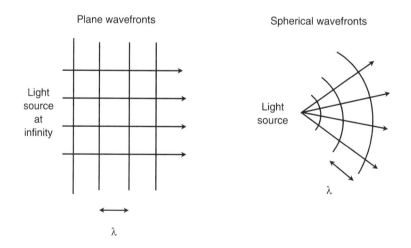

Figure 4.1   Plane and spherical wavefronts.

travels in straight lines and can be represented by rays. Starting with Section 4.7, we will consider optical glass fiber. The above assumptions will no longer hold true. We will see that the wave nature of light must be considered.

There are two types of reflection, specular and diffuse. Specular reflection occurs from a good mirror and makes the same angle from which it came. Simply stated, the angle of incidence equals the angle of reflection. The angular measurements in this case are taken from a line normal to the reflecting surface to the light rays. Figure 4.2a shows this relationship. The second type of reflection, called diffuse reflection, occurs when the individual rays of light strike different portions of an irregular surface as shown in Figure 4.2b. The law of reflection still holds true here since each particular ray may have different normal lines resulting in reflected rays that are not parallel. In this chapter, we will only consider specular reflection.

Next, we consider an air-water interface as shown in Figure 4.3, rather than a plane mirror, for the reflection of light. We see in this figure that the incident light ray makes an angle $\angle\theta_1$ at the air-water interface. The lines on the ray represent plane wavefronts. When this incident ray hits the air-water interface, part of the light ray becomes reflected while the other part becomes refracted. The reflected ray makes an angle $\angle\theta_1$' with the normal. The refracted ray makes an angle $\angle\theta_2$ with the normal. All angles are measured with respect to the normal line. This example can be used to state the laws of reflection and refraction mathematically.

1. $\theta_1 = \theta_1$' (Law of Reflection)
2. $n_1\sin\theta_1 = n_2\sin\theta_2$ (Snell's Law of Refraction)

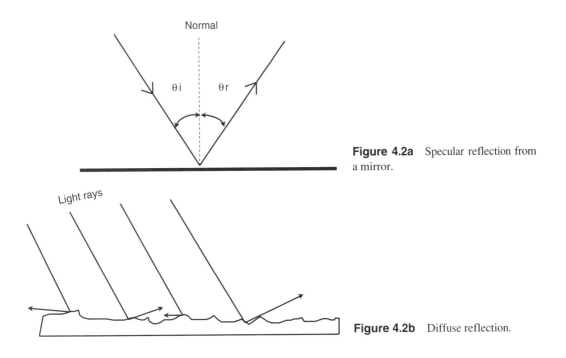

Figure 4.2a   Specular reflection from a mirror.

Figure 4.2b   Diffuse reflection.

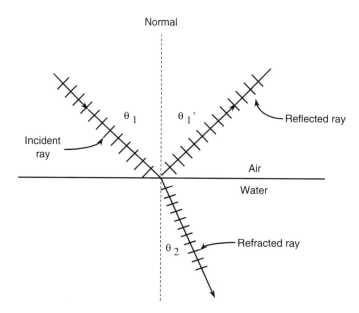

Figure 4.3   Reflection and refraction at an air-water interface.

Equation 2 is known as Snell's law. In this equation, n_1 is a dimensionless constant known as the index of refraction of medium 1. Following the same reasoning, n_2 is a dimensionless constant known as the index of refraction of medium 2. The term refractive index is also used for this constant. As we discussed in chapter 3, the index of refraction of an optical medium is the ratio of the speed of light in a vacuum to the speed of light in the optical medium. This equation is given below:

$$n = v_1/v_2$$

In this equation, v_1 is the speed of light in a vacuum, and v_2 is the speed of light in the optical medium. If, for example, the index of refraction in medium 1 is 1.0, then according to Snell's law, the index of refraction of medium 2 (n_2), can be found by taking the ratio of the sine of the angle of incidence to the sine of the angle of refraction for the light ray. We will show how to calculate this in Example 4.1 below. In Chapter 3, Table 3.2 gives the index of refraction for some common materials used in optical work. If we use glass as our example, the table tells us that it has an index of refraction of 1.5. After rearranging the above equation, we can quickly calculate the speed of light in glass. The steps are detailed below.

Since $n = v_1/v_2$, then $v_2 = v_1/n$. If we substitute in the known values, we can find the speed of light:

$$v_2 = v_1/n = 3 \times 10^8/1.5 = 2 \times 10^8 \text{ m/sec}$$

A third law can be stated that involves the light rays used in our example. The reflected and refracted rays lie in the same plane created by the incident ray. A numerical example follows illustrating these laws.

Example 4.1

A light ray travels from air into water as shown in Figure 4.3. Part of this light ray gets reflected while the other part gets refracted at the interface. (a) If the index of refraction of the water is 1.33, find the angle of refraction when the incident angle is 20°. (b) What is the velocity of light in the water?

To solve for (a), we use Snell's law with n for air equal to 1.0, thus:

$$n = \sin\theta_1/\sin\theta_2$$

Rearranging terms we get:

$$\sin\theta_2 = \sin 20°/1.33 = 0.257$$

which means that $\theta_2 = 14.9°$.
To find the velocity of light in water, we use:

$$n = v_1/v_2 \rightarrow v_2 = v_1/n = 3 \times 10^8/1.33 = 2.26 \times 10^8 \text{ m/sec}$$

To help us understand how refraction occurs at the air-water interface, we will use plane wavefront analysis. In Figure 4.4, we show an enlarged portion from Figure 4.3 detailing two wavefronts as they enter the water. The wavefronts have been extended for the purpose of illustration.

The incident wavefronts are separated by a wavelength λ_1 in air. We find at this point the need to make a basic assumption here so that we may continue with this example. Assume that the speed of light in air is greater than that in water, or $v_2 < v_1$. We will now continue with our analysis of Figure 4.4 using this information.

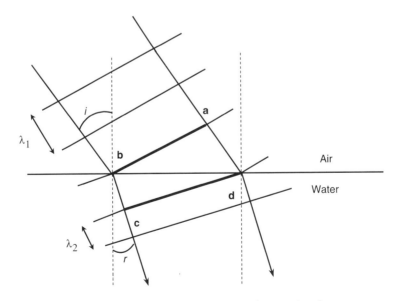

Figure 4.4   Wavefront diagram of an air-water interface.

As the wavefront **ba** approaches the air-water interface, it makes an angle i as shown. The wavefront at point **a** travels to point **d** with velocity v_1 in the same amount of time that the wavefront at point **b** travels in the water to point **c** with velocity v_2. The new wavefront **cd** in the water, now travels at the refraction angle r.

Specifically, it takes a time λ_1/v_1 for the wavefront to move from point **a** to **d**. In this same amount of time, light moves from point **b** to point **c** which is a shorter distance. This distance is given by:

$$\lambda_2 = \lambda_1 v_2 / v_1$$

Rearranging this equation, we get the following result:

$$\lambda_1/\lambda_2 = v_1/v_2 = \text{constant} = n \text{ (index of refraction of water)}$$

Of course, the above equation assumes that the index of refraction for air, n_1, is 1.0 or unity.

4.3 TOTAL INTERNAL REFLECTION

Now, let us see what happens when the rays of light travel from a more dense optical medium to the less dense one. Will the same laws of reflection and refraction apply? To answer this question, we consider the same air-water interface as before. As shown in Figure 4.5a, the incident ray originates in water and the refracted ray enters the air. When the refracted ray exits the interface, it will bend away from the normal since it encounters a less dense medium. A similar case as the one shown in Figure 4.4 demonstrates that this

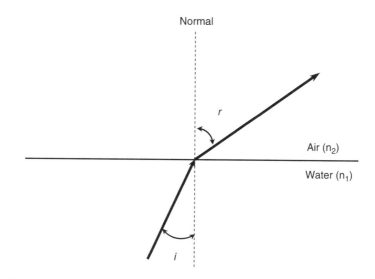

Figure 4.5a   Refraction from a more dense medium to a less dense medium.

will occur. We see in Figure 4.5a that the angle of incidence, *i*, has the smaller value. We expect this to be the case because of the reverse direction from our last example.

Now suppose that we increase the incident angle as shown in Figure 4.5b. As angle i increases, angle r becomes closer to 90°. When this angle equals 90°, light can no longer exit the denser medium. At an angle determined by the indices of refraction of the optical media, there exists an angle θ_C or critical angle for which the incident ray is refracted at an angle of 90°. This relationship can be expressed mathematically by using Snell's law:

$$n_1 \sin \theta_C = n_2 \sin \theta_L$$

In the above equation, $\sin\theta_L$ is the 90° angle of refraction, and $\sin\theta_C$ is the critical angle. This equation can be rearranged to yield an expression for the critical angle as the ratio of the two indices of refraction:

$$\theta_C = \sin^{-1} n_2/n_1$$

We must remember that the sine of an angle cannot exceed 1.0 or unity, thus $n_2 < n_1$. This means that for total internal reflection to occur, the incident light must be in the medium of higher index of refraction. At angles in excess of the critical angle, the light becomes totally reflected at the interface and no refracted ray exists. A practical application of total internal reflection can be found when using optical fibers. In optical fibers, the phenomenon of total internal reflection accounts for the light staying within the glass. This makes possible light transmission along very long lengths of the fiber. The glass fiber usually contains a core through which the light travels. An external cladding layer of a slightly less refractive index surrounds the core to assist in maintaining total internal reflection. We will consider optical glass fiber in this chapter starting with Section 4.7.

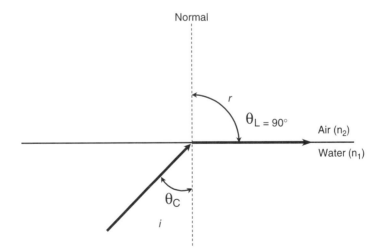

Figure 4.5b A light ray is incident at the critical angle. Total internal reflection will occur for rays greater than θ_C.

Example 4.2

(a) Find the critical angle at an air-water interface. Assume that water has an index of refraction of 1.33. (b) A glass prism is found to have a critical angle of 43° in air. What is its index of refraction?

In (a), we use our last formula for critical angle, thus:

$$\theta_C = \sin^{-1}n_2/n_1 = \sin^{-1}1/1.33 = 0.752$$
$$\theta_C = 48.7°$$

For (b), we rearrange the same formula to find the index of refraction of the glass:

$$n_1 = 1/\sin 43°$$
$$n_1 = 1.47$$

4.4 REFLECTION USING SPHERICAL WAVEFRONTS

In Section 4.2, we considered reflection and refraction of plane waves. This happens to be the special case when the light rays originate at an extremely large distance away. The wavefronts start out in a spherical shape but end up effectively straight or with a curvature approaching that of a straight line. When the light source is close to the object that it encounters, the wavefronts are spherical in nature as shown in the example on the right in Figure 4.1. In this case, the light originates from a point source, and the rays extend outward from this point. It then becomes necessary to consider the more general case of spherical waves. This will lead to a better understanding of mirrors and lenses.

Plane Mirrors

When we last considered reflection, the incident light rays were all parallel to each other simulating a light source at infinity. When moving an object from a large distance to a position close to the mirror, the light rays from that object will no longer be parallel to each other. We assume here that the object has a finite surface area. It can also be considered as a light source since the light rays originate from it. These rays will hit the plane mirror at different angles, as shown in Figure 4.6a. The diagram shows that the light rays originate at a distance D_O from the mirror. A point of intersection can be found by extending the reflected rays to the back of the mirror. This point of intersection is labeled as point I, and corresponds to the image location. The distance D_I is the distance from the mirror to the image. If we measure these two distances relative to the mirror, we will find that they are equal. Therefore, the image of this object appears to be behind the mirror by the same distance as the object distance:

$$D_O = -D_I$$
The object distance = the image distance

The minus sign is arbitrary, and in this case it means that the image appears to be on the opposite side of the mirror. The type of image formed in this case is virtual because no light energy exists at this point behind the mirror. A virtual image cannot be projected

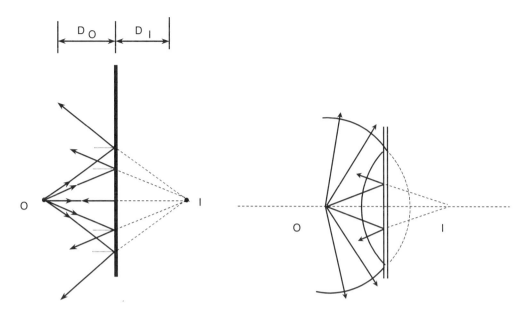

Figure 4.6a Reflection in a plane mirror.   **Figure 4.6b** Reflection of a spherical wavefront from a point source.

onto a screen. The spherical wavefront diagram shown in Figure 4.6b details the rays associated with the object at distance D_O from the mirror initially shown in Figure 4.6a. The spherical wavefront hits the mirror and then gets reflected. We use the diverging rays from the reflected wavefront to find the image distance D_I. When you look at yourself in a plane mirror, your image appears to be the same distance behind the mirror as you are in front of it.

Spherical Mirrors

A real image can be formed using a mirror by having the light energy converge to a common point. This means that a real image can be projected onto a screen. This can be accomplished by using a geometrical shape that directs the rays hitting the mirror to a common point called the focal point. A concave spherical reflecting surface is sufficient to do this. Figure 4.7 shows a spherical mirror using the same object as in Figure 4.6a at distance D_O from the mirror. The line drawn through object O and the center of the mirror establishes the optical axis. Finding the location of the real image requires a minimum of two light rays. Point R is the location of the mirror's center of curvature. The first ray leaves object O along the optical axis and then becomes reflected at the mirror's surface to pass through I. The second ray leaves object O, then travels to the mirror. It makes an angle α with the optical axis. We apply the law of reflection here by drawing a normal line (dashed) at the surface of the mirror where the ray strikes. This normal line also inter-

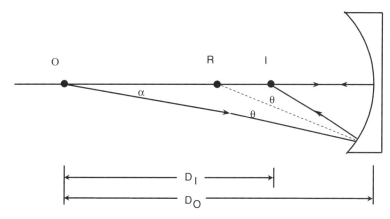

Figure 4.7   Reflection of light from a spherical mirror.

sects point R on the optical axis. Since the angle of incidence equals the angle of reflection, we find that this reflected ray intersects the optical axis at point I. Thus, point I
shows the location of the real image. When only a small portion of the mirror's curvature
is used in its construction, a relatively small angle α will result. In this case, the following
equation holds true:

$$1/D_O + 1/D_I = 1/F$$

This equation is known as the mirror equation. The reciprocal of the object distance, D_O,
plus the reciprocal of the image distance, D_I, equals the reciprocal of the mirror's focal
length. The focal length in this case relates to the radius by $F = r/2$ where r is the mirror's
radius.

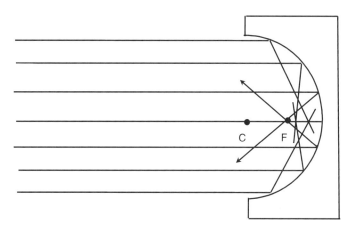

Figure 4.8   Spherical aberration in a concave mirror.

When a large portion of the mirror's curvature is used in its construction, a relatively large angle α will result. In this case, the focal point can only be approximated because not all incident rays lie close to the optical axis. As Figure 4.8 shows, the rays of light striking the mirror near its edges are not reflected to the focus, but end up at points nearer the mirror. This effect is known as spherical aberration. Note that the rays close to the optical axis end up at the focus, F. In this case, a better image resolution can be obtained by using a parabolic shaped mirror as opposed to the spherical shaped one shown in the figure. The mirror equation still holds true for parabolic mirrors. Parabolic mirrors find important applications in automobile headlights and in telescopes.

4.5 IMAGE FORMATION USING CONCAVE MIRRORS

We will next consider five special cases for image formation using concave mirrors. The object used will be a black arrow. The image formed is shown by a gray arrow. A minimum of two rays will be required to construct and locate the image. Refer to Figure 4.9 for ray constructions for Cases 2 and 3 below.

> *Case 1: Object is at infinity.*   The rays in this case are parallel to each other. Using the law of reflection, all of the rays will intersect at the focal point upon reflection. The image formed in this case is a point at the focus of the mirror. This method is sometimes used to find the focal point of a concave mirror. The reverse case works here also. When placing a light source at the focal point, the emitted rays will be parallel.
>
> *Case 2: Object is close to the mirror but beyond the center of curvature.*   This ray construction can be found in Figure 4.9. We will use an arrow located at the object distance. To find where the image will be formed, we need a minimum of two reflected rays from the object. Ray 1 is chosen such that it travels parallel to the optical axis to strike the mirror. Using the law of reflection, this ray should continue by passing through the focal point F as shown. Ray 2 is chosen such that it passes through the focal point before being reflected. The law of reflection requires the reflected ray to be parallel to the optical axis. The intersection of reflected rays 1 and 2 defines the location of the arrow head. We can now make a sketch of the image. The image is real, inverted, and reduced in size from the original object dimensions.
>
> *Case 3: Object is located at the center of curvature.*   In Figure 4.9, we use the same ray construction technique as in Case 2. We find that the image is real, inverted, and the same size as the original object. The object and image locations will also be the same.
>
> *Case 4: Object is located between the center of curvature and the focal point.*   After carefully looking at this case, we find it to be the reverse situation of Case 2. Rays 1 and 2 are drawn in the same fashion as in Case 2. The result is a real, inverted, and enlarged image. The closer the object gets to the focal point, the larger

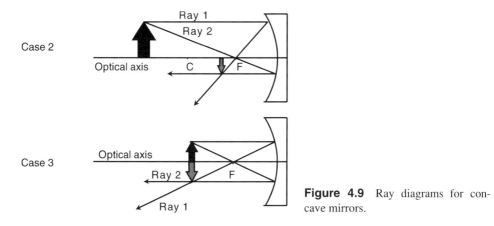

Figure 4.9 Ray diagrams for concave mirrors.

the image will become. In the next case, we will determine the image formation for
the object at the focal point.

Case 5: Object at the focal point.   This is the reverse situation of Case 1. When
placing the object at the focal point, the reflected rays will be parallel to the optical
axis resulting in no image formation.

Note: Ray constructions for Cases 4 and 5 are left as an exercise for the reader.

Example 4.3

A concave mirror has a radius of 20 cm. If an object is placed in front of the mirror along its
optical axis at a distance of 50 cm, (a) Where does the image form? (b) Describe the orientation of the image.

(a) Using the mirror equation we get:

$$1/D_O + 1/D_I = 2/r \rightarrow 1/50 + 1/D_I = 2/20$$

Solving for D_I produces the image distance:

$$D_I = 12.5 \text{ cm}$$

(b) Case 2 in Figure 4.9 describes this example. The image is real, inverted, and reduced
in size.

4.6 THIN LENSES

Lenses find many applications in such devices as cameras, telescopes, microscopes, and
CD players. They are usually made of transparent material such as glass or plastic. One
function of a lens is to produce a desired pattern from incident light. In a CD player, an
objective lens shapes the beam of light into a very small spot that can then "read" the dig-

itally recorded data on the optical disc. This beam of light originates from a diode laser that produces a divergent beam of light having an oval shape. A collimating lens and circularizing optics placed after this diode laser converts the oval shaped beam into a more uniform beam pattern.

When using lenses, many parameters must be considered. One important parameter is the index of refraction that gives the lens the ability to change the shape of the incident light beam. The weight of the lens may also be important in some applications. Returning to our example of the CD player, plastic lenses are an important consideration in this application. Anyone who wears plastic eyeglass lenses knows that they weigh less than their glass counterparts of the same size and thickness. When plastic lenses were first used in CD players, they soon became the better choice over glass for keeping up with the tracking speed during operation. The lighter plastic lens provides a smaller amount of inertia. Another parameter to consider is its geometry or shape. The geometry and the index of refraction of a lens work together to determine how it affects incident light. The two basic lens geometries that we will consider in this chapter are converging and diverging. We will also develop ray diagrams to help us understand how these lenses affect incident light.

Converging Lenses

Figure 4.10a shows a typical converging or convex lens. This lens is commonly known as a double convex lens due to both surfaces being curved outward. The converging lenses that we will consider in this section will all have both surfaces curved outward by the same amount. This type of lens is thicker in the middle than at the edges. Light rays passing through the center portions of this lens are slowed more than at the edges. This occurs because the rays in the center portion of the lens spend more time in the high index mate-

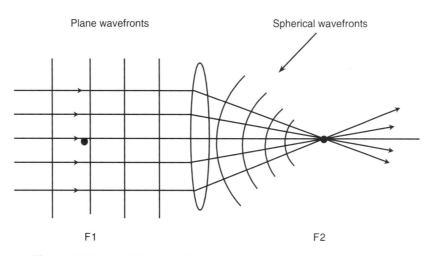

Figure 4.10a Parallel rays of light converge at real focus, F2, when using a double convex lens.

rial than the rays at the edges of the lens. The net result is that all rays converge as shown in the figure. Figure 4.10a shows incident plane wavefronts on the left side of the lens. Notice that the lens surface the light strikes first has its center of curvature on the right side as shown. After passing through the lens, these wavefronts become curved due to this effect. We see that the wavefronts on the right side of the lens are curved such that a center of curvature also exists on the right side. This results in the rays meeting at a common point on this side. The point of intersection for the rays of light is known as the focal point. A real image will form at the focal point.

Diverging Lenses

Figure 4.10b shows a diverging or concave lens. In this case, the edges are thicker than the middle portion. When light strikes the left side of this lens as shown in the example, it will be slowed more at the edges than in the middle. Using the same wavefront analysis as in the last example, we find that the wavefronts on the right side of the lens are curved as shown in Figure 4.10b. There are two important differences between the diverging lens and the converging lens. First, the center of curvature for the first surface that the light rays encounter is located on the left side of the lens. Second, the rays diverge as they exit the lens. We find that the associated rays will meet at a point on the left side of the lens when extending the lines from their points of divergence. The light rays do not actually pass through this point, but we use it to complete the wavefront diagram. The point to the left of the lens is called the virtual focus, and for this reason, no real image will form here with the lens set up as shown. Instead, a virtual image will form on the same side of the lens as the virtual focus.

Ray Diagrams

Spherical lenses usually have two centers of curvature that form the two curved inward surfaces as shown in Figure 4.11. This figure shows a double convex lens with its optical axis passing through the two centers of curvature. For simplicity, we will assume that the

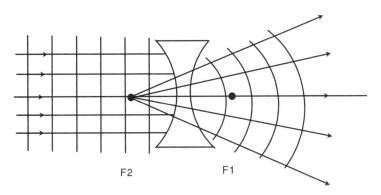

Figure 4.10b Virtual image formation using a diverging lens.

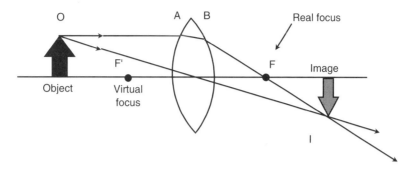

Figure 4.11   Image construction using two rays.

center of curvature coincides with the focus for each side. In this case, the left side of the lens is the virtual side and the right side is the real side. A real image forms on the real side when light passes through the real focus. When locating the object too close to the lens, its image will be on the virtual side of the lens. In this manner, a virtual image forms.

By constructing a ray diagram, the location, size, and orientation of the image, if any, can be defined. We used ray diagrams with mirrors by applying the law of reflection. The ray diagrams to be used with thin lenses will apply to the laws of refraction. To be consistent with the previous diagrams, an arrow will be used as the object. These ray diagrams will require at least two rays to define a point of intersection so that the image location can be determined. In Figure 4.11, ray OA starts off parallel to the optical axis. It then experiences refraction at point A and then again at point B. This ray then continues to intersect point F, the focus, and then beyond. The second ray can be drawn very easily by considering the law of refraction. When a ray passes through the center of a lens, it does not experience refraction, and thus continues in a straight line (the light ray can be considered parallel to the normal at that point). This assumption can only be made if the thickness of the lens is assumed to be negligible. In these constructions, the lens thickness is exaggerated for the purpose of illustration. With this in mind, we draw a line from point O through the center of the lens until it intersects the first ray. The intersection of these two rays defines the image location, I.

The actual location of the focal point on the optical axis depends upon the index of refraction of the lens and the curvature used in its construction. For a double convex lens constructed from crown glass, the focal point and the center of curvature are just about equal. Constructing this lens from an optical material of larger refractive index will result in the focal point being located closer to the lens than the center of curvature.

Image Formation Using Convex Lenses

As with mirrors, we present here some practical examples of image formation using a double convex lens. Two basic cases are given in Figure 4.12. Of course, many other variations exist. In these cases, we will use ray diagrams to specify the image size, location, and orientation.

Case 1: Object at Infinity.   In this case, the incident rays will be parallel to the optical axis. Applying the laws of refraction shows that the image formed in this case will be a point located at the real focus. This method is sometimes used to find the focal point of a lens by using light rays from a distant source such as the sun. Of course, great care must be taken when focusing the sun's rays, as a large amount of energy could be concentrated at one point.

Case 2: Object located at twice the focal length from the lens.   Using two rays to complete this construction as outlined in Figure 4.11, we find that the image in this case is real, the same size, inverted, and located at twice the focal length on the right side of the lens.

Case 3: Object located at the focal point of the lens.   The two rays constructed here show that no image can be formed because the refracted rays are parallel to each other. An application of this lens arrangement can be found in flashlights and car headlights. This is the reverse of Case 1.

Case 4: Object is within the distance of lens to focal point.   The object is moved closer to the lens from the focal point. A real image cannot form on the right side of the lens due to the divergence of the light rays. To complete this ray diagram, we must extend the rays to the left until they intersect. This point of intersection determines the image location. The image formed is virtual, erect, enlarged, and on the left side of the lens. See Figure 4.12 for this ray diagram. Applications of this lens arrangement can be found in a simple magnifying glass.

As with mirrors, a simple equation can be used to determine the relationship between the object and image distances. It applies only to situations where the light rays are mostly parallel to the optical axis and the lens thickness can be assumed to be negligible.

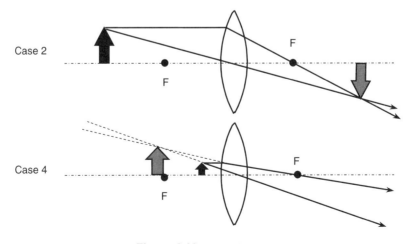

Figure 4.12   Ray diagrams.

$$1/D_O + 1/D_I = (n - 1)(1/r_a - 1/r_b)$$

In this equation, D_O and D_I are the object and image distances respectively. The radius of curvature of the surface that the light strikes first is r_a. Thus, r_b is the radius of curvature of the surface that the light strikes second. You can now see from this equation that the index of refraction and the lens curvatures determine the image location.

We will now derive a more useful form of the above equation. If we place the object at infinity, the incident rays will be parallel to the optical axis. Referring to Case 1 described above, we find that the image will form at the focal point. Substituting an infinite value for D_O, we get the following result:

$$1/F = (n - 1)(1/r_a - 1/r_b)$$

This equation, known as the lens maker's equation, can be used for computing the focal length of a lens using only its index of refraction and radii of curvature of the surfaces, r_a and r_b. When using this equation, some mathematical sign conventions must be observed to obtain a correct result. Next, we will explain the application of these sign conventions.

If the center of curvature lies on the R side of the lens, that distance must be positive (+). If the center of curvature lies on the V side of the lens, that distance must be negative (−). A practical example using the lens maker's equation and these sign conventions is given next.

Example 4.4

Using the lens in Figure 4.11, compute its focal length. The index of refraction of the lens is 1.55. Both lens surfaces have radii of curvature of magnitude 30 centimeters.

Using the lens maker's formula, we substitute the above values using the established sign convention:

$$1/F = (n - 1)(1/r_a - 1/r_b) = (1.55 - 1)(1/30 \text{ cm} - 1/(-30) \text{ cm})$$
$$= (.55)(2/60)$$
$$= .018$$
$$F = +55 \text{ cm}$$

A positive focal length tells us that the image forms on the real side or R-side of the lens. The image will form at this location when parallel incident light converges after being refracted by the lens.

4.7 OPTICAL FIBER

After the development of optical light sources such as LEDs and lasers in the 1960s, it was determined that glass could be used as an optical transmission medium for these sources. For this to be practical, the transmission losses in the glass medium or waveguide had to be less than 20 dB/ Km (the dB or decibel is a measure of relative power that will be discussed in this section). This amount of minimum loss was required for optical fiber to be competitive with copper transmission of electrical current.

The transmission loss problem was solved by Corning Glass Works in 1970. In that year, they announced the development of an optical waveguide fiber having a transmission loss of 20 dB/Km. While rather high by today's standards, this achievement paved the way for further developments in optical fiber. Much lower transmission losses and higher information throughput or bandwidth became a reality. Optical fiber offers many advantages over copper wire and coaxial cable as a transmission medium. Its smaller size and greater bandwidth are the obvious advantages. Another important advantage of optical fiber is its immunity of electromagnetic interference from outside sources. This characteristic allows for efficient transmission in electrically noisy environments, and provides for a more secure transmission link. In later sections of this chapter, we will discuss the important properties of three optical fiber types. We will also consider some of the mechanisms responsible for the macroscopic effects such as attenuation losses and signal distortions.

This is the natural point in this book to introduce the basics of optical fiber since the laws of reflection and refraction are important in its operation. But, as we said in the beginning of this chapter, we must observe at least one variation. The requirement that the aperture size be very large compared to the incident wavelength will no longer apply when we consider optical fiber. This means that the wave nature of light will become important in these discussions.

Our previous discussions on refraction involved just one homogeneous optical media of a given refractive index at an air interface. Before we begin our study of optical fiber, it will then be instructive to consider an example using two homogeneous optical media of different refractive indices with air. The example given in Figure 4.13 shows two layers of glass and incident light rays at two different angles hitting the edge of one glass piece. This example will serve as an introduction to optical fiber due to the two different layers of glass involved. We will next analyze the path of each ray separately as it travels.

Using what we already know about the refraction of light, the ray represented by the dashed line starts from air (n = 1.0) and makes an incident angle ϕ at the air-glass inter-

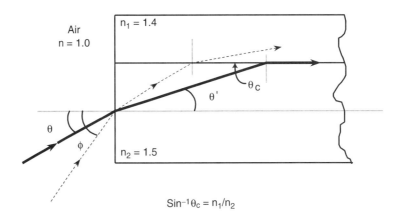

Figure 4.13   Internal reflection using two different glass media.

face. Since the glass has an index of refraction of 1.5, this ray will be bent or refracted toward the normal at this interface. This ray will then continue to travel in a straight line until it encounters the interface between the two glasses. At this point, we must draw a new normal line for this new location. As this ray strikes the interface between the two glass types, it will be refracted away from the normal line since it travels from a more dense optical medium to a less dense one. This ray then travels in a straight line through that glass piece until it strikes the glass-air interface. It should be noted that the incident angle for this ray, $\angle\phi$, has been chosen such that total internal reflection does not occur at the interface of the two glass pieces.

We next consider the second ray represented by the solid line entering the glass ($n_2 = 1.5$) at an angle of incidence $\angle\theta$ in Figure 4.13. We find that this ray will also be refracted toward the normal line thus making an angle of refraction $\angle\theta'$ within the glass having a refractive index of 1.5. The incident angle for this ray has been chosen such that total internal reflection occurs at the interface of the two glass types. For this to happen, $\sin \theta_C$ must be equal to the ratio of n_1 to n_2. In this case, the critical angle can also be shown to be $90° - \theta'$. At this angle, light will be totally reflected instead of refracted at this interface. Any ray entering the lower glass piece (higher index of refraction) at an angle less than θ will also be totally reflected at some point provided that the glass pieces are long enough. Since $n_1/n_2 = 1.4/1.5 = 0.9333$, we find that the critical angle in the glass of higher index of refraction is about 69°. With the understanding of this example, we can now begin our discussion of fiber optic waveguides. An understanding of the procedure used to solve this last example will help us as we develop the initial concepts associated with optical glass fiber.

Going back to Figure 4.13, imagine now a structure where the glass of refractive index 1.5 is formed into a long tubular strand or fiber. This structure will be called the core. Next, the glass of refractive index 1.4 is layered around this long tubular core for its entire length. This layer will be called the cladding. We will next increase the refractive index of this cladding layer slightly to optimize its performance. Later on, we will show why we must do this. Figure 4.14 shows the geometry of this core-cladding structure with the appropriate changes made to Figure 4.13 to describe a typical optical glass fiber. Notice that the cladding now has an index of refraction of 1.49 while the core has an index of refraction of 1.50. Internal reflection will still occur within this structure even with this small difference in refractive index of .01. Looking at Figure 4.14, you can see that there exists an angle θ specified outside of the structure (acceptance angle) for which light entering the core will be internally reflected as it hits the cladding layer of the structure. Using Snell's law, we can find the acceptance angle θ for which there exists an angle θ' within the glass that meets the requirement of internal reflection. The angle θ' is known as the confinement angle in this case. Light rays entering the core at an angle greater than θ will be refracted out of the core. But light rays entering the core as shown in the figure having an angle of incidence less than θ will be internally reflected at the core-cladding interface. The ray represented by the dashed line enters with an angle of incidence less than θ, thus satisfying the requirement for internal reflection. After becoming internally reflected, this rays will also experience an internal reflection upon striking the next core-cladding interface because the requirement for internal reflection is still met at that loca-

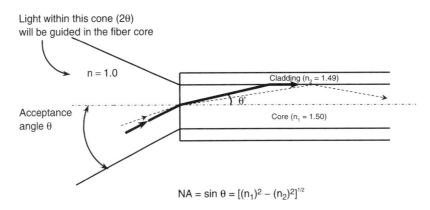

$$NA = \sin \theta = [(n_1)^2 - (n_2)^2]^{1/2}$$

Figure 4.14   Core-cladding structure of optical fiber.

tion. In this way, rays are guided within the core-cladding structure. We can now summarize this relationship below using Snell's law:

$$n \sin \theta = n_1 \sin \theta'$$

In the above equation, n is the refractive index of air (n = 1.0) and n_1 is the core's refractive index (n_1 = 1.5). Now, we will look at the angles involved with the rays. The angle θ is the maximum angle for light entering the core from air that satisfies the requirement for internal reflection at the core-cladding interface. This results in the angle θ' or confinement angle, discussed above.

Since the structure we are now considering is a glass cylinder, a three dimensional representation of the acceptance angle can be considered as a cone of angle 2θ. This cone of acceptance defines the numerical aperture or NA of the fiber, a dimensionless quantity that describes the light gathering ability of a fiber. Typical values for NA range from 0.14 to 0.50. The equation below defines the numerical aperture in terms of the refractive indices of the core and cladding:

$$NA = [\,(n_1)^2 - (n_2)^2\,]^{1/2} \approx \sin\theta$$

In the above equation, n_1 is the refractive index of the core, n_2 is the refractive index of the cladding and n is the refractive index of air (n = 1.0). We will next use the numbers given above to determine the NA in this example.

Example 4.5

A glass fiber has a core refractive index of 1.50 and a cladding refractive index of 1.49. Find the numerical aperture and maximum acceptance angle for this fiber.

Using the above equation and Figure 4.14 we can substitute the above values to get:

$$NA = [\,(1.5)^2 - (1.49)^2\,]^{1/2}$$
$$NA = .17$$

Using Snell's law, we can find the confinement angle for this fiber. Thus:

$$NA = .17 \approx n \sin \theta$$

where θ is the acceptance angle for light outside of the fiber. Solving for θ we get:

$$\theta = 9.8°$$

This means that light entering the fiber at an angle equal to or less than 9.8° will be guided within the fiber due to internal reflection.

The cladding layer of glass surrounding the core serves to improve the performance of the fiber over that of uncoated glass. This cladding layer prevents light from being coupled from one fiber to another when many fibers are used in an assembly. It also prevents the loss of light at the air-glass interface.

4.8 OPTICAL FIBER TYPES

In this section we will discuss three basic fiber types that find their use in a variety of applications. The fiber types for consideration here were developed for the need of increased transmission distance and bandwidth. Other parameters such as input wavelength, cost savings, and fiber strength play an important role. The three fiber types are shown in Figure 4.15. The first type shown, multimode step-index fiber, has a core of constant refractive index. This core can have a diameter of 50 to 200 μm. The second type, multimode graded-index fiber, has a core that varies in refractive index as shown. The core size in this fiber can be from 50 to 85 μm. The third type of fiber, single mode step-index fiber, is designed such that only the lowest order or fundamental mode propagates within its core.

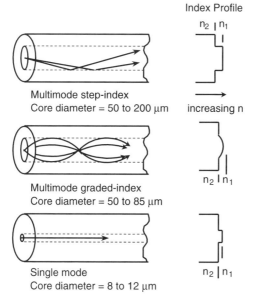

Figure 4.15 A comparison of multimode and single mode fibers.

This fiber has a relatively small core diameter of 8 to 12 μm and finds its use in long haul optical communications systems using diode lasers to launch power into the fiber. Optical fiber size is given in terms of the diameters of the core and cladding. For example, a multimode optical fiber with a core diameter of 62.5 μm and a cladding diameter of 125 μm has the size 62.5/125.

Multimode Step-Index Fiber

This optical fiber type has the simplest construction since only a separate glass core and cladding of uniform refractive index are used. As the term suggests, "step-index" means an abrupt change in refractive index that occurs at the core-cladding interface. The difference in refractive index depends upon a particular fiber design but usually measures less than about 1%. As we saw in our previous example, this difference results in a particular numerical aperture and also aids in confining the light within the fiber. The relatively large core size allows the use of inexpensive light sources such as LEDs. We will see that this advantage has the drawback of limited signal propagation distance and transmission bandwidth.

Returning to Figure 4.14, we see that when light enters the fiber at an angle of θ or less, it will be transmitted through the fiber, assuming no losses. Since the wavelength of the incident light in this case is much less than the fiber's core size, we can use ray diagrams as we did with geometrical optics. This will yield a good approximation when using multimode fiber having a relatively large core-to-wavelength ratio. Besides using a simple ray diagram to explain the reflection and refraction of light in a multimode fiber, we must also consider the wave nature of light. A standing wave must keep a phase condition requirement in this case. This requirement results in only a limited number of paths for electromagnetic waves to propagate such that the field distribution within the fiber remains stable. These stable paths are called modes and satisfy Maxwell's equations. If we considered light only as a ray, there should be an infinite number of paths available. Since we are dealing with a relatively small core size, the condition that these modes must meet can be described by a homogeneous wave equation. A discussion of this is beyond the scope of this book. Any other path not satisfying this condition results in a mode that cannot propagate within the fiber. The number of such modes depends upon the core diameter, numerical aperture, and the wavelength of the light transmitted in the fiber. This relationship can be summarized by the following mathematical expression:

$$M = 1/2(2\pi a/\lambda)^2[(n_1)^2 - (n_2)^2]$$

In this equation, a is the fiber radius, λ is the wavelength of the emitted light, and n_1 and n_2 are the refractive indices of the core and cladding respectively. This equation shows the wavelength dependence on the number of stable modes allowed in the fiber. As the wavelength decreases, more modes will be allowed to propagate within the fiber. Next, we will use this mathematical expression to determine the number of modes allowed to propagate within a typical step-index fiber.

Example 4.6a

Find the number of modes at 880nm and 1.3μm in a step-index glass fiber having a core diameter of 50 μm, $n_1 = 1.48$, and $n_2 = 1.46$.

Using the above formula for the number of modes, we substitute the parameters listed in the given example. We get the following solution for the number of modes at 880 nm:

$$M = (2)(3.14)^2(25 \times 10^{-6})^2/(880 \times 10^{-9})^2 \ [(1.48)^2 - (1.46)^2]$$
$$M = 938 \text{ modes}$$

The problem also asks us to find the number of modes when the wavelength of the light source is changed from 880 nm to 1.3 μm. This longer wavelength happens to be commonly used in optical fiber communications systems. Substituting the values in the equation we get the following solution for the number of modes at 1330 nm.

$$M = (2)(3.14)^2(25 \times 10^{-6})^2/(1.3 \times 10^{-6})^2 \ [(1.48)^2 - (1.46)^2]$$
$$M = 429 \text{ modes}$$

The results from this example confirm that when we increase the wavelength while keeping all other parameters constant, the number of modes in step-index fiber will decrease. As you can also see from the results, there are several hundred modes in the fiber described by the above example when using a light source of 880 nm. To accommodate these modes, each one must make a slightly different angle with the plane of the waveguide or fiber. A term used to describe this parameter for the different modes is called the mode order. The steeper the angle that the ray makes with the plane of the waveguide, the higher the order. High order modes spend more time near the cladding layer of the fiber and less time in the center of the core due to their relatively large angle with the plane of the waveguide. This means that the higher order modes travel a farther distance through the fiber. The low order modes spend most of their time in the center of the core due to their relatively small angle with the plane of the waveguide. These modes travel a shorter distance through the fiber due to the smaller number of reflections experienced at the core-cladding interface. The different distances traveled by each mode contribute to an effect that limits the usefulness of multimode step-index optical fiber. After the next example, we will see how multimode graded-index fiber compensates for this effect.

Example 4.6b

If n_2 is changed to 1.40, what happens to the number of modes at 880 nm?

Using the same formula to determine the number of modes as we used in Example 4.6a we change n_2 to 1.40 (index of refraction of the cladding layer).

We get the following result:

$$M = (2)(3.14)^2(25 \times 10^{-6})^2/(880 \times 10^{-9})^2[(1.48)^2 - (1.40)^2]$$
$$M = 3672 \text{ modes}$$

You can see that when decreasing the refractive index of the cladding layer (NA increases), more modes will be allowed to enter the fiber. In this case, the amount of modes and optical energy entering the fiber increase drastically. We will see from upcoming discussions that this will result in severe pulse spreading due to dispersion. This spreading out of the light energy in the pulse will reduce the effective communications distance. After much research, it has been found that the difference in refractive index between the core and the cladding must be on the order of 1% or less. This relatively small difference results in the best trade-off between the amount of light entering the fiber and pulse spreading due to dispersion.

Multimode Graded-Index Optical Fiber

The multimode step-index fiber as previously described has a basic limitation on the maximum communications distance. Optical digital communications systems use pulses of light in a coded fashion. When a digital pulse of light enters the fiber, many modes will be set up, all of which contain a portion of the optical energy of the pulse. As Figure 4.14 and Example 4.6 show, the stable modes began their journey by entering the fiber at different angles. Of course, these angles must be less than θ to satisfy the condition of internal reflection. First, think of a mode as a directed electromagnetic wave traveling at a particular angle θ_n to the optical axis of the fiber. Since many allowable modes exist in the fiber, there must also be many allowable angles θ_n, where n represents a particular mode. It can easily be seen that each of these modes travels a different path length within the fiber. This difference in path length means that it takes a different amount of time for each mode to travel a given length in the fiber since light travels at a finite speed. This time difference results in the different modes reaching their destination at different times. When looking at all of the mode orders making up the pulse, we see that the pulse of light energy will broaden as it travels further down the fiber. The further the pulse of light travels in the fiber, the wider the pulse becomes as its energy becomes dispersed. The mechanism of attenuation also accounts for a reduction in pulse height. Figure 4.16 shows how this pulse broadening effect and attenuation progressively increase with distance. Modal dispersion limits the pulse rate that can be supported in multimode step-index fiber. You can see from the figure that adjacent pulse energies can overlap or run into each other given enough distance. Another type of dispersion called material dispersion can also cause the same effect. Material dispersion is responsible for the colors that you see when passing white light through a prism. Both types of dispersion will be discussed later in this chapter. Digital signals used in optical fiber consist of step input pulses of optical energy emitted into the fiber from an optical source. These pulses begin with a defined spacing between each adjacent pulse. As Figure 4.16 shows, if these pulses travel far enough in this multimode step-index fiber, the mechanisms of dispersion and attenuation become apparent. The digital pulses will overlap to a point where they cannot be distinguished from one another. We will consider mechanisms of attenuation and dispersion in more detail in the next two sections.

A neat trick can be applied to solve this problem using the index of refraction of glass. We will see how to slow down the lower order modes so that they do not get ahead of the higher order modes. The net effect reduces modal dispersion. Designers of optical fiber realizing this basic limitation in multimode step-index glass fiber have designed a multimode graded-index fiber to address the problem of pulse broadening. This type of fiber has a glass core of refractive index that decreases continuously with radial distance from the center of the fiber. The refractive index profile from the center of the fiber out to the cladding region changes gradually instead of in one step. This parabolic index profile

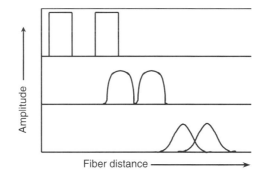

Figure 4.16 Pulse distortion within an optical fiber. This figure shows the relative shapes of two pulses at three different locations in the fiber as they travel further distances. The upper left view shows the two pulses as they leave the transmitter. The middle view shows the two pulses at a much farther distance. The bottom view shows the pulses as they travel even farther. The result above shows that as the optical rise time increases, the bandwidth becomes restricted.

can be seen in Figure 4.15. As you may remember, the velocity of light relates to the refractive index by the following expression, V = c/n, where V is the velocity of the light in the glass fiber of refractive index n. The velocity of light, c, is 3×10^8 meters/second. Applying this relationship to graded-index fiber, you can see that the higher order modes will tend to keep up with the lower order modes within the fiber core medium. This parabolic refractive index distribution will tend to minimize the arrival time differences of every mode. The lower order modes slow down and the higher order modes speed up more. This action will reduce modal dispersion. For example, a mode spending most of its time in the center of the core (low order mode) will travel more slowly than a mode spending most of its time near the core's edge (high order mode). Since the value of n is larger in the center than near the edge, the speed of light in this fiber depends upon the mode order. This gradual decrease in the refractive index has been predetermined such that the modes will ideally end up at the same place for a given length of fiber. This design reduces signal distortion and thus provides for a wider bandwidth.

Single Mode Optical Fiber

In Examples 4.6(a) and 4.6(b), we saw that the number of modes for one particular wavelength depends upon the core diameter and the indices of refraction associated with the core and the cladding. We could still use step-index fiber for long distance optical communications if the modal dispersion effects could be reduced by changing the fiber geometry. We will now look at some different ways in which we can reduce the number of modes. Examining the modal equation, we see that there are three ways to reduce the number of modes in a step-index fiber: (1) reduce the acceptance angle or NA; (2) reduce the core diameter; or (3) increase the input optical wavelength. The first two suggestions have some serious consequences. The reduction of NA or the core diameter will result in less light entering the fiber. Less light means that there will be less optical signal to work with. This may require the use of expensive repeaters to amplify the signal at some point along the length of the fiber. After much testing and evaluation on this issue, it was found that the best way to solve this problem lay in reducing the core diameter and then using a light source of the longest practical wavelength. A typical single mode fiber will have a core diameter of about 9 μm. This fiber has been optimized for a light source such as a diode laser emitting optical radiation in the 1.3 μm region. The advantage of this fiber over the previous types is that only one mode will be supported. This virtually eliminates modal dispersion. We will also see in the next section that light in this wavelength region experiences less loss in the fiber due to absorption.

   We can immediately see a real disadvantage to using this size fiber. Going from a core size of about 100 μm to 9 μm puts very tight tolerances on the positioning of any light source to be used. A more powerful emitter may be required to compensate for the optical power loss due to inefficient coupling. This means that special connectors must be used to assure that as much light as possible gets coupled into the fiber. But today, these problems have been solved with the development of efficient connectors just for this purpose. Single mode fiber is widely used in optical communication systems. It is capable of supporting digitally transmitted optical signals at 2.4 Gbits/sec or more for distances of about 50 kilometers without the use of a repeater or amplifier.

   Since the core diameter approaches that of the wavelength of light used, we cannot use ray diagrams in an analysis with single mode fiber. We must instead use the homogeneous wave equation. As stated earlier, this is beyond the scope of this book. Single mode fiber will later be considered as a comparison with the other types of fiber for its value in reducing signal distortion.

4.9 SIGNAL ATTENUATION IN OPTICAL FIBER

In the last section, we introduced three types of optical fiber and their associated structures. We found that the geometrical design of these optical waveguides addressed various factors involved when optical radiation travels through this medium. For example, the problem of modal dispersion in multimode step-index fiber was basically eliminated by reducing the core size so that, ideally, only a single mode could be transmitted through the fiber.

   As light energy propagates within an optical fiber, it becomes affected in various ways by the glass medium itself. In this section, we will consider one of these properties of glass fiber known as signal attenuation. Special attention must be paid to signal attenuation by not allowing the optical signal energy level to become too low. It must then be amplified to an acceptable level by a repeater or equivalent device to be retransmitted through the fiber. Keeping optical energy losses under control then becomes important to the overall efficiency of an optical communications system.

   Before we can discuss the factors involved when an optical signal becomes attenuated in an optical medium, consideration must be given to the measurement units for the signal power level. It is usually convenient to express optical power in terms of a reference level. This unit of measure, known as the decibel or dB, is logarithmic in nature. Power levels in dB can be calculated as a power ratio and thus provide no indication of absolute power level. The mathematical expression for power level in dB is given below:

$$\text{Power} = 10 \log P_2/P_1 \text{ dB}$$

In this expression, P_1 is the input power level and P_2 is the output power level. The attenuation or loss of power can be calculated by comparing the two power levels and then using the above formula. Sometimes we would like an indication of the absolute power level. In optical fiber communications systems, it is convenient to use the reference optical power level, P_1, as 1 milliwatt (mW). When using 1 mW as a reference level, the unit for the decibel level now becomes the dBm. Thus, the above formula can be changed to yield values in dBm. This new formula can be given as:

$$\text{Power level} = 10 \log P/1\text{mW dBm}$$

In this expression, P is the optical power in milliwatts. An example is given below.

Example 4.7

Find the optical power level of .05 mW expressed in dBm.

Using the above formula for dBm, and then letting P_1 be 1 mW, we get the result below:

$$\text{Power} = 10 \log .05/1 = -13 \text{ dBm}.$$

This means that the power level of .05 mW is 13 dBm below that of 1 mW, so we must use a minus sign to indicate this. A quick calculation of the reference level 1mW gives a value of 0 dBm. Thus, the unit of dBm gives you an absolute power measurement where the unit of dB does not. When given power in dBm, it can easily be converted to an absolute power level by solving for P_1 or P_2 depending upon the application. Table 4.1 gives some examples of dBm units and their corresponding power levels in mW.

Example 4.8

An optical signal experiences a 50% power loss in 2 kilometers of fiber. Find the power loss in dB.

Using the mathematical expression for power level in dB, we can solve this problem as shown below. We use the ratio of power P_2/P_1 as 2/1, which describes a 50% decrease:

Table 4.1   Examples of dBm Levels

Power (mW)	Power Level (dBm)
0.05	−13
0.01	−10
0.5	−3
1	0
2	3
10	10
20	13

$$\text{Power} = 10 \log P_2/P_1 \text{ dB} = 10 \log 2/1 = 10 \log 0.5 = -3 \text{ dB}$$

The minus sign means that this is a power loss.

We are now ready to begin our discussion of attenuation of optical energy. The attenuation of optical energy in glass fiber results from three basic mechanisms. These mechanisms are absorption, scattering, and radiative losses. We shall next consider these mechanisms separately.

Absorption

As light propagates through a glass medium, it will interact with the molecules and atoms of the medium. This involves the interaction of matter and radiation. It turns out that the amount of interaction depends upon the wavelength of the optical radiation. One result of this interaction is the absorption of photons. In glass fiber, this absorption can be caused by the impurities present and the fiber material itself. Impurities such as water ions (OH) present in the fiber due to the manufacturing process have absorption peaks in the visible and infrared regions of the spectrum. Above 1.2 μm, absorption loss in glass fiber becomes a dominant characteristic. This becomes an important consideration since many of the light sources used in optical fiber communication systems emit in this wavelength region. Modern manufacturing techniques have greatly reduced the amount of OH content in glass fiber to such a point that its presence results in a loss of less than 1 dB/Km for some wavelengths. With the long lengths of fiber in some networks, this must still be taken into account.

Scattering Losses

Light becomes partially scattered in many directions within a glass fiber due to very small variations in refractive index. Defects such as gas bubbles, crystallized glass regions, and other impurities may also cause scattering to occur. The sizes of these defects and impurities are typically much smaller than the wavelength of the light source used. Scattered light results in a reduction in the power level of the received signal at the end of a given length of fiber. The mechanism at work here is known as Rayleigh scattering. The physics of Rayleigh scattering is complex and beyond the scope of this book but we will present the result here. The losses due to this type of scattering follow a λ^{-4} dependence. This means that the effect due to scattering increases dramatically with decreasing wavelength. For example, for wavelengths shorter than about 1 μm, Rayleigh scattering in a typical optical fiber becomes a dominant loss mechanism. Above this wavelength region, Rayleigh scattering becomes less pronounced but infrared absorption as discussed previously becomes a dominant mechanism.

Since absorption and scattering losses are wavelength dependent, we can summarize the combined effect in Figure 4.17. This graph shows attenuation in glass fiber in dB/Km as a function of wavelength for both graded-index and single mode types. This curve clearly shows why optoelectronic devices that emit light at 1.3 and 1.55 μm are very important for use in fiberoptic communication systems. Fortunately, there are optical receiver devices sensitive to the same wavelengths. These devices and their applications will be discussed later in this book.

Radiative Losses

Even when keeping absorption and scattering losses to a minimum, there can be significant losses due to the optical fiber geometry itself. These losses usually result from the installation of the fiber. Specifically, great care must be taken not to create bends with small radii of curvature when installing the optical fiber. Light energy can actually leave the fiber core if a radius of curvature becomes small enough. Up to this point in our discussion of optical fibers, we considered fiber oriented in a straight line. In real world applications, optical fiber must be placed in building raceways with other electrical wiring. The placement of this optical fiber may require several bends before reaching its final destination. We can easily see that if an optical fiber has a bend in it, there exists a higher probability for the higher order modes to leak out of this fiber portion than when straight. As you may remember from previous discussions, these higher order modes just barely met the condition for internal reflection as they entered the fiber (the condition here being that the angle of incidence had to be less than or equal to θ_c). If we bend the fiber, the angle at which the higher order modes meet the core-cladding interface will increase. The amount of this increase depends upon the radius of the bend. This means that for some of the higher order modes, the condition of satisfying the angle of incidence for internal reflection no longer exists. The modes no longer satisfying this requirement will exit the fiber at the bend. As the radius of curvature becomes smaller, more light will be coupled out of the fiber.

There are two types of bends in optical glass fiber that must be controlled: (1) bends having radii large compared to the fiber core as described above, and (2) microbends that

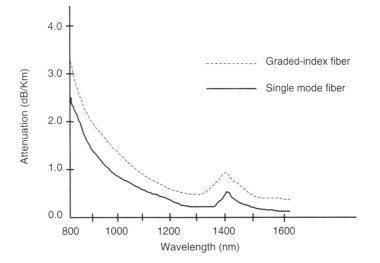

Figure 4.17   Attenuation versus wavelength in optical fiber.

result from small changes in radii of curvature along the fiber length. For example, pressure may be applied at a point along the fiber. This pressure point will cause a small depression that changes the straightness of the fiber itself. A large number of these microbends can cause a significant attenuation loss.

We will now present an example of power loss in optical glass fiber.

Example 4.9

The initial optical power level launched into a glass fiber from a diode laser is 0.5 mW. After travelling 3 kilometers in the fiber, the received power level is 20 µW. What is the power loss in dB? in dBm?

Using the same formula that was used in Example 4.8 we get:

$$\text{Power (dB)} = 10 \log .02/.5 = -14 \text{ dB}$$

Now, using 1 mW as a reference power level to obtain dBm, we get:

$$\text{Power (dBm)} = 10 \log .02/1 = -17 \text{ dBm}$$

4.10   SIGNAL DISTORTION IN OPTICAL FIBER

Of key importance in an optical communication system is the integrity of the signal as it propagates through the glass fiber. As an optical signal propagates through glass fiber, it will eventually become distorted from the original input signal. In the case of digital signals, the pulses will become broadened with increasing propagation distance. At least two different mechanisms cause this distortion: modal and intramodal dispersion. In this section, we will discuss these two mechanisms and their importance in the three types of optical fiber.

Modal Dispersion

In the last section, we briefly mentioned modal dispersion as a distinct limiting factor to bandwidth in multimode step-index fiber. We saw how the placement of the various modes in the fiber resulted in pulse spreading due to the difference in arrival times of these modes. This caused a spreading out of the light energy in the pulse as it propagated through the fiber. When the fiber can only support one mode, pulse spreading due to this mechanism is virtually eliminated. This situation occurs in single mode fiber. In multimode graded-index fiber, the refractive index profile can be selected in such a way that it compensates for this spreading out or broadening of the pulse energy for a specific wavelength. Since the refractive index varies with the wavelength of the light in the glass fiber, modal dispersion can also become a factor in multimode graded-index fiber when using certain light sources.

Figure 4.16 shows the combined effect of attenuation and pulse broadening in optical glass fiber. As the pulse travels through the fiber, it loses some of its energy due to the various attenuation mechanisms. This causes a reduction in the pulse height. The pulse also experiences distortion along the way. In the extreme case, this distortion can cause two adjacent pulses to overlap due to the light energy in each pulse spreading out. When digital pulses of light overlap in this way, the receiver cannot distinguish between the individual pulses, and errors will result. Thus, the distortion mechanisms in the optical fiber limit the available bandwidth.

Intramodal Dispersion

In the last section, we considered the combined effect of all modes in a light pulse as they traveled through the glass fiber. Next, we will consider what happens to a particular mode in the fiber as it travels. In our discussion, we will consider two main causes of intramodal dispersion: (1) material dispersion and (2) waveguide dispersion. Material dispersion results because the refractive index of the fiber is a function of wavelength. This causes the various spectral components of the light energy to travel at different group velocities in the fiber. As we know, material dispersion occurs when white light travels through a glass prism. The various components of white light are dispersed into the pattern of the visible spectrum of colors that our eyes are sensitive to. Waveguide dispersion results because the optical energy in glass fiber propagates in the cladding as well as in the core. For multimode step-index fibers, the amount of energy in the cladding is usually very small compared to that in the core. But, for single mode fibers, an appreciable amount of optical energy may propagate in the cladding. As mentioned before, we cannot use rays to represent the propagation of light within a single mode fiber. If we use the ray approach, we will get the wrong result because the size of the core approaches the wavelength of the light used. The wave nature of light must be considered here. Since the use of the homogeneous wave equation is beyond the scope of this book, we will only consider the result. The actual amount of light that propagates in the cladding depends upon the wavelength used and the core-cladding design. We will now consider both of these dispersion effects.

Material Dispersion.   The variation in refractive index of the glass core with wavelength causes the light energy in a typical light pulse to spread out or disperse as it propagates. In a real world optical communications system, this effect must be taken seriously. Light sources used in transmitters do not emit optical radiation at just one discrete wavelength but over a range of wavelengths. To illustrate this, we will consider an infrared LED or IR LED light source. A typical near IR LED emits optical radiation at a peak wavelength of 880 nm. It also has a spectral bandwidth of about 40 nm. This means that most of its optical energy can be found in the wavelength region from 860 to 900 nm. The strongest emission occurs at 880 nm (the peak wavelength). Emissions gradually decrease on both sides of this 880 nm peak emission point. A graph of output emission vs. wavelength can be defined by a Gaussian distribution function. In Chapter 10, Figure 10.3 shows the spectral distribution for a typical IR LED light source. The different wavelengths suffer dispersion within the glass fiber very similar to the way white light becomes dispersed as it passes through a prism. This problem can be partially solved by using a diode laser as a light source. A diode laser has a spectral width in the range of about 1 to 2 nm. This means that the optical energy contained within a light pulse from a diode laser will have far fewer spectral components than from an IR LED. The tighter spectral width displayed by the diode laser will result in less dispersion. Figure 4.18 is a graph of refractive index vs. wavelength in a typical glass fiber, and shows why this occurs. From this graph, you can also see that this relationship is not linear. Velocity differences caused by the variation in the refractive index with wavelength causes the spreading out of the various spectral components. Material dispersion can be reduced by using light sources having a tighter spectral width such as diode lasers or by operating at longer wavelengths. Since even diode lasers have a finite spectral width, light output will also become dispersed if given enough propagation distance within the fiber.

Many studies have been done to characterize pulse spreading in terms of the distance that the light travels and the wavelength used. The plot given in Figure 4.19 shows how material dispersion varies with wavelength in a typical glass fiber. You can see that at a wavelength of about 1.3 μm, material dispersion effectively goes to zero. Thus, a light source emitting at this wavelength has the potential for providing a much greater transmission bandwidth than a light source emitting at 880 nm. Of course, we assume here that each source has the same spectral width.

Since dispersion affects an optical signal in all types of fibers, most fiber manufacturers have found it convenient to specify this parameter as a measure of information capacity. The units given to this measurement are specified by the product of the bandwidth and distance or MHz · Km. For example, for a step-index fiber, this may be only 20 MHz · Km, while for graded-index fiber it may be as high as 2.5 GHz · Km. This bandwidth limit results from the distortion of the input optical pulse. Figure 4.16 shows how these pulses become distorted with the distance traveled in the fiber.

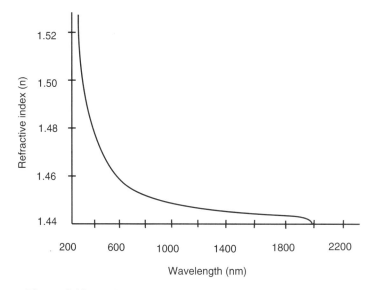

Figure 4.18 Refractive index versus wavelength for optical fiber.

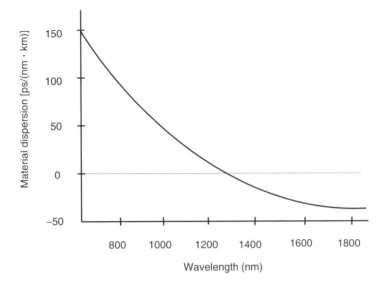

Figure 4.19   Material dispersion versus wavelength in optical fiber.

The amount of time required for the signal pulse to achieve its maximum amplitude is a very important parameter. As the optical pulse spreads out due to dispersion, it will take longer for the leading edge of the signal to rise to this maximum value as seen on an oscilloscope. This effect restricts the bandwidth of the fiber due to this increase in rise time. To determine the amount of distortion in a fiber, measurements must be taken on a series of short light pulses after traveling a given distance in the fiber. We see from Figure 4.16 that the general shape of the optical pulse after it suffers pulse broadening is Gaussian. In this case, the time required for the pulse to reach its half-maximum amplitude, τ_{FWHM}, determines the 3 dB optical bandwidth. This time will increase as the pulse broadens due to traveling a further distance in the fiber. In digital communications, the rise time of the pulse is important since a relatively square pulse shape must be received for error free communications. When using multimode fiber, the relatively long rise time or distortion results from a combination of both modal and material dispersions as previously discussed.

Later in this book, we will consider the rise time for pulses in a fiber optic system. The system rise time can be calculated by taking the root-sum-square of all elements in the system causing the distortion. These elements include the transmitter, the dispersion in the optical fiber, and the receiver. From this time measurement, the bandwidth limit for the entire system can be determined. We will see that bandwidth restrictions occur in both the optical and the electrical realms. Just as the photons in an optical fiber experience dispersion, electrons in a conductor also experience a rise time requirement that distorts the original signal. Capacitance affects an electrical signal similar to the way that dispersion affects an optical signal in a glass fiber. In later chapters, we will see that capacitance can increase the electrical rise time.

Waveguide Dispersion. This type of dispersion results from the wavelength dependence of the modal characteristics of the glass fiber. The effect usually becomes a significant factor when using single mode fiber and thus can generally be neglected when using multimode fiber. In single mode fiber, the optical energy has a significant portion distributed within the cladding. This is due to the energy distribution of the fundamental mode or lowest order mode transmission within the fiber. As the wavelength of this fundamental mode increases when using a particular single mode fiber, more energy will extend into the cladding layer. As a result, the magnitude of the waveguide dispersion will change with wavelength depending upon the difference between the core and the cladding indices of refraction. This can be kept to a minimum by manufacturing single mode fiber with core-cladding differences as small as possible. The core size must also be optimized for a particular wavelength of operation. A core size made too small for the particular wavelength used will result in excess optical energy being distributed into the cladding layer. When the core size becomes too large, then multimode operation may result.

SUMMARY

If the size of the optical system components used, such as mirrors and lenses, is very large compared to the wavelength of light, then we can use rays to approximate the electromagnetic wave. This is known as the study of geometrical optics.

In the general case, when light strikes the boundary between two optically transparent media of different indices of refraction, part of this light will be reflected while the other part will be refracted. The reflected beam will obey the law of reflection. This law states that the angle of incidence will be equal to the angle of reflection. If there is a refracted beam, it will obey Snell's law. This law is given below:

$$n_1 \sin\theta_1 = n_2 \sin\theta \quad \text{(see Figure 4.3)}$$

A special case of Snell's law can occur to produce no refracted ray. In this case, the light ray experiences total internal reflection. For total internal reflection to occur, the light ray must propagate from a more dense optical medium to a less dense one. The light ray must be incident upon the less dense medium at angles less than the critical angle. The following equation can be used in this case:

$$\theta_C = \sin^{-1} n_2/n_1 \quad \text{(see Figure 4.5b)}$$

When using a spherical mirror of focal length F to image an object at a distance D_O from the mirror, the following equation can be used to find the image distance D_I:

$$1/D_O + 1/D_I = 1/F$$

Snell's law of refraction can be applied to construct glass lenses that direct rays of light in a controlled fashion. We use here the example of a converging glass lens. This glass lens has two curved surfaces that extend outward. The following equation can be used to find the image or object distances:

$$1/D_O + 1/D_I = (n - 1)(1/r_a - 1/r_b)$$

In the above equation, D_O is the object distance, D_I is the image distance, n is the index of refraction of the glass, r_a is the radius of curvature of the surface that the light strikes first, and r_b is the radius of curvature of the surface that the light strikes second.

Snell's law is also displayed in the operation of optical fiber. The phenomenon of total internal reflection is responsible for guiding the light energy from one end of the optical fiber to the other. An important parameter of optical fiber known as its numerical aperture or NA defines the cone of acceptance for light rays entering the fiber from a light source. Numerical aperture is a dimensionless quantity, and can be defined in terms of the indices of refraction of the core and the cladding. This formula is given below:

$$NA = [\, (n_1)^2 - (n_2)^2 \,]^{1/2} \approx \text{Sin } \theta \text{ (see Figure 4.14)}$$

The three basic types of optical glass fiber are multimode step-index, multimode graded-index, and single mode. They are described in Figure 4.15. As a light signal propagates within an optical fiber, it will experience attenuation. The signal power levels at two locations can be used to determine the attenuation of the signal between those points. The unit known as the dBm is usually used for this purpose. Mechanisms responsible for attenuation include absorption, scattering, and radiative losses. The power level in dBm can be calculated by using the formula below:

$$\text{Power level} = 10 \log P/1\text{mW dBm}$$

In this expression, P is the optical power in milliwatts.

As a light signal propagates within an optical fiber, it will also experience signal distortion. Mechanisms responsible for signal distortion include modal and intramodal dispersion. Modal dispersion can be a distinct limiting factor to usable bandwidth in multimode step-index fiber. It is basically the spreading out of light energy in a pulse as it propagates within the fiber. To reduce this effect, the refractive index profile of the fiber can be selected in such a way that it compensates for this spreading out of light energy. Multimode graded-index fiber is constructed in such a way that the lower order modes are slowed down in velocity so that they don't get ahead of the higher order modes. This is done by using a parabolic index profile. The result is that, ideally, all modes reach their destination at the same time. Intramodal dispersion concerns what happens to a particular mode in the fiber as it propagates. It also results in pulse spreading. The two main causes for this are material and waveguide dispersion. Material dispersion occurs because the refractive index of the fiber is a function of wavelength. This causes the various spectral components of the light energy in the pulse to travel at different group velocities. Waveguide dispersion results from the wavelength dependence of the modal characteristics of the glass fiber. This is generally a problem that must be addressed when using single mode fiber. In single mode fiber, only the fundamental mode becomes transmitted in the fiber. The energy distribution of this fundamental mode results in a portion of the light energy traveling in the cladding. Thus, the core and cladding sizes must be optimized for a particular wavelength.

5

Interference

5.1 GENERAL REMARKS

In the last chapter, light was represented by using light rays. We could do this because the dimensions of the lenses and mirrors were very large compared to the wavelength of the light used. This area of study is known as geometrical optics. But, when the dimensions of the barriers or apertures become comparable to the wavelength of the light used, then we must consider light's wave nature. This new area of study, called physical or wave optics, will be introduced in this chapter.

Since the energy in an electromagnetic wave spreads out continuously in space and time, two or more waves can interfere with each other. This can be thought of as light interacting with light. The other phenomena that we have already studied such as reflection, refraction, and scattering, describe how light interacts with physical objects or media. The phenomenon of interference is based upon the principle of superposition. This means that for electromagnetic waves, the resultant electric field component at a certain point in space can be found by the vector sum of the separate sources in question. Practical devices that work on this principle include lasers, anti-reflection coatings, and filters.

The phenomenon of interference can be demonstrated using one of two general methods. Each of these methods requires that the light source be monochromatic. The first method called wavefront division occurs when two or more wavefronts overlap producing new wavefronts. Thomas Young's classic double beam interference experiment first performed in 1801 demonstrates this principle. We will consider this method in the next section. The second method, called amplitude division, occurs when a wavefront becomes divided into two or more wavefronts by the use of a beam splitter or similar device. Each portion of the original wavefront is then directed along a different optical path to eventually recombine. In this way, the phenomenon of interference occurs. We will consider this method later in this chapter.

Historically, it was the phenomenon of interference that was used to prove light's wave-like nature. The wave theory of light was opposed by the supporters of Newton's corpuscular theory. We will next consider how the interference of light was demonstrated so that the wave nature of light could be accepted.

5.2 INTERFERENCE BY WAVEFRONT DIVISION

In 1801, Thomas Young demonstrated light's wave nature when he performed his classic experiment on the interference of light. He did this by constructing an apparatus that showed the interference effects produced from two light waves. From the results of his experiment, Young determined the wavelength of the light he was using. This provided evidence for the wave nature of light. In his experiment, bright sunlight was used as the light source. (A source such as a tungsten light bulb could also be used.) A schematic diagram of his experiment is shown in Figure 5.1. We will next describe how he performed the experiment.

Light on the left side of Screen 1 passes through a pinhole, S_0, producing spherical wavefronts. This light then illuminates another barrier labeled 2 that has two more pinholes A and B. The light then passes through both pinholes A and B to a white screen labeled 3 where the interference effects can be observed. As you can see, this diagram shows both diffraction and interference effects. The spherical wave fronts from pinholes

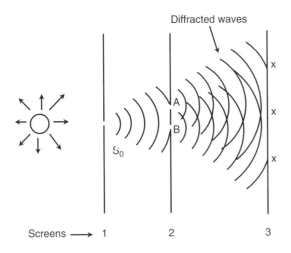

Figure 5.1 Young's experiment. Maximum illumination occurs at each "x" location on Screen 3. Between these locations, minimum illumination occurs.

S_0, A, and B spread out as shown due to diffraction. This occurs when a wave hits a barrier causing some of its energy to spread out behind the obstruction. In this case, plane waves become spherical waves while maintaining their same wavelength. Diffraction effects can be observed when listening to sound emanating from another room. The sound waves will spread around a barrier such as a door opening. Figure 5.2a shows plane waves approaching a barrier that has a slit opening of width a approximately 3 times that of the wavelength of these waves. As the wave encounters the slit, the diffracted wave on the right side of the barrier will diverge outward as shown. If we decrease the width of the aperture to about twice the wavelength, the diffracted wave will diverge with a greater angle. Figure 5.2b shows what happens when the slit width decreases. We see by comparing Figure 5.2a to Figure 5.2b that as the aperture width decreases, diffraction becomes more pronounced. Diffraction will be briefly considered here because it results when using light with small apertures as in Young's experiment. A more detailed discussion of the phenomenon of diffraction will be presented in the next chapter. Since Young's experiment as described in Figure 5.1 does not meet the condition of geometrical optics, a new area of study called wave optics must be used to examine his experiment. We will begin our study of wave optics in this chapter.

Since we have spherical wavefronts, they will overlap at certain locations in the space between the barrier and the screen. The effect of the energy in these two waves overlapping shows up as alternate light and dark narrow bands of light on Screen 3. This interference effect could only result if the light energy were a wave phenomenon. Next, we will take a closer look at the results from this experiment.

The pinholes in the design shown in Figure 5.1 act as sources of spherical wavefronts that expand outward. Next, for simplicity, let us assume that the light consists of one wavelength. In Figure 5.3a we detail the space between Screens 2 and 3 for the purpose of developing a ray diagram. When light enters from the left, the waves at both pinholes A and B are initially in phase since the light was produced from the same wavefront. These two waves are considered to be mutually coherent. We will select a point M

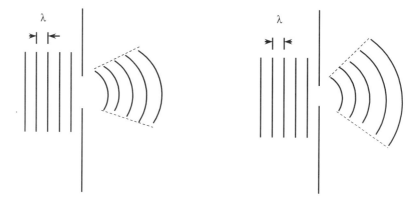

Figure 5.2a   Aperture opening a = 3λ.   **Figure 5.2b**   Aperture opening a = 2λ.

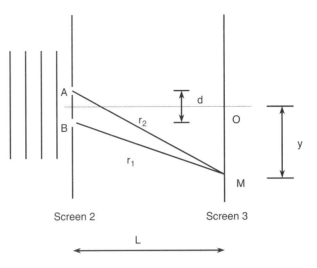

Figure 5.3a   Ray diagram showing optical paths of rays r_1 and r_2.

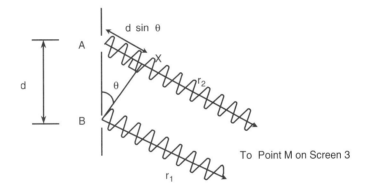

Figure 5.3b   Detail of slit area in Figure 5.3a. We can see that the length AX $= d \sin \theta$.

on Screen 3 where constructive interference occurs. At this point, the energy from each wave adds up because both waves are in phase. The constructive interference of these two waves produces an area of maximum light illumination (maximum). Rays r_1 and r_2 represent optical paths traveled by the light. There are other points on the screen where the two waves end up being in phase. We will see that the location of these points depends upon the spacing d, the wavelength of the light, and the distance L between the screens. We also have areas of minimum light illumination on Screen 3. These areas of minimum light illumination (minimum) result when both waves end up being out of phase with each other. In this case, we have destructive interference and an area of minimum illumination. These alternate maximum and minimum areas are called interference fringes. Of course, the light energy does not become created or destroyed as the terms imply. The interference process just redistributes the light energy over the screen area in this pattern of fringes.

In this experiment, the spacing d must be very small when compared to L. For clarity, this distance has been exaggerated and a detail of this area is given in Figure 5.3b. Now, when the distance d becomes very small compared to L, the light rays r_1 and r_2 are essentially parallel to each other. Making this important assumption will simplify the remainder of this explanation. Looking at Figure 5.3b, we draw a line from B to point X on ray r_2 such that it makes a right angle there. Since the two rays are parallel, the distance BM on ray r_1 equals the distance XM on ray r_2. This leaves the short distance AX on ray r_2 as the extra distance that the light ray r_2 must travel to get to point M. This optical path difference determines the nature of the interference of the two waves leaving pinholes A and B. When the distance AX becomes equal to an integral number of wavelengths, the two waves represented by rays r_1 and r_2 will be in phase at the screen. This region displayed on the screen will be of maximum light illumination. The condition for constructive interference in this case can be summarized by the mathematical expression AX = mλ where AX is the optical path difference, m is an integer representing the number of wavelengths, and λ is the wavelength of the light used. Looking at Figure 5.3b carefully, we can determine another way to express the distance AX. Since the line BX drawn to ray r_2 forms a right angle there, we get another mathematical equality, AX = d sin θ. If we combine these last two expressions we get the condition for constructive interference expressed in terms of d and λ:

$$d \sin \theta = m \lambda, \text{ where } m = 0, 1, 2, \text{ etc.}$$

When m = 0, the central maximum occurs shown by point O on Screen 3. Young used this expression to determine the average wavelength of sunlight. From this he obtained 570 nm. This result is very close to the accepted value of 555 nm.

As Figure 5.1 shows, there are also areas of minimum light intensity caused by destructive interference. This occurs when the two waves are out of phase with each other by 180°(π) or a half wavelength. In this case, we will have a condition of minimum illumination. The length of AX required for this to occur is one half of an integral number of wavelengths. The equation for the condition of destructive interference then becomes:

$$d \sin \theta = (m + 1/2) \lambda, \text{ where } m = 0, 1, 2, \text{ etc.}$$

As you can see, the phase difference between the two waves at particular locations in space determines the interference pattern. We know from Figure 5.3b that the length AX represents the optical path difference between the two waves. This path difference has a definite relationship to phase difference. For simplicity, we will assume that m = 1 for the following discussion. For the condition of minima, AX must be 1/2 of a wavelength in distance to cause the two waves to be 180° (π) out of phase. For the condition of maxima, AX must be one wavelength in distance to cause the two waves to be 0° or 360° (2π) out of phase. This last condition results in both waves being in phase. Putting this together, we have the following mathematical expression relating phase difference to path difference:

$$\text{phase difference} = 2\pi/\lambda \text{ (path difference)}$$

When the optical path difference changes by one wavelength, the phase difference changes by 2π. Using Figure 5.3b and then substituting in the above expression for the optical path difference AX, we get an expression for phase difference:

$$\varphi = 2\pi d/\lambda \sin \theta$$

where φ is the phase difference between the two waves.

Before we can consider some examples using the results from this experiment, the locations of the interference maxima and minima on Screen 3 must be studied. From Figure 5.3a, we can calculate the distance on the screen from O to M by using $\sin \theta$. For small θ, we can use the close approximation of $\sin \theta \approx \theta$. The condition of d being very small compared to L must also hold true for us to simplify the equation for maxima and minima locations:

$$\sin \theta \approx \theta = m\lambda/d, \text{ or } z = m\lambda L/d$$

In the last equation, z is the distance from point O on Screen 3 to a maxima location. As in our previous discussions, m is an integer. Maxima also occur above point O defined by the same equation. Since spherical waves emanate from the pin holes, the areas of maximum illumination show up as concentric rings on Screen 3. A more common configuration uses two narrow slits instead of pin holes. The pattern produced on the screen in this case will consist of alternate bright and dark vertical bands. We will now consider two examples of double beam interference.

Example 5.1

The apparatus detailed in Figure 5.3a is illuminated with monochromatic light having a wavelength of 555 nm. The pinholes are placed a distance of 0.10 mm apart. When Screen 3 is placed 30 cm from the two pinholes, find the angular position of the first maxima.

Using our equation for the condition of maxima, we get:

$$\sin \theta = m\lambda/d = (1) \, (555 \times 10^{-9} \text{ m})/0.10 \times 10^{-3} \text{ m} = .00555$$
$$\theta = 0.32°$$

Example 5.2

Monochromatic red light from a laser having a wavelength of 650 nm illuminates the same design as described above. In an attempt to get more widely spaced fringes, Screen 3 is placed 50 cm from the two pinholes. Find the linear distance on Screen 3 between adjacent maxima. What effect does increasing the wavelength have on this distance? Assuming that this light from the laser is highly coherent, is Screen 1 required to produce interference fringes at Screen 3?

We use the last formula that we developed to determine this distance:

$z_m = m\lambda L/d$ will give us the location of the first maxima when m = 1.

$z_{m+1} = (m + 1)\lambda L/d$ will give us the location of the next maxima.

If we subtract these two distances, we will find the separation between adjacent maxima provided that θ is small. Thus, we have:

$\Delta z = \lambda L/d$ where Δz is the distance between adjacent maxima.

Substituting in for the values we get:

$$\Delta z = (650 \times 10^{-9} \text{ m})(50 \times 10^{-2} \text{ m})/0.10 \times 10^{-3} \text{ m}$$
$$= 3.25 \text{ mm}$$

You can see from the above formula that as the wavelength increases, the distance between adjacent maxima also increases assuming all other variables are kept constant. Also, if a highly coherent light source is used, Screen 1 is not required. The light rays will be in phase at pinholes A and B thus producing the same interference effect on Screen 3.

In the example, we spoke of monochromatic light from a laser illuminating pin holes A and B. The basic requirement for producing well-defined interference fringes on Screen 3 in this example involves a phase difference between the two light waves. If the phase difference at a particular point on the screen is 0 or an even multiple of π, then constructive interference occurs. A bright fringe can be seen at that point. When the phase difference at a particular point on the screen is an odd multiple of π, destructive interference occurs. This results in the formation of a dark fringe at that point. For these interference fringes to occur, the two waves must be mutually coherent. This means that the phase difference between the two waves has to remain constant with time. When the two waves leave pinholes A and B, the peaks and valleys of each wave (represented by a sine wave) line up with each other. Light from a laser provides us with this required condition because, for all practical purposes, the light is composed of one wavelength. The waves are also in phase when they leave the laser. Thus, when the light waves reach pinholes A and B, this phase difference remains the same and interference will occur as described in Figure 5.1.

Effect of Light Source Spectrum

As we know, the light source that Young used in his double beam interference experiment was not monochromatic. We introduced this change in the beginning of our discussion of this experiment to simplify the treatment of this topic. What Young actually saw were diffraction patterns being gradually washed out as the distance increased from the central maximum. When using sunlight to illuminate Young's apparatus, this will occur because of the polychromatic nature of the source. As we know, sunlight contains visible light from red to violet. According to the interference equation $d \sin \theta = m\lambda$, each wavelength will produce an independent interference pattern. Let us find out what happens when we keep the spacing d constant and vary the wavelength of the source.

To accomplish this, let us consider an example using Young's experiment with three different monochromatic light sources. The wavelengths used will be at 400 nm (blue), 500 nm (green), and 600 nm (red). These three sources undergo interference as described in Young's experiment. The result will be three separate interference patterns. This occurs because the optical path difference for each of the three wavelength sources is different. The central bright maximum will be the sum of the three sources since the optical path difference is zero for this point on the screen. As we go further to the right or left of this central maximum, the optical path difference increases by a different amount for each wavelength. For the next maximum, the bright area becomes washed out as the three wavelengths separate from one local area. This occurs because the patterns are out of phase as a result of having different periods. We can calculate the angle of separation for each wavelength using the interference equation. For the three wavelengths, let m = 1 and $d = 0.10 \times 10^{-3}$ m:

Blue $\sin \theta = m\lambda/d = (1)(400 \times 10^{-9})/0.10 \times 10^{-3} = .004$
$\theta = .23°$

Green $\sin \theta = m\lambda/d = (1)(500 \times 10^{-9})/0.10 \times 10^{-3} = .005$
$\theta = .29°$

Red $\sin \theta = m\lambda/d = (1)(600 \times 10^{-9})/0.10 \times 10^{-3} = .006$
$\theta = .34°$

You can see from the above calculations that blue displays the least angular separation while red has the most. This result shows that good visibility of the interference pattern occurs when the spectral width of the light source or optical path difference is small.

5.3 INTERFERENCE FRINGES

In Examples 5.1 and 5.2, we were able to calculate the locations for the maxima and minima of the interference pattern. These locations were the result of the superposition of the two waves. We will next derive an expression that enables us to calculate the intensity of light at these points and all other points between them. This intensity distribution describes the fringe pattern.

To begin our derivation, we will consider only the electric field components of both waves. In the simplest case, if the waves were to travel side by side having the same phase and amplitude, the resultant amplitude E_T can be described by the expression below:

$$E_T = E_1 + E_2$$

But what we have in the situation described in Figure 5.1 is not the simplest case. The two waves have the same frequency and amplitude when they leave the pinholes. They can then acquire a different phase angle between them depending upon the location at which they arrive on Screen 3. Let us consider a more general case. For example, we can express the electric field components of the light arriving at a given point on Screen 3 in Figure 5.1 as:

$$E_1 = E_0 \sin \omega t$$
$$E_2 = E_0 \sin (\omega t - \varphi)$$

where ω equals the angular frequency, and φ equals the phase difference between the two waves. E_0 is the initial amplitude of the wave, and is the same for both waves. In other words, wave E_2 lags behind wave E_1 with a constant phase difference of φ. E_1 and E_2 also have the same wavelength.

To find the resultant of these two waves, we will use vector addition. This method involves using phasors to represent the wave amplitude and angular velocity at a given time. Figure 5.4a shows the phasor representation of the first wave at a given time. This phasor, E_0, has a constant length as it rotates counterclockwise with angular velocity ω. E_1 is simply the projection of this phasor onto the vertical axis. You can see that the length or amplitude of E_1 and the value of ωt will change as the phasor rotates. The mathematical expression for E_1 is given in the first equation above. The second wave has the same amplitude and angular velocity as the first wave. It lags by a phase difference φ. Figure 5.4b shows a representation of this second wave with respect to the first. Again, E_2 is simply the projection of E_0 onto the vertical axis. The phase difference between the two waves results in another wave of different amplitude. Now, we can add these two vectors to find this resultant wave. In Figure 5.4c, we redraw the two phasors and then use the parallelogram method of vector addition. The resultant phasor E represents the sum of the two individual phasors of amplitude E_0 at that instant in time. This phasor now makes an angle γ with the first phasor E_0 that we considered in Figure 5.4a.

To find the amplitude of E in Figure 5.4c, we use the fact that the exterior angle φ of the triangle equals the sum of the two opposite interior angles. This means that $\gamma = 1/2\varphi$. Using trigonometry, we can now find the amplitude of E. We have for this amplitude:

$$E = 2(E_0 \cos \gamma) = 2E_0 \cos (1/2\ \varphi)$$

When we measure the light at a point of the screen, it will be in terms of intensity, not amplitude. Since the intensity of the light is proportional to the square of the amplitude, we

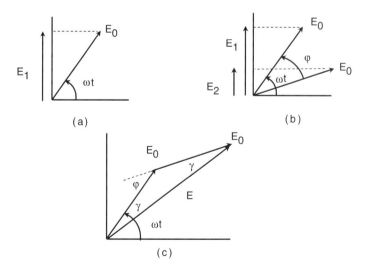

Figure 5.4 (a) Phasor representing a wave of amplitude E_0. **(b)** Two phasors of equal amplitudes, phase difference φ. **(c)** Resultant E vector of both phasors.

can easily make this change to the above equation. To do this, we square both sides of the equation to get a relationship in terms of intensity. We show this operation below.

$$E^2 = 4\,(E_0)^2 \cos^2 (1/2\varphi)$$

and since

$$I \propto E^2$$

we have

$$I = 4I_0 \cos^2 (1/2\varphi)$$

where

$$\varphi = 2\pi d/\lambda \sin \theta$$

In this case, I_0 is the intensity of one of the point sources in Figure 5.1 since both sources are equal. When the phase difference is zero, the intensity will be at a maximum:

$$I_{max} = 4I_0 \cos^2 [1/2(0°)]$$
$$I_{max} = 4I_0$$

When the phase difference between the two waves is 180° (π), we will have minimum light intensity:

$$I_{min} = 4I_1 \cos^2 [1/2\,(180°)]$$
$$I_{min} = 0$$

The plot of the light intensity projected upon the screen vs. phase angle is shown in Figure 5.5. You can see the places of maxima and minima on the fringe pattern calculated above.

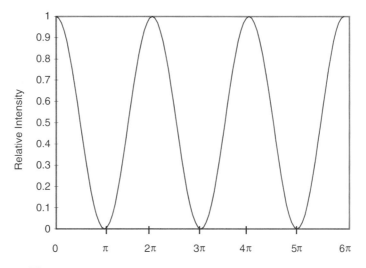

Figure 5.5   Intensity distribution produced by two light sources.

5.4 INTERFERENCE BY AMPLITUDE DIVISION

Up to this point in our discussion of interference effects, we have considered the special case when only two beams of light interfere with each other by wavefront division. We will now consider the more general case of multiple beam interference. Here, interference occurs by a process called amplitude division. This process begins when a wavefront becomes divided into two or more wavefronts. Each wavefront then gets diverted along different optical paths to eventually produce interference effects or fringes. Multiple beam interference finds applications in devices such as the Fabry-Perot interferometer, antireflection coatings, and interference filters. Semiconductor diode lasers operate on the principle of the Fabry-Perot interferometer. We will consider diode lasers in later chapters of this book. Interference filters also operate on this same interference principle. Thus, an understanding of multiple beam interference will be useful in our future studies of these optical and optoelectronic devices.

We saw in Section 5.2 how two mutually coherent light beams could be produced. Figure 5.6 shows one method used to produce a large number of mutually coherent beams. The device in this figure consists of two parallel coated surfaces separated by an air space of thickness d. For simplicity, the thickness of the coated surfaces will be considered negligible. An incident light beam of amplitude A_0, enters the lower left portion of the structure. As this beam enters the structure, we find that it becomes partially reflected and transmitted at the first interface. The angle of incidence θ has been exaggerated for the purpose of illustration, so we will consider this angle to be at near normal incidence. The transmitted beam will then undergo multiple reflections between the two reflective surfaces as shown. Each reflection will involve a certain amount of transmission of light through the structure. This process is known as interference by amplitude division.

Assuming that no absorption occurs within the medium, we can describe the amplitudes of the successive internally reflected rays as A_0TR, A_0TR^2, and so forth. The amplitudes of the successive transmitted rays can also be described as A_0T^2, $A_0T^2R^2$, and so forth. In these cases, R is the coefficient of reflection and T is the coefficient of transmission. The intensity of each successive reflected or transmitted ray decreases due to the amplitude division process. Constructive interference between two successive reflected

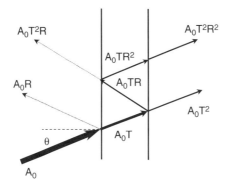

Figure 5.6 Multiple beam interference by amplitude division. The incident beam of amplitude A_0 enters the space between the coated surfaces of negligible thickness where it experiences multiple reflections.

rays will occur when the optical path difference equals an integral number of whole wavelengths. The condition for maximum transmitted intensity is shown below:

$$m\lambda = 2dn \cos \theta$$

In this equation, m is an integer known as the order of interference, d is the distance between the reflective parallel plates, n is the refractive index of the medium between the plates, and θ is the angle of incidence. Destructive interference of the transmitted wavefronts occurs at other wavelengths depending upon the structure's optical parameters. For the condition of destructive interference, the intensity transmitted through the system reduces toward zero as the light reflects back toward the source. Interference fringes will be produced as a result of the interaction of the multiple light beams. For simplicity, we have detailed only one such light beam as it enters the structure described in Figure 5.6.

As in the case with double beam interference, we must now consider the phase relationship involved. For simplicity, we will consider the internally reflected rays within this device to be at normal incidence. When light enters at normal incidence, the plate separation must be a multiple of 1/2 the wavelength for constructive interference to occur in the reflected beam. This means that a multiple of 1/2 a complete wavelength (a phase of mπ radians) must fit into this space between the two plates. When this occurs, the total path difference between any two successive rays becomes one complete wavelength or 2π radians. Thus, we have the following phase relationship for constructive interference.

$$\varphi = 2m\pi$$

In this expression, φ is the phase difference using radian measure and m is an integer. A maximum transmission of light intensity will occur when the transmitted wavefronts undergo an even number of reflections $(0, 2, 4, \ldots)$. This corresponds to a condition of no phase difference between the wavefronts. Figure 5.7 shows the basic requirements for a Fabry-Perot optical cavity. The transmitted light intensity profile from this cavity is

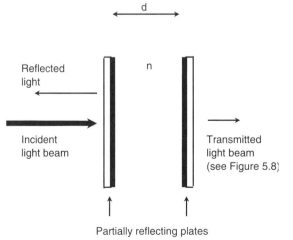

Figure 5.7 Fabry-Perot optical cavity. The transmitted wave must meet the phase requirement $\varphi = 2m\pi$.

shown graphically in Figure 5.8 for 3 different reflectance values of the plates. As you can see, a transmitted intensity maxima occurs at intervals of 2π.

The shape of this intensity profile characterizes the usefulness of a device employing this optical structure. For relatively large values of reflectance, the profile does not follow a sinusoidal pattern. As the reflectance of the two parallel plates increases, the sharpness of the peaks also increases. Only light having the correct phase value will be transmitted. A quality associated with this intensity distribution known as the coefficient of finesse, F, indicates the sharpness of the fringes. The value for the coefficient of finesse depends upon the surface reflectivity of the plates. Assuming that both plates have the same reflectivity, the coefficient of finesse can be expressed mathematically as:

$$F = 4R/(1-R)^2$$

where R is the reflectance of the plate surfaces. The reflectance R will have a value of 1.0 for a 100% reflective surface. Of course, in practice, this does not occur because some absorption occurs at the reflecting surface itself. We will see later in this chapter that relatively large values of R can be obtained by using multiple dielectric films. As the transmission peaks become more narrowed by increasing R, the resolution also increases. This means that the spacing between any two transmission peaks becomes more defined. Another useful measurement used to characterize these peaks is known as the full width at half maximum (FWHM). This can be obtained by measuring the width of the peak at the 50% transmission point as shown in Figure 5.8. Next, we will consider a typical example using multiple beam interference.

Example 5.3

    (a) Two parallel plates of uncoated glass are initially used to make a multiple beam interference set up as previously described. If the reflectance value is 0.04, find the coefficient of finesse:

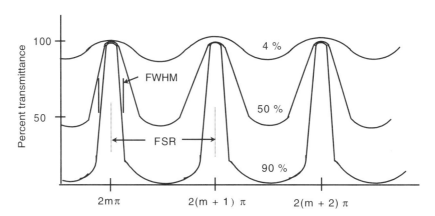

Figure 5.8 Transmitted light intensity profile for 3 different plate reflectance values used in the multiple-beam interference experiment.

$$F = 4R/(1 - R)^2 = (4)(.04)/(1 - .04)^2 = 0.17$$

(b) These same plates are now coated to yield a reflectance value of 0.80. How does this change affect the coefficient of finesse?

$$F = 4R/(1 - R)^2 = (4)(0.8)/(1 - 0.8)^2 = 80$$

In (a), relatively dim broad fringes will be produced. The intensity profile will be similar to the 4% curve in Figure 5.8. There will be poor resolution between these fringes (FWHM will be relatively large). In (b), R is increased to 0.8. The fringe pattern, in this case, will be brighter with more resolved peaks (FWHM will be much less than in the first case). The intensity profile, in this case, will be similar to the 90% curve in Figure 5.8. You have probably noticed another interesting condition associated with increasing the reflectance of the plates. For large values of F, transmission of only specific wavelengths of light occurs. In this way, narrow band interference filters are manufactured that allow only a specific narrow region of the optical spectrum to be transmitted through the device. The use of interference filters instead of prisms or gratings presents an advantage because no dispersion of light occurs.

5.5   THE FABRY-PEROT INTERFEROMETER

Next, we will consider an application of multiple beam interference, the Fabry-Perot interferometer. We will also take a look at the parameters involved with optimizing its operation. As stated in the beginning of the last section, the diode laser uses the Fabry-Perot geometry in its structure, so an understanding of this device will assist us in future discussions involving diode lasers. As we continue our study of the Fabry-Perot interferometer, we will see clear evidence for the wave nature of light. The discussions in this section also serve to initiate our investigation into how the diode laser works. A formal study of the diode laser will begin in Chapter 10. There we will consider the particle nature of light to help explain the emission processes that occur within this structure. As you may have guessed, an understanding of both the wave and particle natures of light is required to correctly describe the operation of the diode laser.

The heart of the Fabry-Perot interferometer consists of two optically flat plates of glass or quartz set up in a parallel fashion as shown in Figure 5.7. The space between the plates, d, can be varied mechanically to obtain the desired interference pattern from a source of light. By using the multiple beam interference formula introduced in the last section, the wavelength of the incident light can be determined. We will see that the information can be obtained from the resulting interference pattern and the distance of separation between the plates. Since most Fabry-Perot interferometers have air between their parallel plates, the optical thickness can be taken to be equal to the physical separation. When a physical medium exists between the mirrors, the refractive index n must be taken into account. In another embodiment, the Fabry-Perot structure becomes a high resolution spectrometer as shown in Figure 5.9. This instrument is useful in determining the spectral makeup of the light emitted by a light source. The lenses shown in this diagram collimate the light from the source and then image the resultant interference pattern onto the screen.

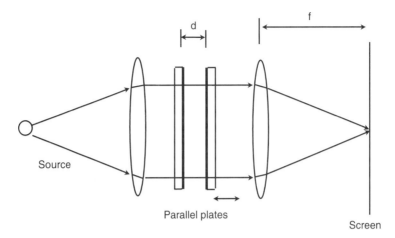

Figure 5.9   The Fabry-Perot interferometer.

When adjusting the plates to a fixed separation d, the device now becomes an etalon or cavity. The plate separation must be fine-tuned to a particular wavelength according to the multiple beam interference formula from the last section. This produces an optical resonator for the wavelengths specified by this formula. The diode laser uses this basic geometrical design in its structure. When the plate separation becomes a multiple of 1/2 the wavelength of the incident light in the medium, this light will be transmitted. Light incident upon the structure not satisfying this wavelength condition will not be transmitted. This separation distance produces an effective optical path difference of an integral number of whole wavelengths. An important application of the etalon can be found in the construction of interference filters. These filters have a high coefficient of finesse thus making them very wavelength selective. Interference filters will be discussed later in this chapter.

Another important parameter of the Fabry-Perot interferometer involves the spacing between two adjacent modes as shown in Figure 5.8. This free spectral range (FSR) of the interferometer can be determined by using the formula for multiple interference. We now present this formula to show how to develop a useful mathematical expression for free spectral range:

$$m\lambda = 2dn \cos \theta$$

In this case, we allow m to be unity since we want to obtain the spacing between two adjacent orders. The angle θ will be consider to be $0°$ (normal incidence). We now get for the condition of maximum transmission at normal incidence:

$$\lambda = 2dn$$

The spacing between two adjacent orders can be obtained in terms of frequency by substituting for λ the very familiar relationship for the speed of light and frequency:

$$\lambda = c/\nu$$

where c is the speed of light and ν is the spacing between adjacent orders in terms of frequency. Doing this substitution, we get an important result:

$$\nu = c/2nd$$

In this expression, ν is the spacing between adjacent orders of interference for our Fabry-Perot interferometer. These adjacent orders are shown in Figure 5.8 as longitudinal modes. The resolution of any particular mode depends upon the coefficient of finesse of the optical cavity. We will now consider examples using the above concepts.

Example 5.4

A Fabry-Perot interferometer has a 1.2 mm air cavity spacing. Light of wavelength 600 nm enters the interferometer at normal incidence. (a) Find the FSR of the interferometer. (b) How many orders m are there between the spacing of the interferometer?

(a) We use the last formula to determine the FSR. Substituting in for the values we get:

$$\nu = 3 \times 10^8/(2)1.2 \times 10^{-3} = 1.25 \times 10^{11} \text{Hz}$$

This is the spacing in frequency (Hertz) between adjacent orders. To find how many orders are present between the spacing of the two plates, we must use the multiple interference formula:

(b) $\lambda = 2d/m \rightarrow m = 2d/\lambda = (2)(1.2 \times 10^{-3})/\, 600 \times 10^{-9} = 4 \times 10^3$ orders

This result shows that if the plates are shifted ever so slightly, there will be a change in the transmission peak output. Thus, the Fabry-Perot interferometer must be constructed such that the plate separation can be adequately controlled. But what happens to the orders or transmission peaks as this distance changes? To find out, we will do a simple calculation in the next example using the Fabry-Perot interferometer from Example 5.4.

Example 5.5

(a) If we now decrease the plate spacing to 0.012 mm in the last example, how many orders exist? (b) Does the FSR change?

(a) Using the same formula as we used in the above example:

$$m = 2d/\lambda = (2)(0.012 \times 10^{-3})/600 \times 10^{-9} = 40 \text{ orders}$$

(b) The FSR now becomes:

$$\nu = 3 \times 10^8/(2)(0.012 \times 10^{-3}) = 1.25 \times 10^{13} \text{ Hz}$$

Thus, when the plate separation decreases, the number of orders also decreases. This result makes logical sense because we now have less distance to fit the transmission peaks. The total number of peaks must decrease to compensate for this decrease in distance. The FSR will also increase when plate separation decreases. Unfortunately, a limit exists as to how much you can increase the FSR. As the FSR increases, there will be less resolution between orders due to an increase in the FWHM of the transmission modes. An increase in FWHM causes adjacent orders to become wider and less resolvable.

If we go the other way by increasing the plate separation, the FSR decreases with a resultant decrease in the FWHM. Thus, for any particular application, a compromise must be determined between free spectral range and the resolution.

5.6   ANTI-REFLECTING FILMS

In the last section, we discussed an application using multiple beam interference where the incident beam experienced a redistribution of its optical energy. Specifically, this energy distribution manifested itself with locations of maximum intensity called maxima. This type of interference is known as amplitude division of the incident beam. When the reflected beams were in phase, a strong response (maximum) was obtained. In this section, we will discuss an application where the multiple reflected beams interfere destructively, with the final result being virtually no reflection. This application of interference can be used with glass lenses and other optical components. A typical glass lens will reflect about 4% of the visible incident light from its front surface. This may sound like a small amount but if there are several lens elements involved, this wasted reflected light becomes significant. Reflection loss can be reduced substantially by applying a thin coating or film layer on the glass surface exposed to the incident light. If this coating has the correct optical qualities, destructive interference will occur resulting in less reflected light. We will now consider just what these optical qualities are and how they cause this condition of destructive interference to occur.

In Figure 5.10, we show an enlargement of a lens area that has been coated with an anti-reflection film of uniform thickness. This layer meets the requirement for the condition of destructive interference of the reflected rays. The indices of refraction of the materials involved are as follows; for glass, n = 1.5, for magnesium fluoride (MgF_2) film, n = 1.35, for air, n = 1.0. This film is used for its durability and its ability to withstand environmental conditions. It can also withstand frequent cleanings without being degraded. As with our other examples, the optical thickness of the dielectric material holds the key to the interaction of the light rays.

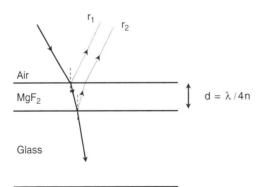

Figure 5.10   Reflections can be reduced by using a thin layer of MgF_2 on glass.

We start our discussion of the example shown in Figure 5.10 by considering only one of the many light rays that initially strike the lens. The incidence angle of this ray has been exaggerated for clarity and can be assumed to be at near normal incidence. When this ray strikes the MgF_2 coating, part of its light energy experiences reflection as ray r_1 and the other part experiences refraction within the MgF_2 medium in accordance with Snell's law. When this refracted ray strikes the MgF_2-glass interface, it experiences partial reflection as ray r_2 and then exits the MgF_2 coating to the air. The remaining light energy exits the coating and then enters the glass. It turns out that there is another reason why MgF_2 was chosen as a coating. This reason involves the propagation of light from a medium of less refractive index to a medium of greater refractive index. In this case, both rays r_1 and r_2 experience a phase change of $180°$ (π) upon reflection.

When a light wave experiences reflection at a boundary, there may also be a phase change involved. The boundaries considered in this chapter are assumed to be dielectric materials such as glass. The phase change upon reflection with such dielectric materials can be only $0°$ or $180°$ (π). Now, there are only two ways in which the light wave may travel in this situation. In the first case, a wave traveling from an optically less dense medium to an optically more dense medium (air to MgF_2 coating) will experience a phase change of $180°$ (π) upon reflection. In the second case, a wave traveling from an optically more dense medium to an optically less dense medium will not experience a phase change upon reflection. In both cases, the transmitted portion of the wave will not experience a phase change.

Now, we can consider the two reflected rays r_1 and r_2. Looking at the ray construction in Figure 5.10, we see that upon reflection, both rays will experience a phase change of $180°$. This will occur since the rays travel from a medium of smaller index of refraction to experience reflection at the interface of a material of greater index of refraction. Thus, there will be no net phase change between the two rays in this case. The phase shift required for destructive interference must be the result of the optical path difference between rays r_1 and r_2. For destructive interference to occur, this distance must be a minimum of a half wavelength. In the example of Figure 5.10, this distance must be equal to 2d where d is the thickness of the thin film. We must also keep in mind that the light wave in the MgF_2 coating has a shorter wavelength than it does in air. The wavelength in an optical medium can be calculated by using the simple equation:

$$\lambda_n = \lambda/n$$

where λ_n is the wavelength of light in the thin film and n is the index of refraction of the thin film. In this case, n = 1.35. Putting this together, we can now present a mathematical expression for the condition of destructive interference in this case:

$$2d = (1/2) (\lambda/n) \text{ which reduces to}$$
$$d = \lambda/(4n)$$

The factor of 1/2 in the first line of the above expression satisfies the wavelength requirement for the distance 2d. This will give us the thinnest possible film layer. Thicker layers must be an odd number of quarter-wavelengths to achieve the correct phase cancellation.

Before we present a practical example using the above expression, we may ask the question, "When we use an anti-reflection coating, what happens to the 4% of light that normally experiences reflection when the glass does not have such a coating?" To answer this question, we must go back to our original discussion involving reflected and transmitted rays. If we neglect any losses due to scattering and absorption, the sum of the intensities of the transmitted and reflected rays must always be equal to intensity of the original incident ray:

$$T + R = 1.0$$

This expression basically describes the physical principle of the conservation of energy. Experiments have shown that the reflected energy involved in the destructive interference process will appear in the transmitted beam. This increases the transmitted intensity by a certain amount more than before the coating was applied. We will now present an example using a thin film to reduce unwanted reflections from a glass lens.

Example 5.6

For light of wavelength 550 nm, find the smallest thickness d that can be used as an anti-reflection film on glass. The film to be used is MgF_2 (n = 1.35) on glass (n = 1.5).

Using the above formula, we get:

$$d = 550/(4)(1.35) = 102 \text{ nm}$$

5.7 INTERFERENCE FILTERS

There are some applications requiring the measurement of very narrow spectral regions from broad band light sources. A practical example of this situation includes viewing the emission spectrum of a flame in a narrow spectral band. In this section, we will apply what we have learned about interference to describe how filters are fabricated to meet the requirement for the above application and many others. These filters are very useful in many applications where a particular wavelength interval or band must be isolated from broad band optical energy.

The bandpass filter makes use of the Fabry-Perot structure to select the transmission wavelength of interest. Instead of an air gap, a thin layer of dielectric material is used. This layer has an optical thickness equal to a multiple half-wavelength optimized for the transmission wavelength. Layers of quarter-wave stacks made from dielectric materials make use of the interference of light to block out unwanted light. Figure 5.11 shows the construction of a typical filter. To achieve the desired effect, these quarter-wave stacks are arranged in alternate layers of high (H) and low (L) refractive indices on both sides of the Fabry-Perot structure. In the figure, the Fabry-Perot structure is a high index dielectric material of half-wave optical thickness designated as the spacer. To decrease the half-width or FWHM of the transmission band, a large number of these alternating high-low quarter wave stacks are used at the design wavelength. In this way, the filter's design takes advantage of both destructive and constructive interference. When light passes through the filter, the differences

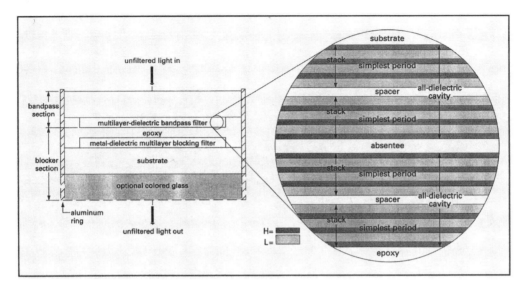

Figure 5.11   Cross-section of a typical two-cavity pass interference filter. The enlarged area to the right shows two interference filters deposited one on top of the other. An "absentee" layer separates these two filters. Reprinted with permission of Melles Griot.

in the refractive indices cause destructive interference at some wavelengths and constructive interference at the design wavelength of interest. These substances are usually deposited in quarter or half-wavelength optical thicknesses to control the reflection and transmission characteristics of the incident light. The optical thickness equals the refractive index times the physical thickness. Table 5.1 lists a few of these substances used as a high or low refractive index component of a quarter-wave stack.

As stated before, as the number of quarter-wave high-low layers increases, the transmission bandwidth becomes narrower. It usually takes at least 3 or 4 of these layers to provide a sufficiently narrow bandwidth for most instrument applications. The spectral profile of the filter includes a measurement of this bandwidth and the center wavelength. The bandwidth is usually measured in terms of the full width at half-maximum (FWHM). Figure 5.12 shows a graph of the change in performance with increasing numbers of cavities. You can see that the FWHM measurement is taken at the 50% transmission point. The center wavelength corresponds to the design wavelength of transmission at normal

Table 5.1   Substances Used for Optical Coatings

Low Refractive Index	High Refractive Index
MgF_2 n = 1.38	ZnS n = 2.3
SiO n = 1.86	Ta_2O_5 n = 2.15
CeF_3 n = 1.60	HfO_2 n = 2.05

incidence. This wavelength usually matches a laser emission wavelength or an atomic emission line such as mercury.

When the angle of incidence changes from the normal, the center wavelength will also change accordingly. This change in incident angle causes an increase in the optical path difference experienced by the light in the film. From our discussions on interference, this means that there will be a spectral shift toward a slightly shorter wavelength. This method can be used to vary the center wavelength of the interference filter.

A typical application of a bandpass interference filter can be found when using a relatively wide band receiver element such as a photodiode to detect a narrow band signal within broad band light emission. Figure 5.13 shows the relative spectral response of the photodiode and the pass band of the expected signal from a transmitter. Usually a certain

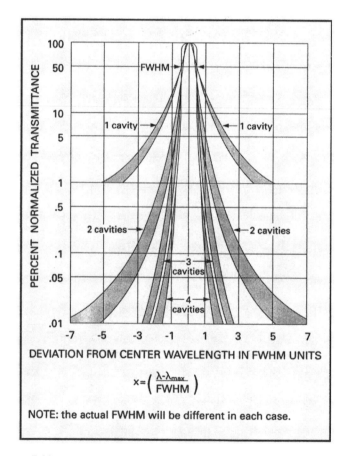

Figure 5.12 Normalized log scale transmission curves for an increasing number of cavities used in an interference filter design. Shaded regions indicate nominal limits of filter characteristics at normal incidence. Reprinted with permission of Melles Griot.

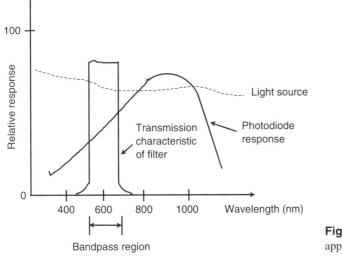

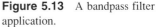

Figure 5.13 A bandpass filter application.

amount of ambient light specified by the line labeled "light source" must be taken into consideration. Without the filter, the amount of ambient light energy could easily exceed that of the signal. The filter used in this case must block out a relatively wide spectral range of optical energy while permitting the optical passage of the signal. This can be accomplished by adding extra layers to the interference filter such as colored glass filters. This technique can produce a filter that sufficiently blocks enough unwanted ambient light so that the receiver can pick up the signal at the specified wavelength. In this case, the specified wavelength of transmission is about 600 nm. As a compromise to achieving this wider spectral blocking range, the optical transmission will usually be decreased from that of a narrower spectral blocking range filter.

SUMMARY

Light is an electromagnetic wave that spreads its energy out continuously in space and time. This can result in two or more of these waves interfering with one another due to the superposition principle. In this chapter, we considered two methods that can be used to cause light to interfere. These methods are wavefront and amplitude division.

During wavefront division, two or more wavefronts overlap in the same space producing a new wavefront. The new wavefront shows up on a screen as alternate light and dark narrow bands called fringes. The light energy in this case has been redistributed; no energy is lost as may be surmised from the dark narrow bands. The light bands meet the requirements for constructive interference. A mathematical expression for constructive interference is given below:

$$d \sin \theta = m \lambda \text{ where } m = 0, 1, 2, \text{ etc.}$$

In this equation, d is the spacing between two slits or pin holes, and λ is the wavelength of light used (see Figure 5.3a and Figure 5.3b).

The dark bands meet the requirements for destructive interference. A mathematical expression for destructive interference is given below:

$$d \sin \theta = (m + 1/2) \lambda \text{ where m = 0, 1, 2, etc.}$$

The same parameters apply to this equation (see Figure 5.3a and Figure 5.3b).

The intensity distribution of these interference fringes on a screen is given by the mathematical expression below (see Figure 5.5 for this function):

$$I = 4I_0 \cos^2 (1/2\varphi)$$

where

$$\varphi = 2\pi d/\lambda \sin \theta$$

This equation can tell the light intensity at all points on the screen, not just at the locations of minimum and maximum illumination.

When interference occurs by the process of amplitude division, a wavefront becomes divided into two or more wavefronts. These new wavefronts becomes diverted along different optical paths to eventually produce interference fringes. This basic process is detailed in Figure 5.6. The Fabry-Perot interferometer operates on the principle of interference by amplitude division. In this device, constructive interference will occur between two successive rays when the optical path difference equals an integral number of whole wavelengths. This condition can be stated by the equation below (see Figure 5.7):

$$m\lambda = 2dn \cos \theta$$

In this equation, m is a whole number, λ is the wavelength of the light used, d is the reflective plate separation, n is the index of refraction of the material between the reflective plates, and θ is the angle of incidence of the initial ray. Understanding the operation of the Fabry-Perot interferometer is important since a typical diode laser uses this geometry to produce a laser beam.

Interference by amplitude division is the basis for many other useful optical devices. A glass lens will reflect about 4% of the visible light from its front surface, assuming that this interface is air to glass. This type of reflection, known as Fresnel reflection, can be reduced or almost eliminated by coating the lens with an anti-reflection film such as MgF_2. If the film layer meets the requirements for destructive interference, the normally reflected 4% of visible light will show up in the transmitted ray (see Figure 5.10). Interference filters that use the Fabry-Perot structure are very useful when the requirement is to isolate different portions of the optical spectrum. These filters use a dielectric material between the two reflective plates instead of air. Layers of quarter-wave plates made in this fashion are stacked in alternate high (H) and low (L) refractive indices on both sides of a Fabry-Perot structure. A large number of these alternating high-low quarter-wave stacks are used to decrease the FWHM of the transmission band. A cross-sectional view of a typical interference filter is given in Figure 5.11.

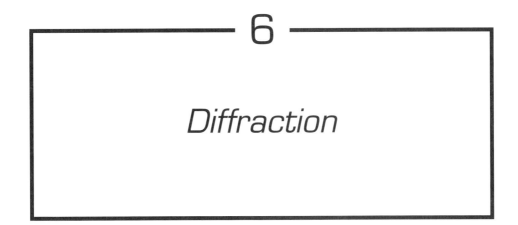

6

Diffraction

6.1 DIFFRACTION—A HISTORICAL PERSPECTIVE

In the last chapter, we saw that light travelling through a small pin hole opening or narrow slit resulted in an image that was spread out. The reason the image was not perfectly sharp has to do with the wave nature of light. Light does not travel in a straight line as would be expected from geometrical optics. This spreading out of light after passing through the pin hole or narrow slit finds its explanation only after considering the wave nature of light. Upon close examination, we see alternate bright and dark areas. A simple experiment can be done to demonstrate this spreading out of light. Take two closely spaced straight edges and then view a distant light source between them. What you will see is a pattern of bright and dark straight lines called a diffraction pattern.

Today, we readily accept this explanation for the diffraction of light as the result of a wave phenomenon. But this was not always the case. When Huygens and Young presented their wave theory of light as described by the interference experiment from the last chapter, they found only a few people who accepted it. Even at that time in history, Newton's corpuscular theory of light was still largely accepted. Newton's book, *Opticks,* had been in print for 115 years. It was not easy to persuade someone that Newton's views on light may require modification.

The general acceptance of the ideas of Young and Huygens on light began in the year 1819. A young military engineer named Augustin Fresnel entered an essay contest offered by the French Academy. A prize would be awarded to the best paper presenting experimental results and an explanation for "diffraction effects." Fresnel, who supported the wave theory of light, would win this contest by disproving the Newtonian corpuscular theory. A member of the committee to judge the papers was a mathematician named Simeon Denis Poisson, who was a staunch supporter of Newton's corpuscular theory. After reading Fresnel's paper, Poisson worked out a very curious prediction based upon Fresnel's dissertation. He reasoned that if Fresnel was indeed correct about the wave nature of light, then a bright spot should appear in the center of the shadow of a circular object exposed to a light source. This result seemed very strange to the supporters of the corpuscular theory and, at first, seemed to offer direct evidence that Fresnel's theory was absurd. Another committee member who was intrigued by this prediction went out immediately to test it. He used a disk with a diameter of 2 mm as the circular object to cast a shadow in direct sunlight. Poisson was proven to be correct in his strange prediction. A central spot was produced by this disk as a result of the diffraction of light waves. Unfortunately, Poisson could find no way to explain this result in terms of the corpuscular theory. Fresnel won the prize. From this point on, the wave theory of light had to be taken more seriously. This unexpected result was the proof required to sway people away from the general acceptance of Newton's corpuscular theory.

In this chapter, we will consider an explanation for diffraction and why it occurs. To help us in this explanation, we will return to Young's experiment. As with our discussions on interference, we will use ray diagrams. The cases involving diffraction will be limited to those where wavefronts initially hitting an aperture or object are parallel to each other. Then later in this chapter, we will consider some practical applications.

6.2 DIFFRACTION OF LIGHT FROM A SINGLE SLIT

As you may remember from Chapter 2 on the historical background of optics, the physicist Christiaan Huygens proposed light to be a wave phenomenon. His principle describes the propagation of light as a wavefront in free space. Every point on this wavefront acts as a point source for a new or secondary wavefront or wavelet. If we know the initial shape of this wavefront at time t, we can find the resulting shape and position at time t + Δt, where Δt represents a small time interval after the formation of the initial wavefront. These secondary spherical wavelets will propagate and expand out at the same speed that the initial wave had. We can draw a line tangent to all of these secondary wavelets to show where the next wavefront or envelope forms. This dashed tangent line is shown in the upper right view of Figure 6.1 as the wavelets leave the slit. To explain just how diffraction of a light wave occurs, Fresnel used the above principle combined with the ideas of Young. This modified view, known as the Huygens-Fresnel principle, is illustrated in Figure 6.1.

Figure 6.1 shows a plane wave moving from the left of a screen with an aperture AB. As the wave reaches the aperture, we must consider the wavefront between points A and B. The detail of the aperture shows that secondary wavelets will be created here, and then propagate through the aperture. These wavelets reinforce or interfere with each other to form a new wavefront. Point P located beyond the screen as shown in Figure 6.1 will receive contributions from the secondary sources or wavelets that originated between the slit. Thus, the waves reaching point P will have different amplitudes and phases due to the fact that each wavelet has traveled a different distance at varying angles to get to point P. If we calculate the light intensity at point P and every other point in this plane, a diffraction pattern will result. We will next develop a way to find out just how much this light wave spreads out to the various locations beyond slit AB. This will be done in a way similar to specifying the locations of the maxima and minima for the interference pattern in Chapter 5.

In Chapter 5, we briefly mentioned the phenomenon of diffraction when introducing Young's double slit interference design. This phenomenon caused the two light waves to spread out beyond the obstructions so that interference could occur. In Young's experiment, we were more concerned with the interference pattern formed when the two waves interacted. To consider the diffraction of light, we will find it convenient to go back to Young's interference set up. For simplicity, we will consider only one of the two slits for this discussion. An enlarged portion of this slit is detailed in the design shown in Figure 6.2.

Figure 6.2 shows the diffraction of monochromatic light waves when passing through a slit aperture having a width a approximately 3 times the wavelength of the light. In this design we have plane wavefronts approaching the aperture from the left. Since the wavefronts can be considered to originate at infinity, they will be parallel. The diffraction

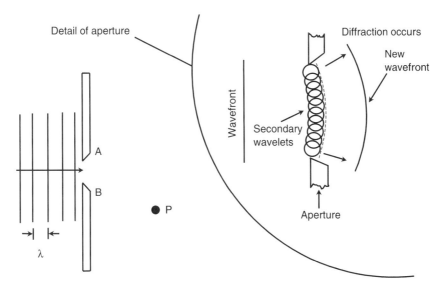

Figure 6.1   The wave disturbance reaches point P due to diffraction.

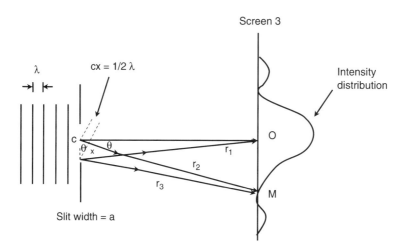

Figure 6.2   The condition for first minimum of the diffraction pattern.

pattern forms on Screen 3. The point labeled c is at the center of the aperture. Point O on Screen 3 lies on the line defined by the optical axis. Thus, the optical axis is along the line labeled cO.

Plane waves approaching the slit from the left are in phase as they reach the slit. We apply Huygens' principle as the wavefront travels through the slit. The formation of secondary wavefronts begins the process of diffraction. The optical path length described by rays cO and ray r_1 are equal. This means that these waves will still be in phase as they reach point O. At point O, these rays will reinforce each other forming a bright area or maximum. Next, we consider the two rays r_2 and r_3. Ray r_2 leaves the center of the slit at angle θ, and then reaches point M. Ray r_3 leaves the bottom of the slit, and then reaches point M also. The angle θ has been chosen such that the distance cx in the figure is 1/2 the wavelength of the incident monochromatic light. This optical path difference causes rays r_2 and r_3 to be out of phase by 180° at point M. Ideally, no light intensity will result at this point. At point M, the contribution from all wavelets between points A and B results in a minimum intensity location. This condition can be stated mathematically as:

$$cx = 1/2 \; \lambda$$

and since cx is also equal to 1/2 a sin θ, we can show further that:

$$cx = (1/2 \; a) \sin \theta = 1/2 \; \lambda$$

Reducing this expression, we get:

$$a \sin \theta = \lambda$$

This is the condition for the first minimum. It can be shown that other minima will occur between adjacent maxima. The same conditions occur at the upper half of the screen. The general formula for the minima condition can then be written as:

$$a \sin \theta = m\lambda \text{ where } m = 1, 2, 3, \dots.$$

A maximum condition will occur approximately 1/2 way between each adjacent pair of minima. The resultant diffraction pattern of maxima and minima has an intensity distribution as shown on the right of Screen 3. This intensity distribution results from the contribution of all wavelets between points A and B.

Example 6.1

Monochromatic light of wavelength 600 nm illuminates an adjustable aperture in a design similar to that in Figure 6.2. (a) When the slit width a is 1800 nm, at what angle does the first minima occur? (b) The slit is now adjusted to a width of 3600 nm while the light illuminates the design. How does the angle of the first minima change? Describe what happens as the slit width increases.

(a) We use the diffraction formula for minima to obtain the answer. Rearranging terms, we get:

$$\sin \theta = m\lambda/a = (1)600/1800 = 0.333$$
$$\theta = 19.5°$$

(b) Using the same formula and then substituting with a larger slit width, we get:

$$\sin \theta = m\lambda/a = (1)600/3600 = 0.166$$
$$\theta = 9.6°$$

We notice that as the slit width increases, the positions on the screen for the minima and maxima come closer together. Figures 6.3a and 6.3b show the intensity distribution for the above two diffraction examples. In the next section of this chapter, we will discuss these intensity distributions in more detail.

Since diffraction is a wave phenomenon, the phase difference becomes important for determining the locations of each minima. We already know that the length cx in Figure 6.2 equals the optical path difference between rays r_2 and r_3. As with the phenomenon of interference, this optical path difference has a relationship to phase difference. For the condition of minima described in Figure 6.2, cx is 1/2 wavelength in distance or π radians. When cx becomes a whole wavelength in distance, θ increases accordingly and the phase difference now becomes 2π. Putting this relationship together mathematically, we get:

$$\text{Phase difference} = 2\pi/\lambda \text{ (path difference)}$$

Using the above relationship and then substituting cx for the path difference, we get:

$$\varphi = (2\pi/\lambda)a \sin \theta$$

where φ is the phase difference.

6.3   SINGLE SLIT DIFFRACTION PATTERNS

We will now consider the two separate intensity distribution plots in Figure 6.3. These plots or graphs were produced from the data obtained using the design in Figure 6.2. Two different ratios for a/λ were used. Graphs (a) and (b) in this figure should be familiar since these are the results from Example 6.1.

In the last section, we determined how to find the locations on the screen where maxima and minima occurred for single slit diffraction. Figure 6.3 shows graphically the relationship of light intensity vs. θ for two separate cases of diffraction with the only difference being the slit width. These intensity distributions can be used to describe the diffraction pattern in each case. Taking a closer look at the graphs in Figure 6.3, you can see that every point on the graph has a corresponding value for θ. The location of points such as the first minimum occurs for different values of θ on both graphs because this angle varies with the slit width according to the formula:

$$a \sin \theta = m\lambda$$

If we want to use one equation to describe the intensity distribution, we must consider the relationship between angle θ in our diffraction formula above and the phase angle φ. The mathematical relationship describing the intensity I of the diffraction pattern is given as:

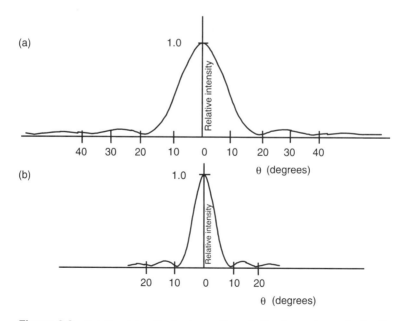

Figure 6.3   Relative intensity for two ratios of a/λ when using a single slit diffraction design.

$$I = I_m [(\sin \beta)/\beta]^2$$

where the angle $\beta = 1/2\varphi = (\pi a/\lambda) \sin \theta$

In the intensity equation, I_m equals the maximum intensity ($\theta = 0°$), β is an angle that equals one half of the phase angle φ and has the relationship to the angle θ as shown. We need to use both of the above equations to determine the variation in intensity of the diffraction pattern at any point. Figure 6.4 shows a plot of this intensity function where the maximum intensity I_m or its maximum value is equal to 1.0. The x-axis displays radian units for angle β. Notice that when $\beta = \pi$, 2π, etc., the relative intensity falls to zero. Conditions of intensity maxima exist approximately halfway between two adjacent minima.

Now, since intensity minima occur when the phase angle is a whole multiple of π (see Figure 6.4), we can also define the mathematical relationship for the condition of intensity minima as:

$$(\pi a/\lambda) \sin \theta = m\pi \quad (m \text{ is a whole number})$$

By rearranging the terms in the above equation, we get:

$$a \sin \theta = m\lambda$$

which is our original mathematical expression to locate the position of minima of the diffraction pattern.

Fraunhofer and Fresnel Diffraction

There are two types of diffraction, Fraunhofer and Fresnel diffraction. The type of diffraction present usually depends upon the distances involved. When the light source and the screen are far apart, $L \gg d^2/\lambda$, Fraunhofer diffraction occurs. Figure 6.2 shows an example of this type of diffraction. In this case, the incident and diffracted wavefronts are effectively plane and parallel.

Fresnel diffraction occurs in the near-field when moving the source and/or the screen close to the slit. Doing this results in wavefronts that are more spherical as they emerge through the slit. This situation makes the calculation of the light intensity at the

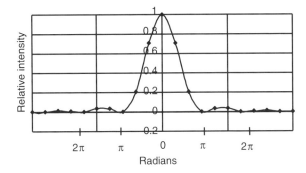

Figure 6.4 Relative intensity as a function of angle β.

points on the screen more difficult. For this reason, only Fraunhofer diffraction will be considered in this chapter. Fresnel diffraction patterns also have little practical use in the kinds of systems we will consider in this book.

6.4 DIFFRACTION OF LIGHT FROM A CIRCULAR APERTURE

In the previous sections of this chapter, we discussed the phenomenon of diffraction using a single slit design. We chose this structure due to its relative simplicity in explaining diffraction as a one-dimensional problem. In this section, we will consider the more common application of diffraction, that from a circular aperture. Circular apertures are found in a large variety of optical devices used for light gathering and light emitting purposes. These apertures may or may not use lens elements. For this reason, it is important to understand the diffraction effects associated with circular apertures.

A diffraction pattern obtained by imaging a distant light source such as a star is shown in Figure 6.5. This figure is a negative; light areas appear dark. You can see that the image of this assumed point source does not show up as an actual point but displays the tell tale signs of a diffraction pattern. This pattern consists of a central bright disk surrounded by several fainter rings or maxima. The central disk is known as the Airy disk. The reason for this pattern has to do with light propagation. Light really does not travel in straight line rays. Instead, light tends to bend around corners and obstructions to a small degree. Geometrical optics only approximates the result. We must consider the wave nature of light in this case. When using a circular aperture, the diffraction pattern produced will be similar to that produced from a single slit. There will be a central maxima in the form of a circle with much fainter rings like maxima as we move away from the center. Since the aperture is a circle, the ratio of its diameter to the wavelength of light (d/λ) determines the diffraction pattern.

The distribution of light energy in the Airy disk and the successive rings is given in Table 6.1. This radial variation of light intensity is an important factor in the design of imaging systems.

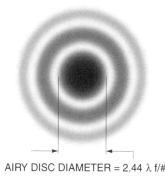

AIRY DISC DIAMETER = 2.44 λ f/#

Figure 6.5 Diffraction pattern for a circular aperture. Reprinted with permission of Melles Griot.

Table 6.1   Distribution of Energy in Diffraction Pattern (Circular Aperture)

Ring	% Energy in Ring
Central Maximum	84.0
1st Bright Ring	7.0
2nd Bright Ring	2.8
3rd Bright Ring	1.5
4th Bright Ring	1.0

As with the single slit design we can also describe the location of the first minimum, or dark ring in this case, by a mathematical expression. For a circular aperture, this expression becomes:

$$\sin \theta = 1.22\ \lambda/d$$

where d is the diameter of the aperture and λ is the wavelength of light used. For comparison, we recall the equation used in the single slit case:

$$\sin \theta = \lambda/a$$

where a is the slit width and λ is the wavelength of light used.

The above equation can be used to determine the location of the first minimum of the diffraction pattern when using a circular aperture. The angular separation defined by this equation produces a condition known as Rayleigh's criterion for resolving two distant objects. To illustrate this condition, we use two distant point sources such as stars. Each of these point sources will produce an Airy disk and successive rings as described in Figure 6.5. When these two points are very close to each other, the diffraction patterns will overlap, making it impossible to distinguish one point from the other. This condition can be seen in Figure 6.6a, where the two points are considered not to be resolved. Figure 6.6b shows the condition for when these two sources are further apart. Here, the angular separation between the two distant point sources barely meets the Rayleigh criterion. In this case, the first minimum of the diffraction pattern of one source falls on the central maximum of the other source. These two objects can now be resolved and have an angular separation, θ_R, given below. The Rayleigh criterion is very important in the study of optics, since it can give an indication of the resolving power of a lens.

$$\theta_R = \sin^{-1}(1.22\lambda/d)$$

Since the angles involved will be very small, the approximation of $\sin \theta \approx \theta$ can be used. Thus, we get an more convenient form of the above equation:

$$\theta_R = 1.22\lambda/d$$

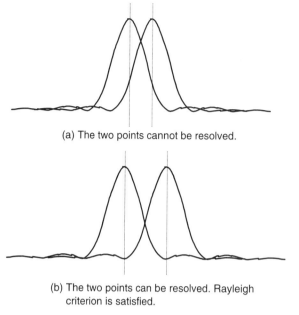

(a) The two points cannot be resolved.

(b) The two points can be resolved. Rayleigh   **Figure 6.6**   Resolving two dis-
    criterion is satisfied.   tant point sources.

We will next consider two examples using the Rayleigh criterion.

Example 6.2

Two distant objects are viewed using a converging lens having a 55 mm diameter. If the wavelength of the light from the objects is 600 nm, what is the minimum angular separation for which these objects can be resolved?

Using the Rayleigh criterion, we get,

$$\theta_R = 1.22\lambda/d = (1.22)\ 6.0 \times 10^{-7}/55 \times 10^{-3} = 1.33 \times 10^{-5} \text{ rad or}$$
$$\theta_R = 2.8 \text{ arc seconds}$$

Example 6.3

A telescope maker needs to know the smallest possible diameter lens that will resolve two point sources at 11 seconds of arc emitting light at 550 nm. Find the diameter of this lens.

We again use the Rayleigh criterion and then rearrange terms to find diameter, d.

$$d = 1.22\lambda/\theta_R = (1.22)(550 \times 10^{-9})/5.28 \times 10^{-5}$$
$$\textit{note:}\ 5.28 \times 10^{-5} \text{ rad} = 11 \text{ seconds of arc}$$

$$d = 12.7 \text{ mm or about } 1/2 \text{ inch}$$

The Rayleigh criterion for diffraction limited performance applies to many other wavelength regions of the electromagnetic spectrum. Table 6.2 lists the diffraction limits

Table 6.2   Examples of the Diffraction-Limited Performance of Radio, Infrared, Optical and Ultraviolet Telescopes

Telescope	Diameter of aperture	Wavelength	Angular resolution λ/D
MPIfRA Effelsberg, West Germany	100 m	74 cm 6 cm	25 arcmin 2 arcmin
James Clerk Maxwell Telescope, Mauna Kea	15 m	1 mm 0.35 mm	14 arcsec 5 arcsec
Infrared Astronomical Satellite (IRAS)	60 cm	100 μm	34 arcsec
ESA Infrared Space Observatory (ISO)	60 cm	10 μm	4 arcsec
UK Infrared Telescope (UKIRT)	3.9 m	10 μm 2.2 μm	0.5 arcsec 0.1 arcsec
Palomar 5 m Telescope	5 m	1 μm 500 μm	0.04 arcsec 0.02 arcsec
NASA-ESA Hubble Space Telescope (HST)	2.4 m	1 μm 500 nm 200 nm 120 nm	0.09 arcsec 0.043 arcsec 0.017 arcsec 0.010 arcsec

Reprinted with permission of Cambridge University Press.

for various telescopes performing in different wavelength regions. These telescopes have angular resolutions for the wavelengths listed. Keep in mind that these values are the theoretical maximum resolutions possible. Adverse atmospheric conditions such as attenuation, dust, and scattering will result in performance somewhat less than the theoretical maximum. Note that the Rayleigh criterion is just one of several ways that angular resolution can be expressed.

Sometimes, we must know the minimum spot size possible when using a particular light source and lens. In this case, the Rayleigh criterion provides the theoretical lower limit to this spot size. Of course, this depends upon the numerical aperture (NA) of the lens and the wavelength of the light being used. The numerical aperture can be defined as the index of refraction of the medium surrounding the image times the sine of the half angle of the cone of illumination, θ. Thus, $NA = n \sin \theta$, where n is usually the index of refraction of air. We discussed numerical aperture in Chapter 4 relative to optical fiber. Another useful parameter of a lens, called relative aperture (f/#), is the focal length divided by the lens diameter. The formula for the minimum diameter of the spot, D, is expressed below when using light of wavelength λ.

$$D = 1.22\lambda/NA \text{ or } D = 2.44\lambda \text{ f/\#}$$

$$\text{where f/\#} = \frac{1}{2}\,NA$$

This means that the spot size varies directly with the wavelength of light used and inversely with the numerical aperture of the lens. Put simply, we can decrease the spot size by decreasing the wavelength or by increasing the numerical aperture. We will next consider a practical example where the spot size is very critical.

Example 6.4

A CD player uses a plastic lens of NA 0.55 to image a beam from a diode laser onto the optical disc to retrieve digital data. This data is in the form of "pits" on the disc that reflect light into a receiver. For effective operation, the size of these pits must be about the same as the spot size of the optical beam. If the diode laser operates at 780 nm, find the lower theoretical limit to the spot size from this beam.

Using the equation for spot size that satisfies the Rayleigh criterion, we get,

$$D = 1.22\lambda/NA = (1.22)(780 \times 10^{-9})/0.55$$
$$D = 1.73 \times 10^{-6} \text{ meter}$$

Example 6.5

A different diode laser is used in the CD player above that operates at 400 nm. Find the theoretical lower limit to this spot size.

$$D = 1.22\lambda/NA = (1.22)(400 \times 10^{-9})/0.55$$
$$D = 8.87 \times 10^{-7} \text{ meter}$$

The results from these two examples tell us that when using a shorter wavelength diode laser, a larger storage density of data can be achieved with CD technology. Most diode lasers used today in CD players emit at a wavelength of 780 nm. A typical 4.75 inch optical disc can store approximately 680 Megabytes of data when operating at this wavelength. By switching to a 400 nm laser, data storage can theoretically be increased to about 2 to 3 Gigabytes for the same size optical disc. Work is currently being done to produce shorter wavelength diode lasers to be used in systems with larger storage capacities.

6.5 DOUBLE SLIT DIFFRACTION

When we introduced Young's double slit interference apparatus in the last chapter, diffraction was required to aid in the interference process. We briefly discussed the phenomenon of diffraction and how the light energy spread out beyond the slits as it emerged. We will now take a closer look at Young's double slit experiment to see the relationship between diffraction and interference. Figure 6.7 shows the double slit design with plane wavefronts approaching the slits. The resultant diffraction pattern forms on Screen 3. For convenience, we make this diagram similar to Figure 6.2 with the exception of using two slits, A and B, instead of one. In this diagram, we consider two sets of rays for the ray constructions. These ray sets, r_1 and r' and r_2 and r'$_2$, are drawn from slits A and B respectively. As in the case of single slit diffraction, the locations of maxima and minima are determined by the optical path difference BX. For the condition of maximum, BX must be an even multiple of a half-wavelength. For the condition of minimum, BX must be an odd multiple of a half-wavelength. If I_m is the maximum illumination as found at the center of Screen 3, the intensity at a particular location on Screen 3 can be expressed by:

$$I = I_m \cos^2 \gamma [(\sin \beta)/\beta]^2$$

$$\text{where } \gamma = (\pi d/\lambda) \sin \theta \text{ and } \beta = (\pi a/\lambda) \sin \theta$$

Right away, we should recognize the $[(\sin \beta)/\beta]^2$ factor as the intensity distribution function for single slit diffraction. For convenience, this distribution is shown again in Figure 6.8a. The other term that we must now consider, $\cos^2 \gamma$, describes the intensity distribution of the interference pattern when using two slits. We should also recognize this $\cos^2$ factor from the intensity equation for interference in the last chapter. This distribution can be seen in Figure 6.8b. These oscillations also appear in Figure 6.8c. The double slit interference pattern becomes shaped by a more gradual intensity variation. The diffraction intensity variation depends upon the width, a, of the slit. To plot the intensity distribution pattern for the double slit interference design previously described, we must take the product of these two terms. Figure 6.8c shows the plot of this intensity distribution function. What we see here is the product of a double slit interference pattern and the diffraction pattern from a single slit. We can see from Figures 6.8b and 6.8c that the actual positions of the interference fringes stay the same. The intensity of each interference fringe depends upon the single slit diffraction pattern.

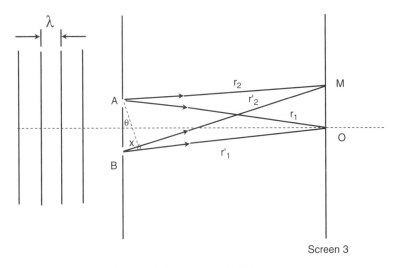

Figure 6.7   Double slit diffraction.

(a)

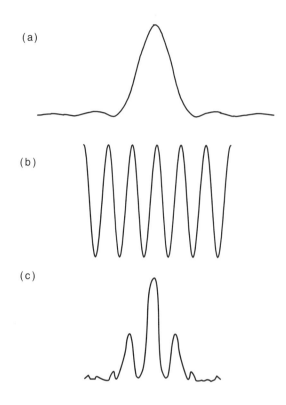

(b)

(c)

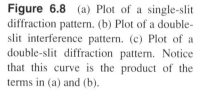

Figure 6.8 (a) Plot of a single-slit diffraction pattern. (b) Plot of a double-slit interference pattern. (c) Plot of a double-slit diffraction pattern. Notice that this curve is the product of the terms in (a) and (b).

Dudley H. Towne summarized the relationship between diffraction and interference in his book *Wave Phenomenon* as follows. The phenomenon is not readily classed as being either "interference" or "diffraction," since both of these effects play an important role. Diffraction is important in that the wave spreads out to a certain extent after it passes through each of the apertures. Interference is important in that destructive and constructive relations exist between the contributions from the individual apertures. Indeed, the distinction between "interference" and "diffraction" is a tenuous one, since the "diffraction pattern" of an individual aperture is the result of interference effects among the various contributions from within the aperture.

6.6 DIFFRACTION GRATINGS—AN IMPORTANT APPLICATION OF DIFFRACTION

We can continue to increase the number of closely spaced slits using ruled lines to produce a diffraction grating. Diffraction gratings usually have thousands of slits or ruled lines per inch. As we increase the number of slits per inch, the fringes become narrower. When we increase the number of slits to 2.3×10^4 per inch, the fringes become sharp enough using this particular configuration to measure the wavelength distribution of the visible light. This type of grating can be used to make a spectroscope for measuring the spectral lines present in certain light sources. To manufacture such a grating, equally spaced parallel grooves are placed on a piece of glass or metal. When the grating is manufactured on a piece of metal, liquid plastic material can be poured onto the surface in a thin layer. As this plastic layer hardens, a transparent finely ruled diffraction grating replica will be produced. In either case, the relationship between the wavelength and the angle of diffraction will be:

$$n\lambda = d \sin \theta, \text{ where n is the order of diffraction and d is the slit separation.}$$

A practical example is given next to illustrate how useful a diffraction grating can be.

Example 6.6

A diffraction grating having 2.3×10^4 grooves per inch is illuminated at normal incidence by a laser beam of emission wavelength 650 nm. Find the angular separation between the normal line and the first order diffracted beam. (25.4 mm = 1 inch)

First, we must calculate the spacing d in MKS units from the information given.

$$d = 25.4 \times 10^{-3} \text{ meter}/2.3 \times 10^4 \text{ grooves/inch}$$
$$d = 11.04 \times 10^{-7} \text{ meter}$$

Since we are interested in the first order diffracted beam, the equation becomes:

$$\lambda/d = \sin \theta$$

$$650 \times 10^{-9}/11.04 \times 10^{-7} = 0.589 = \sin \theta$$

$$\theta = 36.07°$$

Example 6.7

A laser with an emission wavelength of 550 nm is used with the diffraction grating in the last example. What does the diffraction angle change to?

Using the same equation, $\lambda/d = \sin \theta$, we get:

$$550 \times 10^{-9}/11.04 \times 10^{-7} = 0.498 = \sin \theta$$

$$\theta = 29.9°$$

You can see from the results of the last two examples that a change in wavelength of just 100 nm results in a substantial angular difference. This particular diffraction grating is very useful as a dispersive element in a spectroscope. The grating element widely disperses the input visible light, providing for accurate measurements of the spectral components. We will present the details on how to construct and operate a diffraction grating spectroscope in Chapter 9.

Diffraction gratings are also used in certain types of CD players as dispersive elements for the diode laser beam. The laser beam must be split into three separate beams, one zero order beam and two first order beams, for the purposes of data collection, tracking control, and focusing. We will consider this application later on in this book.

6.7 HOLOGRAPHY

Holography involves the application of many of the principles presented in the previous chapters. The discovery of holography in 1948 by Denis Gabor came at a time when the laser did not exist. Gabor was working on a new method to produce a better image from an electron microscope when he made this discovery. At the time, it was known that the image resolution was limited by the spherical aberration of the electron lenses used in the microscope. To demonstrate his new method to those interested, he used an analogy with optical radiation. This method involved making a photograph of an object using two light beams. These two beams of light, one called a reference beam, and the other an object beam, produced phase front information on a photographic plate as a result of the interference between the two beams. He later referred to this method as "wavefront reconstruction."

The light source that he used was not as highly coherent as a laser. Instead, he used the most coherent light source available at the time, a mercury arc lamp. When mercury gas discharges in an arc lamp, many separate wavelengths of light are emitted due to the interaction of radiation and matter. If viewed through a spectroscope, one can see emission lines corresponding to these separate wavelengths. He needed to isolate one of these wavelengths as closely as possible, so he passed the light through a pin hole and then through a filter. This process allowed the use of the emission lines of Hg (mercury), but the coherence length was only about 0.2 mm. The wavelengths used were either violet at 0.44 microns or green at 0.55 microns. Gabor had limited success due to the low coherence quality of the light source that he used.

With the invention of the laser came highly coherent light. The coherent light from the laser provided distinct wavefronts. These wavefronts thus formed sharper interference patterns from the two beams when reproducing Gabor's original experiment. This led to the first successful hologram using laser light in 1960 by Juris Upatnieks and Emmett Leith. Figure 6.9 shows the schematic diagram used to make a hologram.

For high quality holograms, the beam of light from the laser must be spatially coherent. As Figure 6.9 shows, this beam encounters the beam splitter where it becomes separated into two beams. One beam transmits through the splitter, and the other becomes reflected. The transmitted beam, known as the reference beam, goes on to strike the high resolution photographic plate. The wavefronts from this beam are essentially parallel. The portion of the light beam that reflects from the beam splitter goes on to illuminate the object. We have chosen to make a hologram from an apple. For the time being, we will consider only one point on this apple. Diffracted light leaves this point as shown and travels toward the photographic plate. The wavefronts from this point are slightly spherical as they travel toward the photographic plate. When the reference and the object beams meet at the photographic plate, interference occurs. The photographic plate records the interference pattern produced by this interaction. A complete hologram of the object is formed when the contributions from all points diffracted by the object interfere with the reference beam at the location of the photographic plate. Figure 6.10 shows a detail of the wavefronts from the two beams in the vicinity of the photographic plate. The photographic plate will contain information relative to the amplitude and phase of the light due to this interaction. If the object beam was the

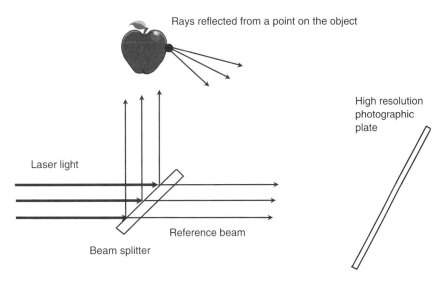

Figure 6.9   Schematic diagram or requirements to produce a hologram.

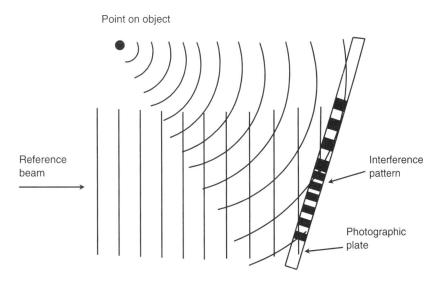

Figure 6.10   Formation of a hologram point object.

only light reaching the photographic plate, then all phase information would be lost. The result would be a typical photograph of the object showing only intensity information.

To reconstruct the image, the developed photographic plate must be illuminated in the original reference beam containing plane waves. A real image can be projected onto a screen by the reference beam. If an observer looks through the transparent photographic plate, a virtual image can also be seen. This image looks like the real object in three dimensions. This three-dimensional quality is the result of the wavefront reconstruction process that produces both intensity and phase information. Thus, as you move relative to the hologram, you will also see the object in a different perspective.

There are some practical considerations that must be observed when making a hologram. The photographic plate must be of high resolution to record the interference patterns produced in this process. During the exposure of the photographic plate, there must be no relative motion or vibrations associated with any of the components used in the design. If relative motion occurs, the interference fringes will not be distinct and may become completely washed out. As for the light intensity, it has been found that if the object beam and reference beam have an illumination ratio of 1:5, this will optimize the reproduction quality. The angle subtended by the reference and object beam at the photographic plate must be carefully chosen. As this angle increases, the interference fringes will come closer together. This means that the resolvance of the photographic plate must increase as this angle increases.

SUMMARY

Diffraction occurs when the aperture or obstacle is comparable to the size of the wavelength of the wave hitting it. To explain how diffraction occurs, the Huygens-Fresnel principle can be used. Every point on a wavefront acts as a point source for a new or secondary wavefront or wavelet. When a wave hits a small aperture, these secondary wavelets will expand at the same speed that the initial wave had. A tangent line drawn to all of these secondary wavelets shows where the next wavefront forms (Figure 6.1).

When waves pass through a narrow slit of width a, a diffraction pattern will be produced. This diffraction pattern will have a location of central maximum illumination surrounded by areas of minimum illumination. These areas of minimum illumination satisfy the equation below (see Figure 6.2).

$$a \sin \theta = m\lambda.$$

Diffraction can also occur when a light wave encounters a circular aperture or lens of diameter d. In this case, the diffraction pattern will consist of a central bright disk surrounded by several fainter rings or maxima (see Figure 6.5). The location of the first minimum or dark ring can be found by using the formula below.

$$\sin \theta = 1.22 \, \lambda/d$$

Diffraction limits the resolution between two distant point sources of light. The resolvability of these two sources can be determined by using Rayleigh's criterion. The angular separation can be calculated by using the formula below.

$$\theta_R = \sin^{-1}(1.22\lambda/d)$$

When using a double slit interference apparatus as describe in Figure 6.7, both interference and diffraction must be taken into account to explain the fringe pattern produced. If I_m is the maximum illumination as found at the center of the screen on which the pattern forms, then the intensity at a particular location on this screen can be expressed by using the following mathematical relationship:

$$I = I_m \cos^2 \gamma[(\sin \beta)/\beta]^2$$

where $\gamma = (\pi d/\lambda) \sin \theta$ and $\beta = (\pi a/\lambda) \sin \theta$

You can see from Figure 6.8 and the above equations that this intensity distribution is the product of a double slit interference pattern and the diffraction pattern from a single slit.

7

Polarization of Light

7.1 POLARIZATION INVOLVES TRANSVERSE WAVES

In Chapter 3, we considered Maxwell's equations for describing how an electromagnetic wave travels through space. This wave consists of electric and magnetic fields mutually perpendicular to each other and changing in intensity as the wave propagates through space. Looking at Figure 3.1 in Chapter 3, the amplitude of the field at each point can be represented by a vector from the z axis to a point on the curve of the electric or magnetic field. In Section 3.2, we considered the equation for the electric field for Figure 3.1 as:

$$E(z,t) = E_0 \sin (kz - \omega t)$$

The electric field oscillates in the yz plane as shown by the E vector, and the vector always points in the y direction. The magnetic field oscillates in the xz plane as shown by the H vector, and this vector always points in the x direction. When the E vector oscillates with a given, constant orientation, the light can be considered linearly polarized. The direction of polarization is traditionally specified by using the electric field orientation. This means that linearly polarized light waves will all have their electric field vectors aligned in the same direction or parallel to each other.

When we studied interference and diffraction effects in the last two chapters, it did not matter whether the waves were longitudinal or transverse. For example, diffraction and interference effects are also observed with longitudinal sound waves. We know from

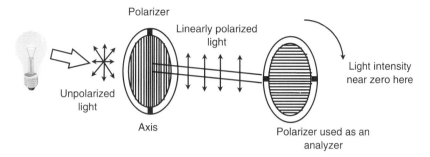

Figure 7.1 Polaroid discs used to analyze linearly polarized light.

Maxwell's equations that an electromagnetic wave has a transverse nature, but how can we prove it? We will describe a relatively simple experiment that proves the transverse wave nature of light and also shows polarization to be a property attributed to transverse waves.

The polarization of a beam of light can be analyzed by using Polaroid discs as shown in the experimental design of Figure 7.1. These discs are made from a material that transmits light in one plane of polarization while absorbing light being polarized in the other planes. A source of unpolarized light such as an incandescent lamp emits light through the first polarizer. The axis of polarization in this case has been set vertically as shown by the mark on the polarizer. This means that, ideally, only light polarized in the vertical plane specified by this polarizer will be transmitted. A second polarizer is used to analyze the light transmitted through the first polarizer. When placing the axis of the second polarizer vertically, the intensity of the light observed will be at a maximum. As we rotate the second polarizer, the intensity of the light observed will gradually decrease to a near zero level when the second polarizer has its axis set horizontally. This effect occurs because the linearly polarized light from the first polarizer cannot be transmitted through the second polarizer that has its axis 90° rotated from the first polarizer. Since light energy propagates through space as a transverse wave, we can see why the intensity drops to near zero in this experiment.

7.2 POLARIZATION CAUSED BY DOUBLE REFRACTION

Another interesting demonstration can be performed to reveal the transverse nature of light. This demonstration involves using a crystal called calcite having two indices of refraction. Christiaan Huygens investigated the polarization of light in 1678 by using a piece of calcite as an optical transmission medium. He caused a beam of unpolarized light to be incident upon the face of a calcite crystal similar to the design in Figure 7.2. Upon striking the surface of the crystal, the beam splits into two parts. This occurs because calcite has two indices of refraction determined by the orientation of the light waves within the crystal. Up to this point, we have studied optical media with the property of being optically isotropic. This means that the refractive index stays the same, independent of the direction in which the light beam travels through the medium. Due to calcite's crystalline lattice structure, light will experience two different speeds, thus making it an optically anisotropic material. Light falling upon the crystal will be split into two rays, one ray called the ordinary ray (o), and the other ray called the extraordinary ray (e). The waves making up the ordinary ray all travel with the same speed and their direction of propagation can be defined by Snell's law. Since the crystal has two indices of refraction, there will be another possible direction of propagation. The waves making up this ray are called the extraordinary ray. The direction of propagation for the extraordinary ray will not obey Snell's law. This fact can be seen from Figure 7.2. The extraordinary ray formed from the incident light does not have a zero angle of refraction as should be the case for a ray at normal incidence. Sometimes, the extraordinary ray will not lie in the same plane as the incident angle. The two beams will thus be spatially separated.

Another interesting result found from the investigation of the two beams concerns the type of polarization. When the two beams emerge from the crystal, they will be linearly polarized at right angles to each other. This allows the generation of linearly polarized light from randomly polarized light.

If we place a thick piece of calcite on a printed page as shown in Figure 7.3, we will see two images, one polarized 90° relative to the other. This causes the printed page under

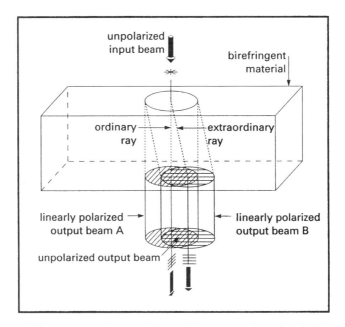

Figure 7.2 Double refraction of a birefringent crystal. Notice that beams A and B emerge linearly polarized and orthogonal to each other. Reprinted with permission of Melles Griot.

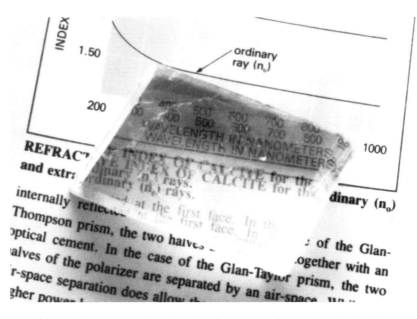

Figure 7.3   A piece of calcite placed on a printed page. Reprinted with permission of Melles Griot.

the crystal to appear double. Calcite and a number of other crystalline materials exhibit this property of double refraction or birefringence.

7.3   CIRCULAR POLARIZATION

For the case of circularly polarized light, the electric field vector no longer oscillates in one plane only as it propagates through space. One way to produce circularly polarized light is to combine two linearly polarized waves together as shown in Figure 7.4. In this figure, we use vectors to represent the magnitude and direction of each electric field. As the figure shows, both waves must have the same magnitudes or $|E_X| = |E_Y|$ and a phase difference of 90°. When one electric field vector reaches its maximum value, the other has a magnitude of zero. From the motion of the two waves, we get a resultant vector that also travels along the z axis. This resultant vector will have the property of a circularly polarized wave. The motion of the tip of the resultant vector in Figure 7.4 makes a helix as it propagates along the z axis. If you could view this motion from a fixed point such as z = 0, the vector tip will form a circle as it propagates away from you.

We can use a piece of calcite crystal to produce light having the property of circular polarization described above. In this case, the calcite crystal must be cut such that its optical axis lies in the plane parallel to the face of the slab as shown in Figure 7.5. We orient the crystal such that linearly polarized light strikes the crystal's plane of vibration at 45°.

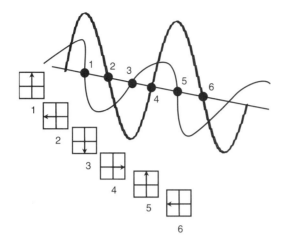

Figure 7.4 Circularly polarized light wave produced as a result of two linearly polarized waves having a phase difference of 90°. The vector at each position shown gives the resultant amplitude and direction.

By doing this, the two rays emerging from the other side of the crystal will have equal amplitudes. They will also be polarized at right angles to each other. Since this material has two refractive indices, the two light waves will travel at different speeds in the calcite. This differential in speed will result in one wavefront being ahead of the other as they propagate through the crystal. We can use this fact to provide the wavefronts with a given phase difference as they exit the other side of the crystal. To accomplish this, we cut the crystal such that its thickness causes the phase of one wavefront to be 90° ahead of the other wavefront as it emerges from the crystal. By doing so, we have just made a quarter-wave plate. A quarter-wave plate, by its thickness and birefringence, causes the two components of the light to be out of phase by 90° for a particular wavelength. The resultant wave will have the characteristic of circular polarization. The orientation of the crystal's

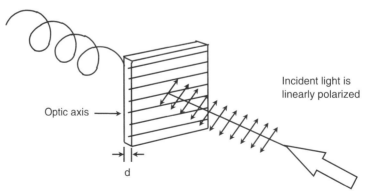

Figure 7.5 A quarter-wave plate can be used to produce circularly polarized light from linearly polarized light.

structure must be such that the higher refractive index, n_1 or the slow axis, is at right angles to the lower refractive index, n_2 or the fast axis, in the plane of the slab. This thickness can be determined by the following mathematical expression:

$$d = \lambda/4(n_1 - n_2)$$

In this expression λ is the wavelength of the incident light in a vacuum. Figure 7.4 shows the two components of circularly polarized light as they leave the calcite crystal. We consider here only the electric field components of each wave. These components have a phase difference of 90° and the same amplitude E_0. Since the waves are also polarized orthogonally to each other and travel in the z direction, the electric field components of the wave are:

$$\hat{i}E_0 \cos (kz - \omega t) \text{ along the x-axis}$$

and

$$\hat{j}E_0 \sin (kz - \omega t) \text{ along the y-axis}$$

Thus, the vector sum of these two fields becomes:

$$E = E_0[\hat{i} \cos (kz - \omega t) + \hat{j} \sin (kz - \omega t)]$$

The above equation describes a single wave having the property of circular polarization. We also use $\hat{i}$ and $\hat{j}$ as unit vectors in the direction of the x and y axes respectively.

The quarter-wave plate finds applications as an optical isolator component and with electro-optic modulators. Sometimes an isolator is required to prevent the reflected portion of an optical beam from getting back into a laser. If this happens, the laser may become unstable. A quarter-wave plate is used in certain types of audio CD players for this reason. In practice, it is used with other optical components to prevent reflected light at the surface of the CD from returning back into the diode laser. The optical isolator allows the transmission of only a specific polarization state. Quarter-wave plates can be used to change the polarization state of a light beam so it cannot not pass back through an optical element after reflection has occurred.

Example 7.1

(a) We wish to make a quarter-wave plate from a piece of calcite. If the crystal is cut such that the two refractive indices are 1.658 and 1.486, what thickness must it be for light of wavelength 600 nm?

Using the above equation for the thickness of a quarter wave plate, we get

$$d = 600 \times 10^{-9}/(4)(1.658 - 1.486)$$
$$d = 8.72 \times 10^{-7} \text{ meter}$$

(b) You can see that the above thickness is rather small. Thus, most quarter wave plates are made from birefringent materials having close values for refractive indices. Mica and quartz are good examples. In this example, we will show why. Mica is usually split in an empirical fashion to find the correct thickness. If the two refractive indices for mica are 1.588 and 1.582, find the thickness needed to make a quarter-wave plate when using light of wavelength 600 nm.

Air space

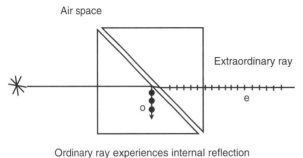

Extraordinary ray

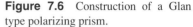

e

o

Ordinary ray experiences internal reflection

Figure 7.6 Construction of a Glan type polarizing prism.

We again use the quarter-wave plate formula for thickness. This time, the two refractive indices are very close to the same values.

$$d = 600 \times 10^{-9}/(4)(1.588 - 1.582)$$
$$d = 0.025 \text{ mm}$$

Calcite can be used as an optical material to produce just one output beam of light having a linear polarization. As you remember, when light enters the calcite crystal, there will be an ordinary and an extraordinary ray. Each ray has an associated refractive index. The crystal can be cut such that only the ordinary ray experiences total internal reflection. In this case, only the extraordinary ray will be left to travel through the crystal. Figure 7.6 shows how this can be done using a piece of calcite. This optical component, known as a Glan type prism, must satisfy the condition below for transmission of only the extraordinary ray.

$$n_E < 1/\sin \theta < n_O$$

The Glan prism is made from two identical prisms of calcite cemented together. Another possible configuration has an air space between them instead of cement. Typical Glan type prisms can separate the extraordinary ray as described above over the wavelength range from ultraviolet to the infrared. This is possible since the index of refraction for the ordinary ray is significantly greater than for the extraordinary ray.

7.4 ELECTRO-OPTIC MODULATORS

We have seen that certain crystals possess a property called birefringence or two refractive indices. This causes light to travel in two possible linear polarized modes at two different speeds. The two rays of light associated with these modes are the ordinary and the extraordinary rays. In these crystals, the refractive indices are fixed.

It has been found for certain crystals that the application of an electric field will cause the refractive indices to vary proportionally to the field's amplitude. This effect, known as the electro-optic effect, can be demonstrated in crystals such as potassium dihydrogen phosphate (KDP) and lithium niobate ($LiNbO_3$). These crystals find a useful application in optical modulators based upon changing the polarization state of the incident light. In practice, the electric field is usually applied normal to the optical axis of the crystal or parallel to it. The effect is relatively small even for crystals with large electro-optic coefficients. For example, a change in refractive index of about 0.01% results when applying an electric field of 10^6 V/m to a lithium niobate crystal. Changing the refractive indices causes differences in velocities between the two wave components. A relatively large electric field must be used to produce a significant refractive index change in bulk type devices. In general, the modulation of light involves inducing changes in one or more properties of the light itself. Variations in intensity, frequency, phase, or polarization are common methods used in communications and many other practical electro-optical devices.

In this section, we will consider two types of optical modulators based upon the electro-optic effect. For the first type, we will explain how this effect produces the desired modulation of a plane polarized light beam. For the second type, we will use these principles of operation to study an integrated-optical modulator. Integrated-optical modulators are used in many high performance lightwave systems.

Bulk Modulators

Figure 7.7 shows the basic components used in the construction of an intensity bulk modulator employing the electro-optic effect. A crystal piece of KDP is placed between two crossed polarizers such that its optical axis lies along the z axis or light path. The two polarizers are orientated as shown in the diagram. The orientation of the crystal birefringent axes must be 45° to the x and y axes. This forms two electrically induced birefringent axes, x' and y', corresponding to the two allowed directions of polarization. Ring electrodes made from gold or silver can be vacuum deposited at the ends of the crystal. This allows light to travel through the crystal without being blocked by typical wire electrodes.

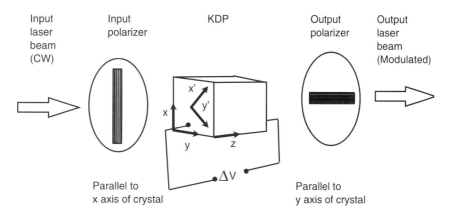

Figure 7.7 Amplitude modulator. The electric field is biased parallel to the optical wave propagation.

Wires attached to these electrodes allow a voltage source to produce a longitudinal electric field within the crystal.

With no applied voltage to the crystal, the light beam passes through the first polarizer to become linearly polarized along the x axis. After this, the light beam passes through the crystal and becomes absorbed by the second polarizer. This condition provides the completely "off " state. If we apply a voltage to the electrodes, an electric field will be generated along the z axis. This field changes the indices of refraction along the two axes of the crystal (x' and y' axes) by different amounts. The amount of this change depends upon the amplitude of the applied voltage. As the electric field increases, this difference also increases. This causes a phase change between the two wave components along the x' and y' axes. The phase change alters the polarization state of the emerging light beam. At the left side of the crystal (z = 0), the x-axis polarization from the first polarizer will split into equal x' and y' components that travel through the crystal at different speeds.

The phase difference at the output of the crystal depends upon the amount of retardation that occurs along the crystal's length. For example, if enough voltage is applied to produce a phase shift of 90° ($\pi/2$) between the two components, the light emerging will be circularly polarized. The output intensity in this case will be 1/2 of the maximum value. This condition can be compared to the quarter-wave plate considered earlier. Finally, when applying enough voltage to produce a 180° (π) phase shift, light linearly polarized in the y direction will occur at the output of the crystal. The output intensity in this case will be the maximum possible since the polarization of the light corresponds to the acceptance position of the second polarizer. The voltage level required for this to occur is known as the half-wave voltage, V_π. This condition provides the completely "on" state. All other voltage levels between zero and V_π will result in a variable output intensity from the input light source. The transmitted power will follow the variations in the input voltage. This process results in amplitude modulation.

The process of amplitude modulation is shown graphically in Figure 7.8. Here we consider the % transmission of light vs. the applied input voltage to the crystal. If we vary the input voltage source in a sinusoidal fashion, the electric field will also vary in the same way. Notice that we only use the linear portion of the curve. This allows for a more faithful reproduction of any input voltage that stays within the vertical dashed lines. To stay within this region, the crystal must be biased at the 50% transmission point. A constant voltage must be applied to the electrodes to keep it at this 50% point. Biasing the input voltage to this point allows intensity changes above and below this reference level. This reference level turns out to be one-half of the half-wave voltage or $V_\pi/2$. Another way to bias the device at the 50% transmission level can be accomplished by placing a quarter-wave plate between the second polarizer and the crystal. This quarter-wave plate provides the required amount of fixed retardation ($\pi/2$). The plate optically biases the light beam without using an applied voltage to the KDP crystal. It must be positioned such that the two principal axes of the plate lie along the x' and y' directions. Specifically, its "fast" axis lies parallel to the x' axis and its "slow" axis lies parallel to the y' axis of the crystal. The use of this quarter-wave plate has the added benefit of lowering V_π by a factor of one half.

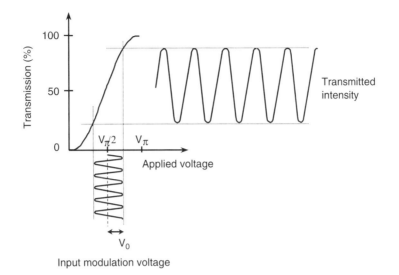

Figure 7.8 Optical power transmission as a function of applied voltage for an amplitude modulator.

As shown, an input sine wave signal voltage, $V_0 \mathrm{Sin}\, \omega_m t$, causes the electric field to vary accordingly. If we operate on the linear portion of the curve as described above, the input voltage, V_0, can be varied electrically to produce a modulated optical light beam. This external modulation of light can provide more dynamic range and linearity than when electrically modulating a diode laser. Another advantage realized by modulating light in this fashion is the elimination of wavelength chirp. Intensity modulation of a diode laser requires the input current level to change abruptly during the modulation process. As a diode laser's input current changes, its output wavelength also changes slightly resulting in more wavelength components. This causes a dispersion penalty in optical fiber. Since the laser emits light continuously during the external modulation process, wavelength chirp does not occur. This provides a more stable modulated light beam that can travel further along an optical fiber due to its narrow spectral width.

Finally, we list the practical limitations of this type of modulator. Photorefractive damage can occur if too much optical power enters the crystal. The optical intensity must be kept below specified limits to avoid damaging the crystal. The frequency response for most bulk type modulators is from DC to about 100 MHz. A primary cause for this upper limit involves the capacitance of the crystal.

Integrated-Optic Modulators

Another device based upon the electro-optic effect finds very important applications in fiber optic communications. The demand for increased information transfer has made this device a practical alternative. Transmission rates of up to 10 Gbits/sec are possible. This device uses lithium niobate in its construction. As with the bulk type modulator discussed previously, an application of an electric field will change the refractive index of this material. Figure 7.9 shows a simplified model of a dielectric waveguide structure consisting of one optical path with electrodes along a given length, L. Applying a voltage to the electrodes will produce an electric field. In this case, light traveling through this material will experience a propagation delay, Δt, based upon the amount of refractive index change. The amount of propagation delay can be determined by using the following equation:

$$\Delta t = (\Delta nL)/c$$

Δn = change in refractive index due to applied electric field
L   = path length traveled by light during interaction of electric field
c   = speed of light

As with the bulk type electro-optic modulator, the propagation delay will cause a phase change at the output when using plane polarized light. This phase change can be determined by multiplying the propagation delay by ω, since $\varphi = \omega \Delta t$. Substituting the expression for Δt into the equation for phase change, we get:

$$\varphi = (\Delta n\omega L)/c$$

One type of integrated-optic modulator is fabricated by placing a waveguide pattern onto the surface of a lithium niobate substrate using photolithography. Figure 7.10 shows an amplitude modulator structure using a Mach-Zehnder interferometer pattern. This structure has been optimized for single mode operation. Thus, single mode fibers attach to the input and output ports of the modulator. Electrodes positioned as shown set up the electric field required to produce the refractive index change. Amplitude modulation occurs as the electric field changes. Since the size of single mode optical fiber is relatively small when compared with a bulk type modulator, the size requirement becomes much less. This smaller size requirement allows for faster switching speeds up to approximately 10 Gbits/sec. To assure that the input plane polarized light will have the correct orientation, polarization maintaining fiber must be used at the optical input.

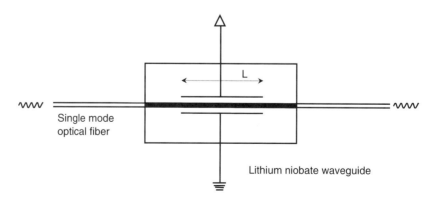

Figure 7.9 A single waveguide structure.

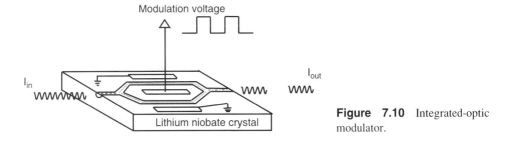

Figure 7.10 Integrated-optic modulator.

Amplitude modulation occurs by controlling the two parallel optical paths with applied voltages at the center terminal. Once plane polarized light enters the device as shown, it splits into two phase modulator sections of equal length. With no applied electric field, the light from each section recombines into a single mode output at the "Y" junction to enter the fiber. The two path lengths contribute to a maximum intensity output since both guided modes are in phase. When applying a voltage to the contacts, the resultant electric field sets up along the path lengths. This electric field will cause a phase change to occur as described by the last formula. A net phase change of 180° between the paths results in total extinction of the light due to destructive interference between the two guided modes. Using this process, the output varies with the applied voltage. In practice, the side electrodes connect to ground. When applying voltage to the center electrode, the resultant electric fields will have opposite polarity across the two optical paths. This results in twice the amount of net phase change from just one path.

SUMMARY

Maxwell's equations describe how an electromagnetic wave propagates through free space at a common speed c. This transverse wave consists of electric and magnetic fields mutually perpendicular to each other and changing in intensity as the wave propagates. The electric field vector can be described by the following equation (see Figure 3.1):

$$E(z,t) = E_0 \sin (kz - \omega t)$$

As Figure 3.1 shows, the electric field oscillates in the yz plane as shown by the E vector, while the magnetic field oscillates in the xz plane as shown by the H vector. If the E vector oscillates with a given, constant orientation, the light can be considered linearly polarized. The direction of polarization is traditionally specified by using the electric field orientation.

The transverse wave nature of light can be demonstrated by shining an unpolarized beam of light onto a calcite crystal. This experiment was done by Christiaan Huygens in 1678. Figure 7.2 shows what happens to this light ray as it propagates through the calcite crystal. The beam splits into two separate parts because calcite has two indices of refraction (it is birefringent). One ray is called the ordinary ray because it obeys Snell's law of refraction. The other ray is called the extraordinary ray because it does not obey Snell's law. When the two emerging beams are analyzed, they are found to be both linearly polarized at right angles to each other.

Circularly polarized light is another state of polarization. The electric field vector for circularly polarized light no longer oscillates in one plane as it propagates. One way to produce circularly polarized light is to combine the two linearly polarized waves together as shown in Figure 7.4. The tip of the resultant vector draws out a helical pattern as it propagates along the z axis.

A piece of calcite can be used to produce light having the property of circular polarization. The crystal must be cut such that its optical axis lies in the plane parallel to the face of the slab. This cut is shown in Figure 7.5. The thickness of the crystal must be determined such that one of the light waves will emerge from the crystal ahead of the other one by a phase angle of 90°. If the crystal is cut in this manner, a quarter-wave plate has just been produced. The thickness of this plate can be determined by the formula below (see Figure 7.5). When linearly polarized light propagates through this plate, it will emerge circularly polarized.

$$d = \lambda/4(n_1 - n_2)$$

Quarter-wave plates find very useful applications in optical isolators and in certain types of audio CD players. The way that the plate works goes as follows. If the circularly polarized light produced from the quarter-wave plate as described above reflects back through the quarter-wave plate, it will become linearly polarized again. This time, the axis of polarization will be 90° from the original wave. Thus, by using a simple polarizing filter or prism, the light is easily prevented from passing back to its source.

8

Light and Thermal Radiation

8.1 MEASUREMENT OF LIGHT

Up to this point, we have been concerned mostly with the propagation of light. We looked at light as a ray, and then as a wave in the areas of study known as geometrical and wave optics, respectively. In Chapter 4, we looked briefly at optical power described in units of milliwatts. Then, in Chapters 5 and 6, we spoke of light intensity in relative units only. Before we can go further in our study of electromagnetic radiation, some measurement terminology must be introduced. There are specific terms used in the measurement of quantities such as light intensity, optical power, etc. In the real world, optoelectronic devices are specified using the terms of measurement from radiometry and photometry. The science of radiometry concerns the measurement of light energy of any wavelength, while the science of photometry concerns light measurement in the visible portion of the spectrum. You might say that photometry is a special case of radiometry; it is concerned with the portion of the spectrum sensitive to the human eye. Since radiometry is the general case, we will consider this area first. The photometric equivalent measurement parameter will then be discussed after the radiometric parameter for each topic section below.

Power is the key parameter used in all radiometric and photometric measurements. The properties of light that will be discussed using radiated or received power include its spatial and angular distributions. Consistent with the traditional study of radiometry, we will treat the propagation of light using the laws of geometrical optics.

Radiant Power/Luminous Flux

This basic radiometric measurement can be thought of as the total radiant power or flux emitted, transferred, or received. Since power is being measured here, the unit of measurement is the watt. Flux implies the time rate of flow of energy. The relationship between power and energy can be summarized by:

$$1 \text{ watt} = 1 \text{ joule/second}$$

The photometric equivalent to radiant power, luminous flux, has the unit of measurement known as the lumen. Luminous flux can be determined by weighing the spectral radiant flux against a visual response function. This function describes the relative sensitivity of the human eye to light within the visible wavelength range.

Now that we have defined power, the common parameter used in all radiometric and photometric measurements, the remaining concepts can be discussed. For example, the radiometric concepts of emittance and radiant intensity can be defined by the density of radiant flux per unit area and per steradian, respectively.

Radiant Emittance/Luminous Emittance

If we measure the density of radiant flux leaving the surface of a radiation source per unit area, we will have its radiant emittance. The units used here are watts/m^2. The equivalent photometric measurement is luminous emittance given in lumens/m^2. You may also find the term exitance used here.

Radiant Intensity/Luminous Intensity

The intensity in these two cases involves the amount of power or flux emitted from a point source within a solid angle. A solid angle can be thought of as the apex angle of a cone formed at the center of a sphere of radius R by an area A on the surface of the

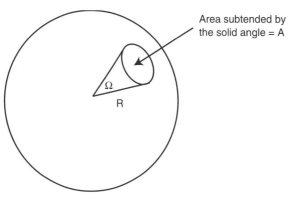

Area subtended by
the solid angle = A

Ω

R

Angle Ω = A/R^2

Figure 8.1 Solid angle measurement.

sphere. As shown in Figure 8.1, the area on the sphere subtended by this solid angle will be a circle of area $A = \pi r^2$. When the area of the circle becomes equal to the square of the radius of the sphere, the solid angle can be defined as one steradian (sr). This corresponds to a cone having an apex angle of about 65.5°. There are 4π steradians of solid angle about the center of the complete sphere. The units of radiant intensity are watts/sr. The photometric equivalent is given by lumens/sr or candela.

Radiance/Luminance

Radiance can be thought of as radiant intensity per unit *projected* area. To define this measurement, we must use both a solid angle and a unit area. Imagine a plane perpendicular to all rays traveling in a particular direction. This light source projects its energy onto this plane. If the plane is at some angle θ from the normal, then the projected area is $A \cos \theta$, where A is the area projected when $\theta = 0°$ (see the figure in Table 8.1). For example, the radiance will be the greatest at 0° than at other angles due to the cosine function. Radiance is the radiant flux transmitted at angle θ to the normal per unit surface area, per unit solid angle. The units of radiance are watts/$(m^2 \cdot sr)$. The photometric equivalent is given by cd/m^2.

Table 8.1

Radiometric Term	Photometric Term	Definition
Radiant Flux (Watts)	Luminous Flux (Lumens)	Total optical power of flux.
Radiant Emittance (Watts/m^2)	Luminous Emittance (lm/m^2) or lux	Flux per unit area from an emitting surface.
Radiant Intensity (Watts/sr)	Luminous Intensity (lm/sr) or candelas	Flux per unit solid angle from a point source.
Radiance (Watts/$m^2 \cdot sr$)	Luminance (cd/m^2)	Flux density per unit solid angle. Emitting surface at angle θ with respect to normal.
Irradiance (Watts/m^2)	Illuminance (lm/m^2) or lux	Incident flux per unit area.

Irradiance/Illuminance

If we now consider the case when the radiation is incident upon a surface, we have irradiance which is measured in the same units as radiant emittance, W/m^2. The unit of illuminance is the lux.

The above measurements are accurate for point sources. Unfortunately, in the real world, point sources do not exist. So, when we apply these measurements to sources such as LEDs, it is an average measurement. Table 8.1 summarizes the radiometric and photometric terms discussed above.

Example 8.1

A light source emits three watts of optical power into a hemisphere. Find the radiant intensity of this light source.

Solution: The problem asks for the radiant intensity that has units of watts per steradian. We know that a sphere has 4π steradians, thus a hemisphere must have 2π steradians. Now we can set up the mathematical solution as:

$$3 \text{ watts (total)}/2\pi \text{ steradians} = 480 \text{ milliwatts/steradian}$$

8.2 BLACKBODY RADIATION

So far, we have studied electromagnetic radiation in the visible and infrared regions of the spectrum by considering phenomena such as refraction, interference and diffraction. These phenomena can be explained with the help of Maxwell's equations that describe the wave nature of light. As you may remember, these equations were discussed briefly in Chapter 3. In that chapter, we also mentioned that light has both wave- and particle-like natures. We deliberately put off discussing light's particle-like nature until we could consider the many examples and situations involving the wave nature of light. Now the time has come to consider the particle nature of light. We will start this study by looking at evidence showing light energy to be quantized or occurring in discrete amounts. This concept directly contradicts the classical wave theory that requires light energy to be continuously variable.

Up until the later part of the nineteenth century, classical physics could explain the observations concerning light phenomena. This changed when scientists were confronted with situations that had no explanation in classical physics. We will consider two such situations that posed a problem for the scientists at the time. The understanding of these situations, plus others that will be presented in the next chapter, helped in the development of the quantum theory in the 1920s.

As we all know, hot objects emit long wave radiation as heat. For example, we can feel long wave infrared radiation emitted from a hot stove. In fact, every object not at the temperature of absolute zero emits electromagnetic radiation. The random motion of molecules contributes to this emission. On an even smaller scale, electrons in motion about the nucleus produce electromagnetic radiation. By the turn of the twentieth century, scientists found a relationship between the surface temperature of an object and the spectral distribution of its radiated energy. This relationship holds true for all objects in thermodynamic equilibrium.

An object can achieve thermodynamic equilibrium when placed in an enclosure having perfect thermal insulating qualities. This will ensure that radiation does not pass through its walls. In this state, the properties of matter and energy simplify considerably. Laboratory experiments performed by placing an object in a special hollow cavity with insulating walls can come very close to this requirement. A small viewing hole placed in this cavity provides access to the emitted energy. The object is heated to a given temperature. When this object reaches thermal equilibrium with the walls of the cavity, a measurement of the emitted energy is taken. Figure 8.2 shows the spectral distribution of the radiation at one particular temperature. This graph was obtained by measuring the intensity of the radiation emitted at various wavelengths. By studying the radiation emitted from an object in thermodynamic equilibrium, we come much closer to understanding the relationship between matter and energy. This relationship is crucial to understanding how optoelectronic devices convert light into electrical energy and visa versa.

The spectral radiance of an object in thermodynamic equilibrium depends upon the object's surface temperature in Kelvins, and the wavelength of the radiation. The characteristic curve shape shown in Figure 8.2 holds true for all matter in thermodynamic equilibrium. More specifically, this relationship holds true for objects emitting as a blackbody. A blackbody is a perfect emitter of radiation. When such an object is heated to a given temperature, it will emit the maximum amount of radiation possible at every wavelength. Starting from the right side of the graph in Figure 8.2, you can see that as the wavelength decreases, the intensity increases for a while. The intensity then peaks at a certain wavelength, and then decreases to a minimum value. To produce a practical blackbody, place a small hole in a closed cavity such as a metal barrel. The radiation streaming through this hole will obey the blackbody radiation relationship if the walls of the container are maintained at a given temperature. When the temperature of the object becomes low enough, such as when it is at room temperature, very little visible light will be emitted. This is why we call this type of radiation blackbody radiation.

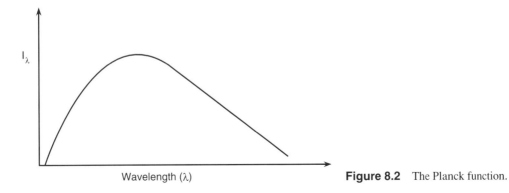

Wavelength (λ)   **Figure 8.2** The Planck function.

Next, consider an object heated to different temperatures in steps. At each step or interval, this object reaches thermodynamic equilibrium. When thermodynamic equilibrium occurs, measurements similar to those performed to produce the curve in Figure 8.2 are taken. We will then have a blackbody curve for each temperature. The general curve shape will be the same for each temperature, but the curve's peak and x axis position will change. Figure 8.3 shows this data plotted for different temperatures. As the temperature increases, the maximum intensity also increases, causing the peak point to move toward shorter wavelengths. The overall shape of each curve stays the same with no intersections occurring as the temperature changes.

A simple example will help to add some insight into what these curves mean. Imagine a piece of iron heated in a fire capable of attaining several hundred degrees Fahrenheit. The iron will absorb heat from the fire. At first, the iron piece will emit only infrared radiation that we can feel but not see. As the temperature rises, the iron piece starts glowing dark red. At this point, it is emitting energy described by a wider spectral distribution that includes the visible red portion. As the temperature increases further, the iron piece changes visibly from dark red to bright red, and then finally to white hot. The white hot piece is now emitting even more energy. This situation can be described by an even wider spectral distribution. The spectral distribution of the white hot piece of iron also shifts its

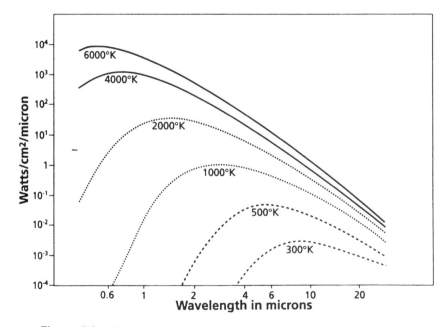

Figure 8.3 Blackbody spectral radiant emittance for various temperatures. Reprinted with permission of Cal Sensors.

maximum intensity toward shorter wavelengths as the temperature increases. This shift eventually causes the iron piece to appear light blue or white. While this example does not describe a perfect blackbody radiator, it illustrates the important concept involving surface temperature, spectral distribution, and the total amount of radiated energy.

8.3 PLANCK'S FORMULA FOR BLACKBODY RADIATION

Unfortunately, the results shown in Figures 8.2 and 8.3 cannot be explained by using the classical theory of a system in thermodynamic equilibrium. In the early 1900s, two scientists, Lord Rayleigh and Sir James Jeans, tried to explain this spectral energy distribution by using the concept of modes. They envisioned the energy in a blackbody cavity consisting of modes or standing waves that exist between the sides of the cavity. Each mode had a specific frequency or wavelength, and could emit or absorb energy continuously. Rayleigh then made the assumption that there must be an equipartition of energy among the modes. In this case, the energy associated with each mode must then be equal to kT, where k is Boltzmann's constant, and T is the absolute temperature. Next, he worked out a mathematical expression to fit the spectral energy distribution curve. In his mathematical expression, the spectral energy increased by the square of the frequency (v^2). This result worked well for the longer wavelengths, but did not agree with the experimental results obtained for the shorter wavelengths where the curve peaks and then decreases in intensity. As a matter of fact, Rayleigh's expression predicted that at very high frequencies, the total energy emitted should be infinite (remember the v^2 dependency). This of course cannot occur, and certainly was not observed. Being a clear contradiction of classical radiation theory, his approach was later known as the "ultraviolet catastrophe." This classical approach, in essence, was based upon the emission of vibrating electrical charges as described by the electromagnetic wave theory.

In October of 1900, the German physicist Max Planck came up with a mathematical relationship that could explain the results of Figure 8.2. He basically combined the result of Rayleigh's mathematical expression for longer wavelengths with another expression developed by a scientist named Wilhelm Wein. Wein's results worked well for shorter wavelengths but only approximated the observations obtained at longer wavelengths. The final equation that Planck obtained fit the curve nicely. The equation that he obtained is given below:

$$I_\lambda = \frac{2\pi hc}{\lambda^5} \frac{1}{\left(e^{hc/\lambda kT} - 1\right)}$$

In this mathematical expression, h is a constant of proportionality now called Planck's constant, c is the speed of light, λ is the wavelength of the radiation, k is Boltzmann's constant, and T is the absolute temperature. The intensity of the radiation at a particular wavelength is I_λ (spectral radiance). We can assign the units of power per unit area per unit wavelength interval to this parameter.

Since this formula was obtained empirically, Planck sought an explanation as to its physical basis. Classical wave mechanics could not give any insight into the strong suppression of the shorter wavelengths displayed by the blackbody distribution curve. This suppression can be seen in all of the curves in Figure 8.3. It starts to occur at the peak in each curve. As we showed before, this peak shifts toward longer wavelengths as the object cools. Consider our previous example of the hot iron piece. If we let the temperature decrease enough, there will be virtually no radiation in the visible portion of the spectrum. The height or intensity of the peak will also decrease, and subsequently, the spectral distribution will also change.

A comparison of the curves in Figure 8.3 shows that a maximum wavelength value, λ_{max}, is obtained for each curve. It turns out that λ_{max} has a direct relationship to temperature. You can see that as the temperature increases, the peak wavelength decreases. Each curve displays a shift to the left of the previous lower temperature curve. This relationship has some very useful applications and works out to be the simple relationship shown below:

$$\lambda_{max} = \frac{2898(\mu m)}{T(K)}$$

When using temperature in Kelvins, the wavelength will be given in micrometers (μm). We will consider two examples using the above relationship.

Example 8.2

Using the Planck function, find the wavelength λ_{max} at which a tungsten filament radiates most strongly when it has an absolute temperature of 1800 K.

Using the above formula, we make the following substitutions:

$$\lambda_{max} = 2898/1800 = 1.61 \; \mu m$$

This wavelength is in the infrared portion of the spectrum.

Example 8.3

The surface temperature of the human body is 35° C or 308 K. Find the wavelength that radiates most strongly at this temperature.

Using the above formula, we make the following substitutions:

$$\lambda_{max} = 2898/308 = 9.4 \; \mu m$$

The above result tells us that λ_{max} occurs in the infrared portion of the spectrum. A very important application utilizing this area of the spectrum can be found in burglar alarms and motion detectors. Passive infrared sensors detect the motion of a human body. They are optimized to respond to electromagnetic energy emitted at this temperature. In theory, they should work fine, but in practice, some precautions must be observed. You can readily see that many other objects also emit strongly in this same portion of the infrared. Convection heat, house pets, and windows heated by the sun can cause the passive infrared sensor to respond to the same signal as that emitted from a human body. To reduce the likelihood of this occurrence, the motion detector usually employs a second type of sensor based upon microwave technology that also detects motion. When using both sensors in one device package, the likelihood of unwanted activation due to other sources can be substantially reduced. Other practical devices that operate on the principle described by Example 8.3 include infrared thermometers and imaging arrays.

The Planck formula will give the intensity of the radiation at one particular wavelength. But, sometimes it may be useful to know the total radiation emitted due to all wavelengths. This can be found mathematically by using the process of integration. Integration, in this case, involves basically finding the area under the curve described by the Planck equation. The interested reader can refer to Appendix C where this integration is done to obtain another useful relationship.

When Planck worked out his radiation formula, he came to the unavoidable conclusion that a blackbody emits energy in discrete indivisible units. He found that each unit had an energy level equal to the frequency of the radiation multiplied by a constant. This constant of proportionality, h, he called "quantum action." It later became known as Planck's constant in his honor. To explain the shape of the blackbody curve, Planck proposed that the atoms of the radiating object could gain or lose energy only in discrete amounts of h multiplied by the frequency of the radiation. This energy increased or decreased only in multiples of these units. Thus, he found this amount to be given by the following expression:

$$E = h\nu$$

In this expression, h is Planck's constant that numerically is equal to 6.626×10^{-34} joule·sec. and ν is the frequency of the electromagnetic radiation. Energy can be exchanged only in jumps or quanta in an amount proportional to Planck's constant.

Planck had mixed feelings about this result. The idea that energy could be exchanged only in discrete steps was contrary to what everyone thought at the time. This exchange of energy should be continuously variable, like the release of energy from a spring. By doing so, it would be consistent with classical Newtonian physics. The idea that this energy was exchanged between the oscillations and the electromagnetic field continuously was disproved by Planck's result. Thus, the stage was set for the eventual acceptance of the quantum theory about 20 years later. Classical Newtonian physics was not adequate enough to explain the blackbody result. Physicists such as Albert Einstein and Niels Bohr did most of the ground work that eventually led to the acceptance of the quantum theory.

The recent discovery of the microwave background radiation confirms the Planck formula or radiation law. The presently accepted theory for the creation of the universe, the "Big Bang" theory, proposes that this microwave background radiation is actually a remnant from the intense heat released during its formation. According to this theory, the universe evolved from a primeval fireball. From this beginning, the universe expanded outward and continues to do so today. It took about 500,000 years after this initial event for the universe to become transparent to electromagnetic radiation. At this point, radiation decoupled from matter as the universe kept expanding while the matter cooled. This remnant temperature from the beginning of the universe can be detected today with special equipment.

In 1965, two Bell Laboratory scientists, Arno Penzias and Robert Wilson, were given the assignment to test a sensitive microwave receiver. When they pointed this receiver antenna to the zenith, a faint background noise level was received that could not be explained. This noise level remained constant with a change in direction. At first, they thought that there was a problem with the equipment. After checking and calibrating the equipment, they continued to receive the same signal. The maximum signal intensity was found to be at a wavelength of 1.1 mm. The signal itself was isotropic and unpolarized with no variation from season to season. This noise signal was later discovered to be the cosmic background radiation, leftover radiation from the initial primeval fireball. A calculation using Planck's radiation law for a wavelength of 1.1 mm gave a temperature of about 3.0 K. After plotting many other points from their data, a characteristic blackbody distribution of radiation intensity around this peak wavelength was discovered. In 1978, Penzias and Wilson received the Nobel prize for this discovery.

A more precise measurement of the spectral distribution of this radiation was obtained by the Cosmic Background Explorer (COBE) launched in November of 1989. This satellite also measured the isotropy of this background radiation. The results obtained from this research project yielded a radiation temperature of 2.735 K. The plot of hundreds of data points displayed a blackbody distribution curve. The cosmic background radiation is a good example of a naturally occurring blackbody spectrum.

8.4 THE PHOTOELECTRIC EFFECT

The second situation that had physicists puzzled was the photoelectric effect. Albert Einstein was the first to explain the physics involved with this effect by applying Planck's discovery. It was known at the time that when X-rays or UV radiation strike a metal plate, electrons will eject from the surface. When making careful measurements using monochromatic radiation, it was found that the energies for the emitted electron never exceeded a certain amount even when the intensity of the radiation was increased. This value was found to be linear with the frequency of the monochromatic radiation.

The apparatus used for this experiment consisted of cathode and anode surfaces to which a controlled voltage could be applied. Light radiation was incident upon the cathode during the experiment. When the incident radiation had enough energy, an electron was ejected from the surface of the metal cathode. Once ejected, it could move to the anode thus producing an electrical current flow in an external circuit. The voltage was made adjustable for the purpose of controlling the motion of the electrons. Thus, the maximum energy of the electron could be determined by varying the voltage between the cathode and anode until the current went to zero. At this point, the voltage level represents the stopping potential, V_0. The kinetic energy associated with this stopping potential can be calculated from the following relationship:

$$K_m = e V_0$$

In this expression, e is the electronic charge (1.602×10^{-19} coulomb).

Einstein assumed, in this case, that the energy quantization used by Max Planck was characteristic of light. In other words, light is composed of small packets of energy called quanta. The individual quantum we now call the photon contains the energy, $h\nu$, where h is Planck's constant and ν is the frequency of the photon. In this experiment, when the photons penetrated the surface of the metal cathode, their energy was completely given to the electrons. In order for an electron to escape from the metal, an amount of energy known as the work function of the metal, $\emptyset$, must be applied to overcome the potential barrier at the surface. The energy K_m now becomes the maximum kinetic energy that an electron can have. The mathematical relationship can be summarized below:

$$K_m = eV_0 = h\nu - \emptyset$$

This equation predicts a linear relationship between the stopping potential V_0 and the frequency of the incident radiation. Figure 8.4 shows data obtained from this experiment using the apparatus previously described. You can see that a linear relationship exists here. Another important result shows that the energy of the emitted electron depends upon the frequency of the light, not its intensity.

When looking at the data from this plot, we find an interesting situation. The point labeled f_0, or threshold frequency, has an important relationship to the work function. At this voltage level, no electrons are ejected from the metal. Thus, the following relationship holds true:

$$\emptyset = h\nu = (6.63 \times 10^{-34} \text{ joule·sec})(4.3 \times 10^{14} \text{ Hz})$$
$$\emptyset = 2.9 \times 10^{-19} \text{ joule or } 1.8 \text{ eV}$$

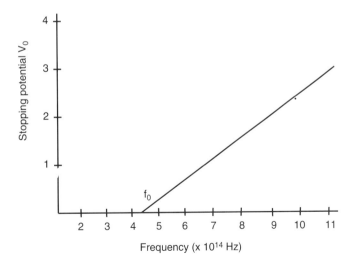

Figure 8.4   Data from the photoelectric effect experiment.

Photons of energy less than this amount will not be able to eject an electron from the metal. The conversion to electron volts can easily be made by using the following relationship:

$$1 \text{ eV} = 1.602 \times 10^{-19} \text{ joule.}$$

The results from this experiment cannot be explained using the wave nature of light. Einstein explained this effect by considering the particle nature of light. The photoelectric effect must be viewed as a collision between the emitted photon of the light radiation and the electron within the metal. Since a collision is usually associated with particles, the photoelectric effect provides proof for the particle nature of light. This photon model helped to explain the photoelectric effect when the wave nature of light could not. We will next discuss two observations that could not be explained using the wave nature of light, but could be explained when using the particle nature.

Results from the Photoelectric Effect Experiment

1. This basic result involved increasing the intensity level to a point where the ejection of photoelectrons occurred from the metal surface. It was found that if this intensity were doubled, for example, the energy level of the ejected photoelectrons would remain the same. If we try to explain this in terms of a wave phenomenon, this result does not make any sense. According to the wave theory, as you increase the intensity of the light, the ejected photoelectrons should also increase their energy. You can see that the last equation does not involve light intensity at all. The result from the actual experiment showed that when the light intensity level was increased, the number of photoelectrons also increased. The energy level of each photoelectron did not change.

2. Another important observation obtained from this experiment was that this effect only occurred after a minimum energy was applied. This minimum energy, or work function $\varnothing$, must be exceeded for the photoelectric effect to occur. The reason why this energy must be applied requires a little knowledge of the structure of matter. The electrons of the copper atoms involved in this process are bound to the metal's surface by a potential barrier. It takes a minimum amount of energy to eject an electron from this surface. When applying an amount of energy less than the minimum required to eject an electron, it remains bound to the metal. If we are dealing with a wave phenomenon only, an electron should be ejected at any frequency, provided that the requirement of intensity is met. We will next consider an example using the results from the photoelectric effect.

Example 8.4

The threshold frequency for potassium metal is 5.32×10^{14} Hz. What is the work function for this metal?

Using the formula for the work function, we get:

$$\varnothing = h\nu = (6.63 \times 10^{-34})(5.32 \times 10^{14}) \text{ joule}$$
$$\varnothing = 3.53 \times 10^{-19} \text{ joule or 2.2 eV}$$

Example 8.5

Now, if light having a wavelength of 450 nm strikes the potassium metal described in the last example, what will be the maximum energy of the photoelectrons leaving the potassium metal surface?

To solve this, we must first find the energy in electron-volts (eV) of the incident light. This comes out to be:

$$E = h\nu = hc/\lambda = (6.63 \times 10^{-34})(3 \times 10^{8})/(450 \times 10^{-9}) \text{ eV}$$
$$E = 4.42 \times 10^{-19} \text{ joule or 2.76 eV}$$

The maximum energy of the photoelectrons can now be found by subtracting this energy from that which binds the potassium electron to the surface of the metal. Thus:

$$K_m = h\nu - \varnothing = (2.76 - 2.2) \text{ eV} = 0.56 \text{ eV}$$

SUMMARY

In this chapter, we discussed two situations that give proof to light's particle-like nature. The shape of the Planck blackbody curve cannot be explained if electromagnetic radiation has only a wave-like nature. The photoelectric effect can only be explained by considering the particle nature of light. Thus, the classical view of light had to be changed to take these observations into account. This helped in developing the quantum theory in the 1920s.

The evidence is clear; light energy is quantized or occurs in discrete energy packets. These discrete energy packets, known as photons, will be one of the main subjects of the next chapter. The exchange of energy in the form of photons is at the very heart of the study of electro-optics. This is why it is very important to first learn about the electro-magnetic energy involved before learning about the circuits used in this exchange of energy.

9

Quanta and Optical Spectra

9.1 THE DUAL NATURE OF LIGHT—PARTICLES AND WAVES

In the first few chapters, we studied the various properties of light such as propagation, reflection, refraction, interference, diffraction, and polarization. These topics concentrated on the wave nature of light. All of these phenomena can be explained using classical physics. In this chapter, we will begin our study of the production of light by studying the basic structure of matter. As discussed in Chapter 3, Maxwell's equations treat the propagation of light, whereas the quantum theory describes the interaction of radiation and matter. The electron plays an important role in this interaction. A combined theory called quantum electrodynamics best describes the modern view of light phenomena by stressing the wave and particle nature of light. As you may have guessed by now, waves and particles have very little in common. Nothing in our human experience can be used to describe a phenomenon that has both a wave and a particle nature. This accounts for the confusion in this area of study. The physicist, Richard Feynman, summed it up beautifully when he described this dual nature of light by saying, "the paradox is only a conflict between reality and your feeling of what reality ought to be."

9.2 THE DISCOVERY OF THE ELECTRON

Before the late nineteenth century, the atom was thought to be the smallest building block of matter. But, by the end of that century, there was some evidence that the atom could be made of still smaller particles. This evidence was obtained by J.J. Thomson in 1897 with the discovery of the first subatomic particle, the electron. He is generally credited for this discovery, but many other scientists were also involved in similar work at the time.

Thomson's experiment involved the use of a special glass vacuum tube. He used high voltage electrodes on both ends of this tube. One electrode was charged negatively (called the cathode) and the other one was charged positively (called the anode). Thomson noticed that when he applied a high voltage to the electrodes, the tube walls glowed with a greenish fluorescence.

He later made more refinements to this vacuum tube, such as the addition of a special screen at the end to make observations easier. Thomson discovered that this green glow originated from the negative lead by traveling in a straight line. He subsequently called this phenomenon "cathode rays." These cathode rays were invisible to the naked eye unless they interacted with another substance such as the glass walls or the special screen at the end of the vacuum tube.

One important result from this experiment was that cathode ray particles are much lighter than normal atoms. This conclusion was brought about after much experimentation with changing the gas type and cathode material in the tube. These changes did not affect the characteristics of the cathode ray particles. He also used a magnet near the trajectory of these particles and noticed that the straight line motion was disturbed by the field. From the effect the magnet had on the particles, he proposed that these particles contained a charge.

With this information, he refined his design for a more interesting experiment. He added two flat parallel metal plates separated by 1.5 cm. A voltage was applied to these plates creating an electric field between them. He used the magnet again to deflect the cathode rays. Figure 9.1 shows the details of his design. The cathode rays passed through the two plates, and then through the field of the magnet. The two forces involved here could be used to deflect the straight line motion of the cathode ray particles. For example, the deflection could be controlled by varying the plate voltage.

With the experimental design described in Figure 9.1, Thomson was able to determine a parameter called the charge-to-mass ratio (e/m) of the electron. To explain his experiment in more detail, we must first consider the forces involved. The first force occurred between the parallel plates when a high voltage was applied. They were positioned such that the cathode ray beam experienced a downward force due to the electric field there. The second force was due to the magnet. It was positioned such that its force was exerted upward. Since it was already determined that these particles possess charge, they will be affected by both the electric and magnetic fields. The amount of force needed to deflect these particles depends upon the amount of charge present. We must also remember that the particles have a certain forward velocity through both fields. In his famous experiment, Thomson adjusted both the electric and magnetic fields to certain values so that

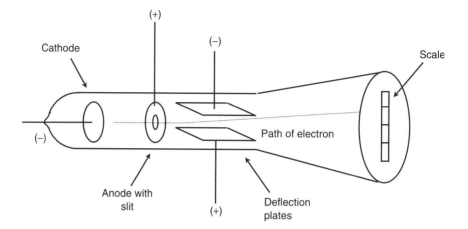

Figure 9.1   J.J. Thomson's design for measuring charge-to-mass ratio.

no deflection occurred. Under these conditions, the two forces resulted in a net zero force. As we know from basic physics, the mathematical expression for the two forces here are as follows:

$$F = eE \text{ (Electric force)} \quad \text{and} \quad F = evB \text{ (Magnetic force)}$$

We can now present the mathematical expression describing the state when the two forces resulted in a net zero force. In this case, both forces are equal. Thus:

$$evB = eE$$

In the above expression, e is the electron charge, v is the velocity of the electron, B is the magnetic field, and E is the electric field. Reducing this expression we get:

$$E = vB$$

Since the particles display a motion similar to that of a projectile fired horizontally in the earth's gravitational field, the vertical displacement or deflection, y, of the electron can be expressed as:

$$y = \frac{1}{2}at^2 \text{ where a is the acceleration of the particle, and t is time}$$

From Newton's second law, we know that F = ma or a = F/m. If we use the mathematical expression for the force caused by the electric field, F = eE, we can continue further with our derivation:

$$y = \frac{1}{2}at^2 = \frac{eE}{2m}\left(\frac{L}{v}\right)^2 = \frac{eE}{2m}\left(\frac{BL}{E}\right)^2$$

In this expression, L is the distance traveled by the cathode rays. Rearranging terms we get the desired result:

$$\frac{e}{m} = \frac{2yE}{B^2L^2}$$

Thomson obtained 1.6×10^{11} C/kg for this ratio. This result agrees closely with the modern accepted value of 1.7589×10^{11} C/kg. When he used different gases in the tube with different cathode metals, he obtained the same result. This means that the cathode ray particles were common to all metals.

Thomson was able to obtain only the charge-to-mass ratio itself. To find the value for either e or m, one of these must be known. At the time of Thomson's experiment, it was not possible to perform an experiment to determine either value. Then, in 1912, the American physicist Robert A. Millikan measured the charge on the electron successfully by using his oil drop apparatus. Knowing the value for the electron charge meant that the electron mass could now be determined. We will discuss this experiment later in this chapter.

Before Millikan's experiment, various attempts were made to determine the charge on the electron. These attempts gave only approximate results. There were disagreements among the physicists at that time relative to the value of the electron charge. Some physicists argued that if all electrons in Thomson's experiment displayed the same e/m ratio, that did not necessarily mean that they had the same charge. Also, up until this point in time, there was no proof that all electrons were identical.

Example 9.1

Let us now consider an example using J.J. Thomson's design as detailed in Figure 9.1. The following physical parameters will apply in this example: the two plates are 5.0 cm long with a separation of 1.5 cm. (a) If these plates are kept at a potential of 60 volts, find the resulting deflection of electrons as they travel the 5 cm path between the plates. The electrons have a kinetic energy of 2500 eV. (b) This deflection can be balanced by applying a magnetic field as described in Thomson's experiment. Find the strength of the magnetic field necessary to cause no deflection of the electrons.

(a) Looking at the equation developed for vertical displacement, y, we find that we are lacking two variables to solve this problem. These variables are the electric field, E, and the velocity, v, of the electrons. We will find each of these variables from the given information.

The electric field, E, between the plates can be found from the applied voltage and the separation distance. Thus:

$$E = V/L = 60/1.5 \times 10^{-2} = 4.0 \times 10^3 \text{ volts/meter}$$

The velocity of the electron can be found from the kinetic energy where K.E. = (1/2) mv^2. When we rearrange the terms, we get:

$$v^2 = 2(K.E.)/m = (2)(2500)(1.6 \times 10^{-19})/9.11 \times 10^{-31} \text{ m}^2/\text{sec}^2$$
$$v^2 = 8.78 \times 10^{14} \text{ m}^2/\text{sec}^2 \text{ (keep in this form for substituting)}$$

Now, we can substitute the values directly into the equation:

$$y = (1/2)eE/m(L/v)^2 = (1/2)[(1.6 \times 10^{-19})(4 \times 10^3)/9.1 \times 10^{-31}](25 \times 10^{-4})/$$
$$(8.78 \times 10^{14}) \text{ m}$$

$$y = 1 \times 10^{-3} \text{ meter or } 0.1 \text{ centimeter}$$

(b) For the condition of no deflection with applied electric and magnetic fields, the following equation can be used:

$$evB = eE \text{ or } B = E/v$$

Solving for B we get:

$$B = 4 \times 10^3/2.96 \times 10^7 = 1.35 \times 10^{-4} \text{ weber/meter}^2$$

We now continue with our historical development by considering the Millikan oil drop experiment, one of the classic physics experiments. In this experiment, Millikan determined the charge on the electron in units of coulombs. His apparatus used charged parallel plates placed inside of a metal box. The top plate had a small hole in the center. The metal box allowed the environmental conditions inside to be controlled. When a voltage was applied across these parallel plates, an electric field was produced between the plates. Using a special microscope, it was possible to view a portion of the volume of air between the plates. Figure 9.2 shows the details of this apparatus.

In his experiment, a light oil mist was introduced into the volume above the parallel plates. The force of gravity caused these oil drops to fall onto the top plate. Since this plate contained a small hole, eventually an oil drop made it through the opening. With no voltage applied across the parallel plates, the oil drop fell with a terminal velocity determined by the weight of the drop and air resistance. With the application of a voltage to the plates, the descent or rise of the drop could be controlled. Millikan could slow it down, make it stationary, or make it rise back to the top.

The control of the oil drop's motion could only be possible if it had an electric charge on it. In this case, the electric field between the plates would exert a force on the drop to counter the gravitational force. Using x-rays, he could add one or more charges onto the oil drop and observe its change in downward velocity. By varying the electric field between the plates, he found that the field strength was related to the charge on the oil drop. After observing thousands of oil drops, he came to two very important conclusions:

1. The value for the charge on the oil drop never fell below a minimum amount.
2. The measured charge always came out to a whole number multiple of this minimum value.

From this, Millikan assumed that this minimum charge on the oil drop was due to only one electron. He found the value for e to be 1.591×10^{-19} C. Today, the accepted value for the electron charge is 1.6021×10^{-19} C. Now, with this value, we can calculate the mass of the electron using J.J. Thomson's result. We show this calculation below:

$$m = (e)/(e/m) = 1.6021 \times 10^{-19} \text{ C}/1.7589 \times 10^{11} \text{ C/kg}$$
$$m = 9.109 \times 10^{-31} \text{ kg}$$

This result showed a very important characteristic about the amount of electric charge present. Electric charge is not continuously variable in nature but occurs only in discrete amounts. Most scientists at that time did not seem very surprised at this finding. They saw this quantization of electric charge to be very similar to the discrete nature of matter proposed by Dalton and Avogadro. In the last chapter, we saw further evidence for this discreteness with the quantization of light energy and the discrete energy states of atoms.

The electron was eventually accepted as a fundamental building block of the atom. However, the scientists at the time were not sure just how this electron integrated into the overall atomic structure. This acceptance also left the question of charge distribution within the atom unanswered. For the atom to be electrically neutral (no net charge), it must also contain the same magnitude of positive charge. How these charges were distributed still remained a mystery at that time. There were a number of theories proposed by scientists to help explain the electron's role in the atomic structure.

One of the most popular theories about atomic structure was called the "plum pudding" model for which J.J. Thomson supported. In this theory, the positive charges were distributed evenly within a spherical volume. The newly discovered electrons were thought to oscillate about fixed centers inside this volume. Ernest Rutherford, a well-known scientist at the time, wanted to test the validity of this and some of the other theories. He eventually initiated a research study to look at the basic structure of the atom.

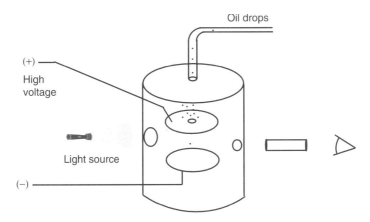

Figure 9.2   Millikan oil-drop apparatus.

9.3   THE THEORY OF ATOMIC SPECTRA

In 1911, Ernest Rutherford conducted experiments that eventually led to the discovery that a typical atom is composed of a relatively heavy nucleus around which much lighter electrons travel. This nucleus has a positive charge while the electrons have a negative charge. A stable atom will have just enough electrons in number (charges) to balance off the positive charges in the nucleus. He arrived at this conclusion after very close examination of his test data.

Figure 9.3 shows Rutherford's design for the above experiments. He used a radioactive source of polonium contained in a lead box to generate the alpha particles. The alpha particle is basically a helium nucleus. It is composed of two protons and two neutrons. This configuration results in a net positive charge of +2. Traveling with a velocity of 1.6×10^7 meters per second, these particles acted as subatomic bullets aimed at a thin gold foil of thickness 10^{-7} meter. Since their mass, charge, and velocity were known, he could determine important parameters about the gold nucleus from this interaction. Fluorescent screens were placed, as shown, to display the locations where the alpha particles ended up after interacting with the gold atoms. After many hours of observations, he came up with some unexpected results. Most of the particles passed straight through the foil to the fluorescent screen on the other side. For this to occur, he reasoned that the gold atoms must have spaces between them. A few of the particles were deflected from the foil by angles that varied from a few degrees to 180°. The few particles that were deflected at large angles caused a problem for the atomic theory at the time.

In the early 1900s, physicists thought that the positive charge of the atom was spread out to a radius of about 10^{-10} meter. The electrons were also thought to move about inside this volume. With this view in mind, Rutherford could not explain how a massive

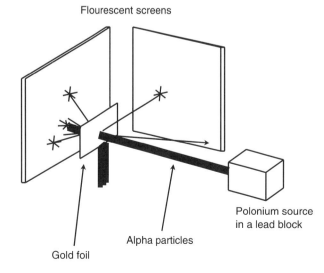

Flourescent screens

Gold foil

Alpha particles

Polonium source
in a lead block

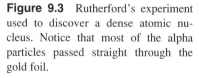

Figure 9.3 Rutherford's experiment used to discover a dense atomic nucleus. Notice that most of the alpha particles passed straight through the gold foil.

alpha particle could be deflected by 180° due to the tiny electrons inside the atom. The mass of the electron was known to be much less than that of an alpha particle. Rutherford was later quoted as saying, "It was quite the most incredible event that has happened to me in my life. It was almost as incredible as if you fired a 15-inch shell at a piece of tissue paper and it came back and hit you."

The solution to the problem required a new model for the atom. Rutherford proposed that the electrons moved around the outside of the nucleus instead of inside of it. As for the positive charges, his model had them confined to a smaller radius of about 10^{-14} meter. Using this model, Rutherford could now explain the observations from his experiments. The number of alpha particles passing through the foil compared with the number that rebounded gave him a clue as to the size of the atomic nucleus. It was found that one out of every 8000 alpha particles were deflected by more than 90°. This deflection resulted when the alpha particle passed through a region of enormous field strength found at the surface of the nucleus. This field results from the positive charges (protons) being confined to such a small volume (the nucleus). Returning back to Figure 9.3, we can see the path taken by the alpha particles as they move through the gold foil. There are a few grazing angles but most of the alpha particles go straight through the foil. Since the nucleus has a net positive charge, the force on the alpha particle will be repulsive. The few particles that were deflected experienced the repulsive force of the gold nucleus.

Now let's look at the physics involved with the force between charged particles. As we know, these charges exert forces on each other. If e_1 and e_2 are both electric charges separated by a distance r, the force that each charge exerts on the other charge in a vacuum is given by Coulomb's law:

$$F = k \frac{e_1 e_2}{r^2}$$

The coulomb force is given in Newtons when using the MKS system. The unit of charge is the coulomb, and the distance between charges is measured in meters. If there are like charges, the force is repulsive. The force is attractive if these charges are unlike.

Next, let's compare the motion of the electron around its nucleus to that of the earth around the sun. Classical physics says that the electron does not fall into the nucleus of an atom because it is orbiting at a fast enough speed. Coulomb's equation above shows the same inverse square relationship with distance that characterizes the gravitational force. If classical physics were correct, the electron should move in a similar fashion to an orbiting gravitational mass. Nevertheless, it turns out that there is another condition that classical physics cannot explain. While the electron is orbiting its nucleus, it is constantly being accelerated toward it. According to classical physics, an accelerating electric charge should radiate energy. This means that the orbiting electron should be constantly losing energy due to this emission of radiation during acceleration. The energy loss should eventually cause the electron to spiral into the nucleus. In fact, calculations using classical mechanics predict that the electron should achieve this state in about 10^{-6} sec. This situation is not observed in a stable atom. How then can we explain the motion of the electron around the nucleus?

Neils Bohr studied the atomic structure by applying classical physics as far as he could. He had to make additional assumptions where necessary to help explain observed phenomena. It was his conclusion that you could not deduce stable positions for the electron by using classical mechanics or electrodynamics. Bohr eventually came up with an empirical model for the hydrogen atom which is applicable to other systems as well. This model will be used later to study the physics involved with optoelectronic devices.

In the last chapter, we discussed the fact that the energy associated with an electron orbiting an atom can be exchanged only in discrete amounts called quanta. Neils Bohr did his famous work on this subject. He was primarily interested in finding a theoretical basis for explaining the spectral or emission lines of hydrogen. He noticed a series of bright lines when the light emitted from excited hydrogen gas was passed through a device called a spectroscope. These bright lines, he reasoned, are a characteristic of a particular atom or molecule. Thus, hydrogen will be different from oxygen in the number and positions of these emission lines. Hydrogen also has the simplest emission spectrum.

To explain why hydrogen atoms emit only discrete characteristic frequencies, Bohr proposed two assumptions or postulates:

1. Electrons can occupy only certain discrete quantized orbits or stationary states of fixed energy. These particular stable orbits or states are nonradiating and obey both Coulomb's and Newton's laws. This means that the angular momentum of each orbit must also be fixed. The lowest or normal state is known as the ground state.

2. When an electron changes from one stationary state to another, the energy emitted or absorbed equals the difference in energy between the two states. The frequency, ν, of this radiation can then be determined by the formula

$$\nu = \frac{\Delta E}{h}$$

where ΔE is the difference between the two energy states and h is Planck's constant. Thus, the energy difference from a higher energy level to a lower energy level can be written as $\Delta E = E_{n2} - E_{n1}$.

This emitted or absorbed energy involves the exchange of photons. The photon's energy is the difference between these two energy states. Thus, each particular emission line in the hydrogen spectrum corresponds to the emission of photons with a particular energy given by the above equation.

Bohr came to the conclusion that the classical mechanical interpretation of the atomic system required modification. We have already seen two instances where this was done. Planck's blackbody radiation theory and Einstein's photoelectric effect required modifications from the classical view. Thus, Bohr assumed that there were only certain electron orbits that did not result in the emission of energy. If electrons stayed in these orbits, the energy remained fixed. When electrons changed from one orbit to another, an exchange of energy was involved. This also meant that the angular momentum of an electron in one of these fixed energy orbits is quantized. An expression for the angular momentum in this case is given below:

$$L = \frac{nh}{2\pi} = m_e vr \quad \text{where } n = 1, 2, 3, \ldots$$

In this expression, L is the angular momentum, n is the particular state of the electron, m_e is the mass of the electron with velocity, v, in a circular orbit of radius, r. Now, we will apply Coulomb's law in this case. When we make the Coulomb force equal to the centripetal force of the electron, we get the following equality (the electron has a charge of $-e$, the proton has a charge of $+e$):

$$F = \frac{e^2}{4\pi\varepsilon_0 r^2} = \frac{mv^2}{r}$$

Next, we substitute for velocity, v, using the formula $mvr = nh/2\pi$. After rearranging terms and simplifying, we get an expression for the allowed radii of the Bohr atom:

$$r = \frac{\varepsilon_0 h^2}{\pi m_e e^2}\left(n^2\right) = r_b n^2$$

where r_b is the first Bohr radius (n = 1). The various orbits can then have the following values, r_b, $4r_b$, $9r_b$, and so forth. Solving for r_b, we get the actual length of this radius:

$$r_b = 5.292 \times 10^{-11} \text{ meter}$$

Close examination of the hydrogen emission lines showed that there are at least three different series or sets of these lines. The first series were discovered by Johann Balmer in 1885. This particular series of emission lines can be found in the visible part of the spectrum. Figure 9.4, a sketch of a spectrograph for hydrogen, shows the three different series of hydrogen emission lines and their location in the spectrum. For the Balmer series, three bright lines can easily be seen through a simple spectroscope while viewing a hydrogen discharge tube. The hydrogen alpha line is located at 656.3 nm (6563 Angstroms). The hydrogen beta line is at 486.1 nm (4861 Angstroms), and the hydrogen gamma line is at 434.0 nm (4340 Angstroms). As you can see, another series of lines is lo-

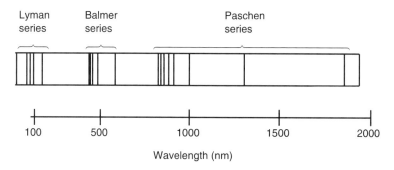

Figure 9.4   Spectrograph of hydrogen showing the first three series of emission lines.

cated in the infrared region and one more is located in the ultraviolet region of the electro-magnetic spectrum.

To explain the formation of these lines, Bohr reasoned that there are only a finite number of allowable stable orbits that an electron can occupy around the nucleus. Each orbit has a different energy associated with it given by the following equation:

$$E_n = -\frac{m_e e^4}{8h^2\varepsilon_0^2 n^2}$$

E_n is the energy associated with the formation of a stable orbit. The mass of the electron is m_e, e is the electron charge, and ε_0 is the vacuum permittivity. In the above formula, n is an integer or quantum number. When n = 1, this is the lowest allowed orbit for the Bohr hydrogen atom. Solving the above equation for the condition when n = 1, we get E = −13.5 electron volts (eV). The negative sign means that the energy given applies to a bound state for the electron. In this case, an excess of 13.5 eV must be applied to this system for the electron to break free of the nucleus. If this happens, an ion will be formed because the lost negative charge of the electron leaves a net positive charge of the nucleus (one proton). This energy is also referred to as the ionization energy.

The energy emitted by an electron as it falls or jumps from a higher level to a lower level orbit can then be defined as:

$$E_2 - E_1 = \frac{m_3 e^4}{8h^2\varepsilon_0^2}\left(\frac{1}{n_1^2} - \frac{1}{n_2^2}\right)$$

The energy released is then given by the following expression where v is the frequency of the emitted photon:

$$\Delta E = E_{n2} - E_{n1} = hv$$

A transition diagram for hydrogen is given in Figure 9.5. The various energy levels for the allowed orbits are represented by horizontal lines. These transitions correspond to the vertical arrows and show up as emission lines. This diagram shows the energy transitions for the Lyman, Balmer, and Paschen series. Each series has a common lower energy level. The spectral lines are obtained by the various combinations of integers for n_1 and n_2. The energy associated with each line can be calculated by using the above formula. As stated previously, the Balmer series of spectral lines were the first to be discovered. When Johann Balmer was presented with the observations of the four visible emission lines of hydrogen, he was able to come up with the portion of the formula contained in the brackets. Balmer had an interest in solving numerical puzzles, so he tried various numerical combinations. By setting the lowest value to n = 2 and then using n = 3, 4, 5 ,6 for the higher levels, he able to match this result with the four observed wavelengths. In the process, he had to treat all of the quantities to the left of the brackets as one constant.

The Bohr model of the hydrogen atom had at least one serious problem, but it does give essentially correct numerical results. In the Bohr model, electrons move around the nucleus in fixed, stable orbits. This is impossible according to classical physics because a charged particle, such as an electron, moving in an electrostatic field should radiate en-

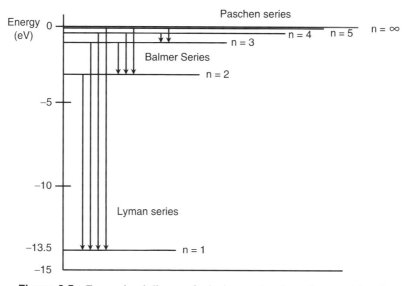

Figure 9.5   Energy level diagram for hydrogen showing a few transitions in each series.

ergy. As this energy radiates from the charged particle, its energy level should gradually decrease, causing a smaller orbital radius. This means the electron should, according to classical physics, eventually fall into the nucleus of the atom. The Bohr theory was eventually superseded by the modern quantum theory to explain this discrepancy. The quantum theory uses wave or state functions to describe the atomic system. A study of these wave functions is beyond the scope of this book. Let us next consider an example using Bohr's model of the hydrogen atom.

Example 9.2

Looking at the diagram in Figure 9.5, we see the Balmer series of emission lines for hydrogen. Find the wavelength of the emitted photon as an electron makes a transition from the energy state at n = 4 to n = 2.

To solve for the wavelength of the photon, we must first find its energy. Then, from its energy, we can determine its wavelength using the Planck relationship:

$$E_4 - E_2 = \frac{m_e e^4}{8h^2\varepsilon^2}\left[\frac{1}{(n_2)^2} - \frac{1}{(n_4)^2}\right]$$

When we substitute the values into the right side of the equation, we find:

$$E = (9.11 \times 10^{-31})(1.6 \times 10^{-19})^4/(8)(6.63 \times 10^{-34})^2(8.85 \times 10^{-12})^2\,[1/2^2 - 1/4^2]$$

$$E = 4.06 \times 10^{-19} \text{ joules or } 2.54 \text{ eV}$$

Now, since E = hv, we can now solve for the wavelength by finding the frequency:

$$v = E/h = 4.06 \times 10^{-19}/6.63 \times 10^{-34} = 0.6124 \times 10^{15} \text{ Hz and}$$

$$\lambda = c/v = 3 \times 10^8/0.6124 \times 10^{15} = 4.898 \times 10^{-7} \text{ meter or } 489.8 \text{ nm (blue light)}$$

9.4 ELECTRON WAVES

In 1924, Louis de Broglie predicted the existence of matter waves. He proposed that this wave-particle nature of light could also apply to any particle. The relationship that he came up with represents a direct violation of the classical view. That is, an electron acts as a wave phenomenon. He used the expressions below to prove his theory:

$$\lambda = \frac{h}{p}$$

and

$$E = hv = pc$$

In both expressions, p is the momentum of the particle of matter. The dual nature of light can be described using the above two equations. Each equation contains a variable for the wave concept (λ and v) and the particle concept (p and E). According to deBroglie, this relationship should hold true for any particle having a linear momentum, p. Looking at this relationship, we can see that as a particle's momentum increases, its wavelength will decrease. Since h is a very small number, this dual wave-particle nature of matter becomes apparent only in the microscopic realm of atoms and subatomic particles. And, it is probably a good thing that h is very small. Imagine what this world would be like if Planck's constant were significantly larger than it is now! With the value of h as it is, electron size particles should display wave properties such as diffraction, but macroscopic objects, such as baseballs, should display properties of a particle. About three years after deBroglie proposed this relationship, the wavelike nature of the electron was displayed by an electron beam.

In 1927, C. J. Davisson and L. H. Germer proved that the electron can be described as a wave phenomenon when they obtained diffraction patterns from reflected electrons. Davisson was working on the problems associated with vacuum tubes at AT&T when he noticed the scattering of electrons from a nickel crystal target. With the help of Germer, he researched this effect further by allowing an electron beam to hit this nickel crystal target at normal incidence. They also used a detector sensitive to electrons (an ionization chamber). When they made a plot of scattered electrons versus the angle of collection, they obtained a diffraction pattern with maxima and minima. Applying the deBroglie formula to find the electron wavelength, they found close agreement with the classical diffraction formula, $n\lambda = d \sin \theta$. In this case, the spacing, d, of the atomic planes in the nickel crystal was 2.15×10^{-10} meter. X-ray diffraction studies provided this distance. The angle of diffraction for which a strong maximum occurred was found to be 50 degrees. We solve this expression below:

$$n\lambda = d \sin \theta \text{ where } n = 1$$
$$\lambda = (2.15 \times 10^{-10})(.766) = 1.65 \times 10^{-10} \text{ meter}$$

The atomic planes of the crystal performed like a diffraction grating producing the observed diffraction pattern. This physical design satisfies the Bragg condition that is also observed in many other instances involving other types of electromagnetic radiation. Using the deBroglie wavelength formula, they obtained 1.67×10^{-10} meter (see Example 9.3 for this solution). Thus, there was very close agreement between both methods. Now we see why the wave properties of the electron were not discovered until this time in history. From our earlier discussions on diffraction, recall that the slit width or aperture size must be on the order of the wavelength to observe this effect. When the aperture becomes much larger, the effects of diffraction and interference are not observed. In the above experiment, the spacing provided by the atomic planes met this requirement.

In 1937, Davisson and G.P. Thomson, son of J.J. Thomson, received the Nobel prize for their work that showed the wave nature of electrons. We will next present interesting examples that involve calculating the deBroglie wavelengths for two very different sized objects.

Example 9.3

In Davisson's experiment described above, he used electrons having an energy of 54 eV. Find the deBroglie wavelength in this case.

Since we know its energy, we can find the electron's velocity from the kinetic energy formula. We need to calculate its velocity to determine its momentum, p. Thus:

$$\text{K.E.} = K = \frac{1}{2}mv^2 \text{ or } mv = p = (2Km)^{1/2}$$

$$p = [(2)(54)1.6 \times 10^{-19})(9.1 \times 10^{-31})]^{1/2} \text{ kg m/sec.}$$
$$p = 39.7 \times 10^{-25} \text{ kg m/sec.}$$

Now, we can use the deBroglie relationship to find the wavelength:

$$\lambda = \frac{h}{p} = 6.63 \times 10^{-34}/39.7 \times 10^{-25} = 1.67 \times 10^{-10} \text{ meter}$$

This result compares very closely to what Davisson found when using the classical diffraction formula and his observations.

Example 9.4

We will next consider an example from the macroscopic world. Find the deBroglie wavelength of a 100 gram golf ball traveling at 30 meters per second.

$$\lambda = \frac{h}{p} = 6.63 \times 10^{-34}/(.10)(30) = 2.21 \times 10^{-34} \text{ meter}$$

This wavelength is extremely small, and for this reason, we do not think of macroscopic-sized objects as having a wave-like nature.

With his wavelength formula and the condition for angular momentum, deBroglie came up with an explanation as to why electrons are only found in certain acceptable orbits around the nucleus. To illustrate this, we use the analogy of a standing sound wave in a long gas-filled tube. We can also use an example of an electromagnetic wave in a waveguide. In both of these cases, there exists a particular length at which the wave resonates within the cavity producing a standing wave.

Figure 9.6 shows the application of a standing electron wave adjusted in wavelength to fit an integral number of times, n, around the circumference or orbit of the Bohr atom. This concept, while regarded as oversimplified, can be used to help us understand why the electron can occupy only discrete orbits or states. Any other orbit that does not contain an integral number of wavelengths in circumference is not allowed, and thus does not occur. These orbits are also known as stationary states because they have a defined energy level. The mathematical expression for these acceptable orbits can be given as:

$$n = \frac{2\pi r}{\lambda} = \frac{2\pi r}{h/p} \quad \text{where } n = 1,\ 2,\ 3,\ldots$$

Thus, we have seen that electrons and photons possess particle and wave-like natures. This means that they can display normal wave properties such as diffraction and interference. Light phenomena, which we normally think of as wave motion, can also display a particle-like nature. This was shown in our discussion on the photoelectric effect. There is a way to help us understand this dual nature of electromagnetic radiation. When considering the propagation of light, it will behave as a classical wave. When energy exchange is involved, it will behave as a classical particle. However, nothing can behave as a classical wave and a classical particle at the same time.

The exchange of energy using quanta is of primary importance in the study of electro-optics. When using optical sources, photons are generated in the interaction of matter and energy. Semiconductor light sources will be studied in the next chapter. These semiconductor light sources emit photons at a particular energy determined by the relationship, $E = h\nu$. The emitted photons behave as particles just before they leave the surface of the source. When they propagate through media such as air, space, or a fiber optic

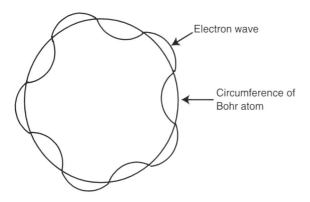

Electron wave

Circumference of
Bohr atom

Figure 9.6   An electron wave around a Bohr atom. It can fit only an integral number of times around it.

waveguide, wave-like properties apply. Maxwell's equations can be used to describe the electromagnetic field in this case. Next, as these photons strike the surface of an appropriate optical receiver, we must again consider the particle-like properties of the photon. The process involved here is very much the reverse of the emission process that produced the photons.

The absorption of a photon by an atom of the receiver material results in an electron of the receiver material becoming excited to a higher energy level. For this to occur, the energy of the incident photon, $E = h\nu$, must be equal to that required to excite the electron to a higher level (band gap energy). When the electron is excited to a higher energy level, the conduction properties of the atom are changed and electron flow in the bulk material occurs.

In Section 9.3, we discussed the theory of atomic spectra. We learned that a series of bright emission lines can be seen from excited hydrogen atoms when using a device called a spectroscope. In the next section, we will detail how to construct such a device.

9.5 LOW COST SPECTROSCOPE

In our previous discussions, we described how photons were produced in the interaction of radiation and matter. A practical optical instrument, the spectroscope, can be used to analyze light to determine what wavelengths or photon energies are involved in this interaction. The spectroscope accomplishes this by dispersing the light into its individual wavelength components. A glass prism or a diffraction grating can be used for this purpose. Our low cost spectroscope will use a diffraction grating in its construction. It is simple to make, and provides for very accurate optical measurements in the visible portion of the spectrum. The description of how to make this spectroscope originally appeared in *The Physics Teacher* magazine, April 1967, page 173. To make this spectroscope, you will need the following materials:

1. A box measuring approximately 9" long by 6" wide by 2" deep with a removable or hinged cover (A cigar box or video tape box will work nicely)
2. A diffraction grating having 23,000 lines/inch (This can be obtained from Edmund Scientific, part number C40,267)
3. Plastic centimeter ruler
4. 3" × 5" Index card (or an optional slit from Edmund Scientific, part number C39,522)
5. Razor blade

To construct this spectroscope, refer to Figure 9.7.

1. Make a 1" hole in the box centered as shown approximately 1½ inches from the right edge. This hole will be needed to mount the diffraction grating later on in the construction.

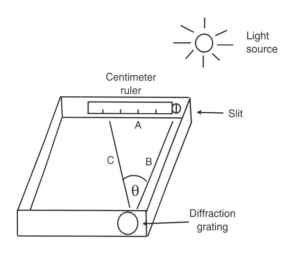

Figure 9.7 A low cost spectroscope.

2. Directly opposite the line of sight of this hole, make another hole for the slit. This hole must be about ½" in diameter. It is very important that the center line for the holes line up with the side of the box as shown.

3. To make the slit for the light to enter, cut into the index card with a razor blade. The slit should be about 12 mm long and no more than 1 mm wide. You may need to practice on another piece of paper to get this right. As an option, a mounted slit can be used.

4. After successfully making the slit, trim the index card and tape the slit over the ½" hole. The slit must be oriented vertically and then centered over the hole.

5. Carefully cut a piece of diffraction grating to be placed over the 1" hole. The ruled lines must also be oriented vertically. Use tape to keep it in place.

6. The centimeter ruler must be fastened to the inside of the box as shown. The edge or zero point of this ruler should line up with the slit. This will make future measurements easier.

The operation of the spectroscope can easily be checked by aiming the slit at a light source such as a fluorescent light. When looking into the end with the grating, bright emission lines on a continuous spectrum background can be seen. Other lamps, such as mercury vapor and quartz halogen, should give similar results but with a different set of emission lines. Looking closely at these lines, you can also see the centimeter ruler. We will use this ruler to determine the diffraction angle for a particular emission line. If the diffraction angle and slit spacing are known, we can easily determine the wavelength of the emission line.

As you remember from Chapter 6, the mathematical relationship that we can use in this case is the diffraction formula:

$$n\lambda = d \sin \theta.$$

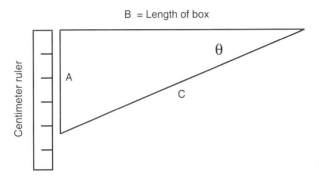

Figure 9.8 Measurement of θ using the spectroscope.

In this equation, n is the order number (one in this case) and d is the spacing between the lines on the grating. To determine the angle θ, we must know the lengths of two sides of the triangle shown in Figure 9.8. Substituting into the above equation for sin θ, we get:

$$n\lambda = d \sin \theta = d \ A/C$$

If you use this method to determine the diffraction angle, θ, you will need to make a scale model of the triangle to determine the lengths of A and C. These measurements must be as accurate as possible. An alternate method is to obtain the fixed measurement of side B with the measurement of side A. From these two sides, the diffraction angle can be determined by using the following trigonometric relationship:

$$\tan \theta = A/B$$

Finally, we must determine the spacing, d, from the lines per inch of the diffraction grating. If we convert this measurement to centimeters, our calculations will be easier. We know that 2.54 centimeters equals one inch. Using this information, 23,000 lines/inch of the grating equals 9055 lines/centimeter. The spacing, d, can now be calculated as follows:

$$d = 1/9055 \text{ lines/cm} = 1.10 \times 10^{-4} \text{ cm or } 1.1 \times 10^{-6} \text{ meter}$$

We are now ready to demonstrate how to determine the wavelength of an emission line from the measured angle of diffraction.

Example 9.5

Using our spectroscope, we find the angular measurements for two particular emission lines. The first one has a measured angle of diffraction of 36°. The second has a measured angle of diffraction of 23.5°. What wavelengths do these emission lines correspond to?

Using the above formula, we get for the first emission line:

$$n\lambda = d \sin \theta = 1.10 \times 10^{-6} \ (.5878) \text{ meter}$$
$$\lambda = 646.6 \text{ nm (red)}$$

The second emission line can then be calculated as:

$$n\lambda = d \sin \theta = 1.10 \times 10^{-6} (.3987) \text{ meter}$$
$$\lambda = 438.6 \text{ nm (violet)}$$

You can see from the above results the accuracy of this instrument due to the relatively large dynamic range available between red and violet wavelengths.

SUMMARY

In this chapter, we have seen the important role that the electron plays in the interaction of radiation and matter. Electrons can occupy only certain discrete quantized orbits or stationary states of fixed energy. This energy is not continuously variable as classical physics would predict. This means that when the electron changes from one orbit or state to another, the energy emitted or absorbed equals the difference between these two states. This is why distinct emission lines can be seen when using a spectroscope to view excited hydrogen atoms. Hydrogen atoms were considered because they are the simplest to use in an explanation for how this energy exchange happens. We will see later that this interaction of radiation and matter also occurs in solid materials.

In the next chapter, we will continue with a discussion of this interaction of radiation and matter as we consider solid materials such as semiconductors. This discussion will involve the particle-like nature of light. We will see why semiconductors make very useful light-emitting materials.

PART 2

10

Semiconductor Light Sources

In the last chapter, we considered the behavior of light as a particle. We will continue with our discussion of light's particle nature as we consider the interaction of radiation and matter within semiconductors. These discussions will bring us to understand why semiconductors make good candidates for light emission materials. As anyone who has worked with LEDs knows, by varying the semiconductor alloy, a different emission color or wavelength will result. The LED has many useful optical and electrical characteristics. These will also be discussed in detail. Next, we will study the operation of the diode laser. To understand this operation more fully, we must consider both the particle and wave natures of light. The particle nature will help us to understand the process of photon production. The wave nature will help us to understand how a coherent beam of light is produced within the laser's optical resonant cavity. As with the LED, we will consider the diode laser's useful optical and electrical characteristics. Finally, we will conclude this chapter by comparing some of the trade-offs involved with using either the LED or diode laser as a semiconductor light source. These trade-offs will involve parameters such as bandwidth, spectral width, modulation rate, light output, and source coupling to optical fibers. Manufacturers' data sheets on specific devices will be used in these discussions.

10.1 EMISSION PROCESSES IN SEMICONDUCTORS

When we consider electronic applications, the use of materials that conduct electricity becomes important. We know that some materials are good conductors while others are good insulators. When considering these materials, we may wonder why a metal such as copper conducts electricity while a substance like rubber does not. To investigate the reason this occurs requires some knowledge of the basic structure of matter.

In conductors you will find the free motion of electric charges or electrons, while in insulators this motion does not occur. The electrons in the outer orbital of the copper atoms are involved with this motion through the conductor. A property of matter describing this ability or inability to conduct free electrons is known as resistivity. Table 10.1 lists some materials and their resistivity. Notice that an insulator such as diamond has a much larger resistivity than a metal such as copper. The higher the resistivity, the better that material will perform as an insulator.

In the last chapter, we considered individual hydrogen atoms and the discrete energy levels of its electron. When atoms come very close together to form a solid, these discrete energy levels broaden into bands of allowable energies. This close spacing also allows neighboring atoms of the solid to share their valence electrons. Valence electrons are located in the outermost orbital. In doing so, bonding occurs between the atoms. For conductors, there are two energy levels to consider, the valence and conduction bands. When an electron is in the valence band, it occupies a lower energy state and stays bound to the atom. An electron can be placed in the conduction band by the addition of energy such as that from an electric field. In the conduction band, the electron is free to move within the conductor.

The first example that we will consider is the metal sodium. Sodium has one valence electron that can easily be raised to a higher energy state by an electric field. This becomes true for sodium atoms brought very close together to form a solid. The energy levels of the neighboring atoms' outer electrons actually overlap each other. This situation makes it very easy for the outer electrons of sodium to wander from one atom to another with the addition of an electric field. The ability to carry charge through the material makes it a conductor. This same situation also exists for substances such as copper. The presence of closely spaced copper atoms results in orbital energy levels that overlap. This

Table 10.1   Resistivity of Some Common Materials

Material	Resistivity (ohm·meter)
Diamond	10^8
Silicon	3.0×10^3
Sodium	4.7×10^{-8}
Copper	1.7×10^{-8}
Aluminum	2.8×10^{-8}

makes it relatively easy for the one valence electron to be raised to the conduction band level.

When we consider an insulator, the situation becomes quite different. The outer electrons associated with insulators are not free to wander within the solid. In this case, the energy levels do not overlap. As a matter of fact, a relatively wide energy gap exists between the valence band and conduction band. This means it takes a relatively large amount of energy to move an electron of an insulator material across this energy gap. Diamond for example has an energy gap of 5.4 eV making it a very poor conductor.

Semiconductor materials are neither good conductors or insulators. Their resistivity lies somewhere between that of conductors and insulators. The element silicon is an example of a semiconductor. Silicon's place in the periodic table is in the fourth column, making it a group IV material. This means that silicon has four electrons in its outer orbit available to form covalent bonds with neighboring atoms. In Table 10.2, we list some of the most common elements used to construct semiconductor compounds. These elements can be found in the periodic table of the elements. For convenience, each element is listed in columns by group along with its symbol and atomic number. Many of these elements will be used in the remaining chapters of this book. In future discussions using semiconductors, the elemental symbol will be used in place of its name for simplicity, unless otherwise stated.

When two or more of these compounds are used to manufacture a semiconductor material such as GaAs, it can be referred to as a group III-V compound. Some other examples of group III-V compounds are GaP and GaAlAs.

The element of silicon cannot be considered a conductor or an insulator. Its conduction properties can be better understood by considering both the energy band and lattice spacing diagrams for pure silicon as shown in Figure 10.1(a). Silicon has a total of 14 electrons in its atomic structure; four of these are the outer valence electrons occupying the valence energy band. The complete occupation of this band by electrons means that they cannot wander freely within the solid. These four valence electrons are used for covalent bonding with other silicon atoms, and form the silicon lattice structure. Since the energy levels do not overlap, an application of at least 1.1 eV is required to move one of

Table 10.2   A Partial List of Elements in Groups II to VI

Group IIA	Group IIIA	Group IVA	Group VA	Group VIA
^{12}Mg	^{13}Al	^{6}C	^{7}N	^{8}O
Magnesium	Aluminum	Carbon	Nitrogen	Oxygen
^{20}Ca	^{31}Ga	^{14}Si	^{15}P	^{16}S
Calcium	Gallium	Silicon	Phosphorus	Sulfur
	^{49}In	^{32}Ge	^{33}As	
	Indium	Germanium	Arsenic	
			^{51}Sb	
			Antimony	

these electrons from the valence band to the conduction band. In the energy level diagram given below the figure, 1.1 eV represents this band gap energy. Besides using an electric field, this energy can be applied optically or thermally to the valence band electrons as well. In doing so, an electron leaves behind a hole or unoccupied state in the valence band as it moves to the conduction band. With the application of enough energy, this hole will then be filled by another electron in the valence band. As this electron moves to fill this hole or vacancy, it leaves behind another hole in the process. In this fashion, the movement of both electrons and holes occurs within the solid. This process makes it appear that the hole is moving and thus participating in current flow. The hole may serve as a carrier of electricity comparable to that of a free electron. We will see in our discussion of the emission process in semiconductors that both electrons and holes must be considered.

As you can see, pure silicon does not make a very good conductor. It is considered to be an intrinsic semiconductor since it contains no foreign atoms. But fortunately it turns out that the current carrier type and density in the silicon can be controlled by adding trace amounts of suitable atoms into the lattice structure. This doping process greatly improves the usefulness of the semiconductor material. To see how this can be accomplished, let's consider a group of silicon atoms such as those shown in Figure 10.1a. For simplicity, we represent the lattice structure here in two dimensions. Since each atom of silicon has four valence electrons, they will be shared covalently with four other silicon atoms, thus forming the lattice structure. The valence electrons are represented by dots in the diagram. To initiate the doping process we add trace amounts of dopants with the correct chemical properties. In this case, phosphorus is chosen for this dopant since it has a valence of five. For simplicity, we consider one of the silicon atoms being replaced with an atom of phosphorus. The lattice structure for this situation now conforms to Figure 10.1b. Here we can see that four out of the five valence electrons of the phosphorus atom are shared with neighboring silicon atoms leaving one unbound electron available as a carrier of current. This affects the normal energy band gap as shown in the diagram under the lattice spacing configuration. The extra electrons of the phosphorus atoms added to the silicon structure create an energy level slightly below the conduction band known as a donor level. This makes the band gap energy level on the order of 0.05 eV less than it would have been without the addition of the phosphorus atoms. The result of this change in the energy level makes it much easier for the phosphorus electrons to jump the energy gap to the conduction band. If other impurities from the group V elements, such as As or Sb, are added to silicon, extra electrons will also be produced. These group V materials also have five electrons in their outer shell.

The above situation describes a donor level because atoms such as phosphorus basically donate electrons to the conduction band without creating holes in the valence band. Normally, a hole is created in the absence of an electron in the valence band. But, when adding impurity atoms, such as phosphorus to the silicon crystal, we can think of this as just adding more electrons to the conduction band as shown in Figure 10.1b without adding holes to the valence band. A semiconductor manufactured in this way is known as an n-type semiconductor. The "n" means "negative" since the number of negatively charged carriers exceeds the number of positively charged ones. Thus, the electrons in an n-type semiconductor are known as majority carriers while the holes are known as minority carriers in the valence band. As you can see in Figure 10.1(b), the number of holes in the valence band is much less than the number of electrons in the conduction band. Consequently, the dominant carriers in this type of material are the electrons. The amount of dopants or impurities added to the semiconductor controls the density of electrons in the conduction band.

Dopants or impurities from the group III elements such as Al, Ga, or In can be added to pure silicon to create an acceptor site. What happens here is more complicated than with the n-type semiconductor describe above. This new situation is shown in Figure 10.1(c). The group III elements have three electrons in their outer shell for covalent bonding. A group III impurity will allow silicon to accept electrons or produce holes as shown in the figure. The three outer electrons of the aluminum atom, for example, form only three covalent bonds with neighbor atoms as shown. The vacancy that exists in the fourth bond results in a hole. This group III impurity makes positive carriers available since it creates holes that can accept electrons. Thus, the act of accepting an electron will create a hole in the valence band. Looking at the energy band diagram under this lattice configuration, we can see that the energy band gap is decreased slightly by the presence of these holes. An acceptor level is thus created. A semiconductor material made in this way is known as a p-type semiconductor. The "p" means "positive" since the number of positively charged carriers exceeds the amount of negatively charged carriers. Thus, the holes

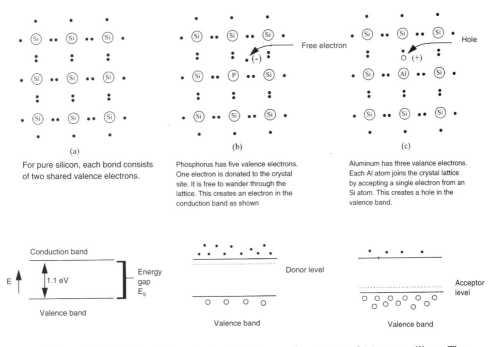

(a)

For pure silicon, each bond consists of two shared valence electrons.

(b)

Phosphorus has five valence electrons. One electron is donated to the crystal site. It is free to wander through the lattice. This creates an electron in the conduction band as shown

(c)

Aluminum has three valance electrons. Each Al atom joins the crystal lattice by accepting a single electron from an Si atom. This creates a hole in the valence band.

Figure 10.1   Silicon lattice structure for (a) pure, (b) n-type, and (c) p-type silicon. The energy band diagram is given below each type.

are the majority carriers while the electrons are the minority carriers. The number of holes in the valence band can be controlled by the concentration of acceptor atoms in the semiconductor.

To make a useful device from these semiconductor materials, both p and n types are needed in a single continuous crystal structure. Using these two materials in this manner creates a pn junction. Applying a negative voltage to the n-type material and a positive voltage to the p-type material results in a current flow from the p side of the junction to the n side of the junction. A semiconductor set up in this way is said to be forward biased. This process occurs in the operation of transistors and diodes.

When we forward bias the junction, electrons and holes will recombine at the pn junction. In other words, the electrons from the conduction band recombine or meet with the holes in the valence band. Energy will be released during this recombination process. The form that this energy takes depends upon whether the semiconductor is a direct or indirect band gap material. In a direct band gap semiconductor material, the transition from conduction band to valence band is a one-step process. This can occur when the electrons and holes have the same momentum. Since momentum must be conserved in this process, the carriers can combine directly across the band gap. The emitted energy during this process shows up as photons. This type of transfer results in radiative emission. For example, if group III-IV materials, such as GaAs, are used together, radiative photons can be produced. When these elements are used in correct amounts to produce radiative photons, the material can be considered to have a direct band gap. The energy emitted approximately equals the band gap level. An important factor involved here is the lattice spacing. This will be discussed later in more detail. In an indirect band gap semiconductor material, a direct transition of the electron from conduction band to valence band is not possible because the electrons and holes do not have the same momentum. Since momentum must be conserved in this process, a third particle known as a phonon must be involved in an intermediate step in order for this transition to occur. When this transition occurs, some of the energy will cause vibrations in the crystal structure. This energy will be in the form of heat when using materials such as pure germanium or silicon, thus making it much harder for indirect band gap materials to emit light. This type of transfer results in non-radiative emission. In the next section, we will see how different semiconductor materials can be combined together to produce a direct band gap material.

When an electron drops from the bottom of the conduction band to the top of the valence band, the energy released shows up as a photon. This photon has a wavelength λ given by the following expression:

$$\lambda = hc/E_g$$

In this expression, h is Planck's constant, c is the speed of light, and E_g is the band gap energy needed for this transition to occur. The above processes describe the interaction of radiation and matter at the quantum level. These processes are fundamental to the operation of all optoelectronic devices since the photon is the basic unit of energy exchange. In the next section, we will describe the process of spontaneous emission. This will initiate our discussions of LEDs.

10.2 SEMICONDUCTOR MATERIALS USED IN LIGHT SOURCES

When combining two or more dissimilar materials to make a semiconductor device, attention must also be paid to the lattice structure of the crystal itself. A parameter called the lattice constant must be considered. If a mismatch occurs between the lattice constants, or the spacing between lattice structures increases beyond a certain amount, non-radiative emission will occur. This parameter basically involves the spacing between the crystal lattice structures within the crystal. The alloys of materials such as GaAlAs and GaAs are used to make semiconductor light sources since their lattice constants are nearly the same. A direct band gap material can be made from these semiconductors capable of producing electrons and holes with identical momentum values, a requirement for the production of photons. Another reason for this combination is the emission wavelength. When these compounds are combined in correct proportions, emission of photons in the near infrared region occurs when electrically excited.

The relatively simple structure described above, known as a homojunction, is made from one compound, typically GaAs. This structure consists of a pn junction from which radiation can either be emitted or absorbed. Unfortunately, these homojunction structures are not very efficient emitters of photons even with the improvements previously discussed. In the late 1960s, a way was found to make efficient emitters using these materials. The result was a double-heterostructure. It consisted of a very thin direct band gap semiconductor layer sandwiched between two thicker semiconductor layers. Figure 10.2 shows the detail of this structure. When this structure is forward biased, holes and electrons from the thicker outer regions enter the thinner active layer. These outer regions are labeled "carrier confinement and light guiding regions." Once the carriers enter this active region, they become trapped between the potential energy barriers created by the thicker layers. This is called carrier confinement. The higher refractive index of the active region compared to that of the surrounding regions provides for optical confinement as well. This causes the photons produced there to become trapped within this region. The net result of the mechanisms of carrier and optical confinement work together to provide a device of high efficiency and high radiance.

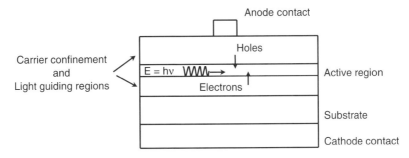

Figure 10.2   Semiconductor double-heterostructure.

As previously discussed, the active layer must be a direct band gap type semiconductor so that radiative recombination of the electrons and holes can occur. This process will produce spontaneous emission that can then be enhanced by using a heterojunction type structure to achieve carrier and optical confinement. Various alloys of groups III and V materials can then be used in different proportions to achieve a desired wavelength emission.

GaAlAs is a very common alloy used to construct semiconductor emitters such as LEDs. When used as a material in the active layer, it produces optical emission in the 800 to 900 nanometer region of the spectrum. The ratio of AlAs to GaAs determines the band gap of the material and thus the emission wavelength. Thus, the desired optical emission can be fine tuned by using various ratios of these two compounds. An equation for this compound can be set up as $Ga_{1-x} Al_x As$. A limit to the value for x (mole fraction) exists for this substance, and thus for the band gap energy. Table 10.3 gives some typical values for x and the corresponding band gap energy. As x approaches 0.37, this material changes from a direct to an indirect band gap compound. Thus, we must consider another alloy if we want a direct band gap compound greater than 1.9 eV.

Another commonly used alloy, $In_{1-x} Ga_x As_y P_{1-y}$, is a semiconductor material having an emission wavelength range between 1.0 and 1.7 μm. The mole fractions of x and y can be varied to offer a wider range of possible emission wavelengths. As we know, this wavelength region is of special interest because of the very low attenuation when used in typical optical fiber.

10.3   LIGHT EMITTING DIODES

The first visible light emitting diodes or LEDs were introduced in the late 1960s. These devices were made from semiconductor materials such as GaAsP with a resultant wavelength emission in the red portion of the visible spectrum. The optical intensity was rather weak, meaning that they could only be used for indoor applications. The structure of these first LEDs consisted of a single n-type layer of GaAsP grown on an n-type GaAs substrate. The pn junction was formed by a Zn diffusion on top. It turned out that the GaAsP

Table 10.3   Band Gap Energy for $Ga_{1-x} Al_x As$ (x = aluminum mole fraction)

x	Band Gap Energy (eV)	Emission Wavelength (nm)
.05	1.47	843.5
.10	1.55	800.0
.20	1.66	747.0
.30	1.80	688.9
.35	1.85	670.3
.37	1.90	652.6

layer was not lattice matched to the GaAs substrate. This resulted in an inefficient photon emitter.

The next development occurred in 1968 when LEDs were made with nitrogen-doped GaAsP on a GaAs substrate. These devices could emit in the green through red portion of the spectrum with a factor of about 10 in increased brightness. They still suffered from the same lattice mismatch.

By the 1980s, heterojunction LEDs became available. These LEDs were based on the GaAlAs heterostructure detailed in the last section. The only visible wavelength that these LEDs could produce was red. Shorter wavelengths are not possible when using GaAlAs materials since the band structure starts to change to indirect at about 630 nanometers. To make LEDs emit efficiently at shorter wavelengths, AlInGaP was used. This material allowed emission in the green and yellow region of the spectrum.

For emission in the blue area of the spectrum, new types of compounds had to be found. The band gap energies of GaAsP and GaP are not high enough to emit these more energetic photons. Materials that can emit at these wavelengths include GaN, ZnS, ZnSe, and SiC. The first blue LEDs were fabricated in the early 1970s using GaN. These first devices did not produce light by the radiative recombination of charged carriers. Instead, a collision-ionization process requiring much higher operating voltages was used. The resultant light intensity was weak due to this less efficient process. The same problem existed for ZnS and ZnSe materials. That left SiC material as a source emitter for blue LEDs. This material is presently used to make blue LEDs, but its performance does not match that of the longer wavelength LEDs due to its indirect band gap structure. Table 10.4 lists many of the semiconductor compounds used to construct LEDs and their peak emission wavelength.

Of primary importance are LEDs that emit in the near infrared. These infrared emitting diodes, or IR LEDs, are used in remote control devices, smoke detectors, communications systems, and many other consumer devices. The emission spectrum of a typical

Table 10.4   Semiconductor Materials

LED Material	Peak Wavelength (nm)	Color
InGaN	450	Blue
SiC	470	Blue
GaP	569	Green
AlInGaP	585	Yellow
GaAsP/GaP	585	Yellow
GaAsP/GaP	635	Orange
AlGaAs	660	Red
GaAsP	700	Red
GaAlAs	820, 880	IR
GaAs	940	IR
GaAlAs InP Alloys	1300–1500	IR

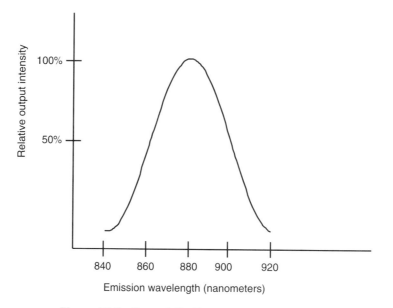

Figure 10.3 Spectral distribution of a typical IR LED.

GaAlAs infrared LED is shown in Figure 10.3. As you can see, photons are emitted at a peak wavelength of 880 nanometers. A decreasing amount of emission occurs on both sides of this peak. The spectral bandwidth is a measure of the width in wavelength at one-half the intensity of the peak emission, also called the full width at half-maximum (FWHM). In the case shown in this figure, the FWHM is 36 nanometers. Another term used for this is the half-power spectral width. Visible LEDs have approximately the same spectral bandwidth as IR LEDs. This specification is included in the data book for any particular emitter selected.

Figure 10.4 shows the basic construction of an LED lamp. The LED chip containing the pn junction structure is mounted on a recessed header. This protects the fragile structure from damage due to the environment and also reflects the light emitted from the sides

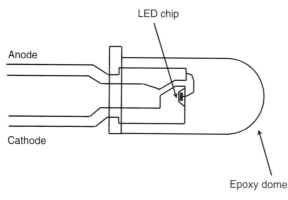

Figure 10.4 Basic construction of the LED.

of the pn junction. The LED package can also incorporate a lens to direct optical power at a specified beam angle.

One popular housing for LED chips is in a plastic epoxy/lead frame T-1 ¾ package. These plastic LEDs have only one optical surface in the lens system since the semiconductor chip is encapsulated in plastic. This results in a beam pattern having more consistency. The placement of the semiconductor chip is very critical in this type of package as this affects the emission pattern. Very close attention must be paid to this detail in the manufacture of this device. If a particular beam emission angle is required with a good consistency from device to device, a separate lens system in front of the LED is recommended when using either type of emitter described above.

As Figure 10.4 shows, the LED chip is placed in a reflecting cavity to increase the forward light output. This cavity sets on top of the cathode lead frame. Light emission that occurs from the sides of the chip reflects from the sides of this cavity. Electrical contact is accomplished by a thin wire bonded to the anode lead frame and to the top of the LED chip. The chip makes electrical contact with the cathode post through the reflector cavity. This entire assembly becomes encapsulated in an epoxy dome lens.

A certain amount of optical loss occurs in any type of the LED due to at least three loss mechanisms: absorption, internal reflection, and Fresnel reflection. When the photon emission process occurs within the semiconductor chip, photons will propagate outward from the chip. The longer this distance, the greater the internal absorption will be. This absorption also occurs within the epoxy. LEDs with chips placed in smaller packages tend to have a reduced optical loss due to absorption within the epoxy. The electrical contacts used on the surface of the chip also absorb a certain amount of optical power.

When photons are emitted at the chip/epoxy interface at an angle greater than the critical angle, they will be reflected back into the chip itself. Typical indices of refraction of the epoxy and the chip are about 1.5 and 3.1 respectively. A quick calculation shows that the critical angle will be about 29° in this case. For LEDs housed in a hermetic metal can, the chip can be coated with a plastic encapsulant to decrease internal reflection. When the photons reach the epoxy/air interface, there can also be a significant amount of internal reflection. The internally reflected rays account for a side lobe beam. In most cases, this beam is unwanted and must be blocked by adding an aperture or similar device. The extent of this reflection depends upon the location of the semiconductor chip relative to the interface. Figure 10.5a shows the rays making up the main forward beam and the side lobe beam. A typical emission pattern produced by a narrow beam LED is shown in Figure 10.5b. Notice that a significant amount of optical energy shows up in the side lobes of the emission pattern. We will consider this type of LED in the next section.

As we know, when light propagates from an optical medium of n_1 to another optical medium of n_2, Fresnel reflection will occur if these two media have unequal refractive indices. By coating the LED chip with an index matching material, the loss due to Fresnel reflection can be reduced to the point where optical transmission becomes greater than 90%. By contrast, an LED chip emitting into air has an optical transmission of about 80%.

Another embodiment of an LED structure uses a lens system in a metal can such as a TO-18 package. The lens in this type of package is usually made from glass formed by melting and reflowing of a cut glass disc. This procedure results in a lens that can have

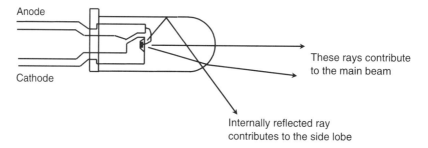

Figure 10.5a   The direction of some rays in a narrow beam LED.

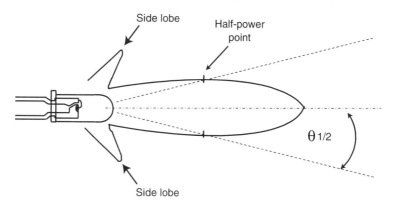

Figure 10.5b   The emission from a narrow beam LED.

considerable variation from unit to unit since the lenses are not precision ground. Thus, with this type of LED, device-to-device optical parameters such as focal point and beam emission angle will have significant variations.

10.4 LED CHARACTERISTICS

In this section, we will look at the characteristics of both visible and infrared LEDs. These characteristics are given in most data books published by the LED manufacturer. Some of these characteristics are given in graphical form, while others are given as numerical values. Each characteristic will be considered separately with a detailed explanation. To assist in the explanation of these parameters, we will use an actual data sheet for a plastic packaged IR LED. This data sheet can be found at the end of this chapter for the Siemens device SFH 484 and SFH 485.

Relative Spectral Emission—The LED produces light by the process of spontaneous emission as described in the last section. This process produces incoherent but relatively monochromatic light. A measurement of the relative light intensity over a wavelength region yields a Gaussian distribution as shown in Figure 3 of the Siemens data sheet. This emitted wavelength range is generally 100 nanometers or less, with the peak output power occurring at the peak spectral wavelength λ_{PEAK}. The useful width of this spectral distribution is measured between the half intensity points on both sides of the Gaussian curve. This width is referred to as the spectral bandwidth $\Delta\lambda$. For LEDs, there is generally 15 to 20 nanometers on both sides of the peak spectral wavelength resulting in a FWHM of 30 to 40 nanometers. The energy band gap of the semiconductor material determines the peak emission wavelength. As discussed earlier, this energy band gap can be changed by varying the mole fractions in the active semiconductor materials. Temperature variations can also change the peak wavelength. The temperature dependence on peak wavelength will be discussed later in this section.

Radiant Intensity vs. Forward Current—LEDs produce a nearly linear output optical power with drive current. Figure 4 of the Siemens data sheet shows this relationship. The maximum current rating for an LED should never be exceeded as this could cause permanent damage to the device. When current flows through the semiconductor material, a voltage appears across the junction. As current increases, more power will be dissipated resulting in a rise in temperature at the junction. The dissipated power can be calculated by multiplying the current by the voltage drop (power = current × voltage). The design and construction of the LED helps to dissipate heat from the device. A heat sink can be used to allow more current to the device. For a typical infrared LED, linear operation can be expected in a range from about 2 milliamperes to 100 milliamperes. Current levels above this will result in non-linear operation with less efficiency. One way to exceed this limit and still retain linear operation is to pulse the LED on and off at a low duty cycle. The average power used in this case is relatively small, thus allowing the device to operate without overheating. This operation will be discussed later in this section.

Figure 4 of the Siemens 484 data sheet gives information for the radiant intensity as a ratio for the vertical axis (I_E/I_E 100mA). The graph has been normalized for a 100mA input current. At this level, the output is unity. We must next look at the specification for radiant intensity in the axial direction. This can be found on the first page of this data sheet. Here, it gives a minimum radiant intensity of 50 mW/sr when I_F = 100mA. To find the minimum radiant intensity for 20 mA, we must consult Figure 4 again (horizontal axis). We find that at 20 mA, the multiplication factor is 0.2. This gives us a value of 10 mW/sr for minimum radiant intensity when using 20 mA with this device.

Characterization of Optical Power—The description of the physical volume that the output beam occupies is a very important parameter. This specification defines the usable optical power emitted within a specified half-angle cone. A measurement of its solid angle must be taken starting from the optical axis of the LED, and then moving out to the side of the beam. The measurement stops where the radiant intensity falls off to ½ of its maximum value. The data sheet gives such curves, one for each device. Figure 1 on the data sheet shows the characteristic curve for the SFH 484, while Figure 2 shows the characteristic curve for the SFH 485. It is important to note that this specification describes the spatial distribution of the emitted power. These diagrams are sometimes referred to as far field emission patterns. The radiant intensity, or power contained within a solid angle, has the units of milliwatts/steradian. This specification varies considerably from device to device, so most manufacturers grade these devices accordingly. The grading of these devices can be found at the bottom of the first page of the data sheet. As an example, SFH 484 is available as 484-1 or 484-2 depending upon the desired radiant intensity in the axial direction. Radiant intensity can also vary with beam angle and input current. The total optical power output is found by integrating the complete volume into which the light is emitted. For emitters, a useful unit of optical power is expressed in milliwatts. This total power output is used to specify LEDs, since this measurement can be reproduced to a high degree of accuracy.

Keep in mind that the disadvantage of specifying an LED only by its total power output is that spatial distribution information is lacking. Not all of this optical power may be useful in some applications. Both infrared emitters specified on the data sheet have the same total output power but very different far field emission patterns. Looking at the polar plot for SFH 485 you can see that this device emits its optical power in a relatively wide pattern. The photons emitted from the semiconductor chip in this device strike the epoxy/air interface at almost normal incidence. Thus, only slight internal reflection losses occur at this interface. The placement of the semiconductor chip relative to the epoxy/air interface is the major factor involved here. Now, let's look at the polar plot for device SFH 484. This plot shows a distribution pattern characterized by a relatively narrow main optical beam with side lobes. Figure 10.5a shows why the location of the emitter chip results in this beam pattern. For most applications, the optical energy in the side lobes cannot be used. To prevent this unwanted optical energy from entering the far field, an aperturing arrangement around the LED may be required. The side lobes result from internally reflected light at the epoxy/air interface. The semiconductor chip must be placed a further distance away from the epoxy/air interface to provide the LED with this narrow emission pattern. Thus, as the half-angle specification decreases to provide increased radiant intensity (mW/sr) in the axial direction, the amount of useful optical power (mW/cm^2) in the main beam will decrease. For example, an LED with a half-angle of 40° has a negligible amount of optical power contained in its side lobes. Designing this same LED with a half-angle of 10° will result in 20 to 30% of its optical power in its side lobes. For applications requiring a maximum output power with a narrow beam, an external lens should be used to redirect the side lobe energy.

Pulse Handling Capability—As previously discussed, the LED can handle more current by pulsing it on and off at a low duty cycle. This increased current equates to an increased optical output. This increase in output means a less sensitive receiver circuit can be used than in the case of an LED that is continuously on. A less sensitive receiver can also provide increased bandwidth resulting in more information throughput. To implement this, the receiver circuit must be optimized to look for short pulses rather than a constant light intensity. With all probability, the receiver circuit will use a photodiode to convert these short optical pulses into short bursts of current. Each current pulse is then converted into a signal voltage pulse to be amplified by an electronic circuit. An application of this type of LED operation can be found in television remote control units. These units are battery operated and provide a long period of service between battery changes. The average current consumption is quite low because the circuit used pulses the IR LED on and "off" for each instruction or command.

Figure 7 in the data sheet shows the relationship between duty cycle and forward current. The duty cycle is the ratio of the "on" time of the LED pulse to the period between pulses. It can be expressed mathematically as $D = t_p/T$ where t_p is the "on" time of the LED pulse, T is the period of time between pulses, and D is the duty cycle. Looking at the curves on this logarithmic graph, the lower line is labeled "DC" for a forward current, I_F, of 10^2 mA. This is the permissible value of the forward current when the LED is continuously on. As t_p decreases and T increases, the duty cycle becomes less, and the average power dissipation also decreases. This fact allows us to increase the current through the LED to provide more light output. This increased light output will occur during the pulse "on" time. For example, according to the graph, the maximum permissible DC forward current is 100 milliamperes. As the current increases beyond this amount, self-heating of the unit causes it to become less and less efficient. This same device can handle over 1000 milliamperes or 1 ampere of forward current for about 10^{-4} second provided the duty cycle is low enough. In this case, the device will still maintain its efficiency. This increased forward current will result in substantially more light output during the "on" time period. This is possible since using this low duty cycle results in the device being "off" most of the time. This greatly reduces the adverse self-heating affect providing a more efficient operation. The specific amount of light output will depend upon the specifications of the LED itself. Next, we present an example of this duty cycle relationship using the SFH 484.

Example 10.1

An IR LED is pulsed on for 150 μseconds every 20 milliseconds. What is the duty cycle?
    Using the above formula, we get the following result:

$$D = 150 \times 10^{-6}/20 \times 10^{-3} = 7.5 \times 10^{-3} = .075$$
The device is "on" 7.5% of the time.

Temperature Characteristics—Changes in temperature will affect the operation of an LED. As the temperature of the semiconductor material increases, the peak wavelength emission, λ_p, will shift toward longer wavelengths. The shift for the SFH 484 is typically 0.25 nanometers/Kelvin. This parameter can be found on the first page of the data sheet as the temperature coefficient, TC_λ. While this effect is relatively small, it still must be accounted for when temperature variations are large and the emission wavelength requirement is tight.

Another characteristic that displays temperature dependence is the maximum permissible forward current. The second page of the data sheet shows a graph of maximum permissible forward current vs. temperature (Figure 5). As you can see, this parameter stays relatively constant until about 25°C. As the temperature is increased above 25°C, the maximum permissible current decreases substantially due to the heating effects of the pn junction. If the device is to be used at elevated temperatures, this parameter will be a very important consideration.

Radiant intensity also decreases with increasing temperature. In some data books this parameter is given as a coefficient. For the SFH 484, this value is -0.5%/K as shown on the first page of the data sheet for I_E or Φ_E. This means that the radiant intensity decreases by 0.5% of its normal value as the temperature increases by 1° C or 1 K. The reverse holds true also. As the temperature decreases, the radiant intensity will increase by the same percentage. This parameter becomes important when large temperature variations occur and you must know how tight a tolerance can be kept on radiant intensity.

Capacitance—This characteristic is important when considering transient behavior such as when modulating the device. For the SFH 484, this capacitance, C_O, is 25 pF at a frequency of 1 MHz. The capacitance depends upon such things as the junction area of the semiconductor, frequency, carrier lifetime, and the applied voltage. The capacitance is one factor that limits the speed at which the device reacts to an input current pulse. A detailed discussion of transient response will be given later in this chapter.

10.5 DIODE LASERS—THE OPTICAL AMPLIFICATION PROCESS

The word "laser" is an acronym for light amplification through stimulated emission of radiation. In this section, we will explore the basics of the light amplification process. This process involves the interaction of radiation and matter. Both the wave and particle natures of light must be considered to successfully explain this process. While we are mainly interested in the diode laser's operation, the amplification process described in this section can apply to other types of lasers such as gas or ruby rod types. Thus, we will keep the discussion of this process more general in this section. Then, in the next section, we will apply these basic principles of operation to the diode laser.

Diode lasers and LEDs are made from the same semiconductor materials. For laser output to occur in a diode laser, internal geometry must be able to keep a relatively large current confined to a small lasing cavity. This cavity must have precise dimensions determined by the desired output emission spectrum of the laser. With the correct dimensions, the cavity will support an optical standing wave so that light amplification can occur within the optical gain medium. Most of us are familiar with the helium-neon laser. In this case, a precise optical lasing cavity composed of helium and neon gases forms the optical gain medium.

To produce laser action or stimulated emission, three basic processes are involved. Each of these three processes are represented in simple two energy diagrams detailed in Figure 10.6. For convenience, the first energy diagram on the left of the figure shows the thermal equilibrium condition. Energy level E_2 is greater than level E_1 by the amount $h\nu_1$. The first process, detailed in Figure 10.6a, involves photon energy absorption. When a photon is absorbed by the active medium of the laser, an electron can then move from its ground state to a higher energy level. The energy requirement of Planck's law must be met for this to occur. This energy must be $E_2 - E_1 = h\nu_{12}$, where E_1 is the ground state energy and E_2 is the excited state energy. When a photon of energy $h\nu_{12}$ enters this medium, it can be absorbed resulting in the movement of an electron up to the E_2 energy level. This places the electron in a higher orbit or energy level, making it unstable. The electron now has a natural tendency to return to the ground state again (E_1) a short time later, and, in the process, emit a photon of the same energy $h\nu_{12}$. Figure 10.6b shows this energy level diagram for the process known as spontaneous emission. Spontaneous emissions have a random polarization.

We learned from the last chapter that, for the case of spontaneous emission, the atoms of the material are in equilibrium with thermal radiation. This condition is described by Planck's radiation law. The energy distribution curve for this equilibrium condition is shown in Figure 10.7(a). As you can see, the number of atoms in the lower energy state E_1 is much greater than the amount in state E_2. We will see how to change this equilibrium condition as shown in Figure 10.7(a) by considering the third basic process required for laser operation.

The third process, known as stimulated emission, requires the initial condition of spontaneous emission of photons. When spontaneous emissions occur, as described above, there will be a certain amount of induced or stimulated emissions. This can occur when an electron makes a transition from level E_2 to E_1. During this process, the medium emits a photon of energy $h\nu_{12}$. This photon then strikes another excited atom also at the E_2 level. In this interaction, another photon will be created by this excited atom having the same energy and phase as the original photon. This stimulated emission process is shown in Figure 10.6c. According to Planck's law, the rate at which stimulated emission occurs is proportional to the energy density of the particular radiation.

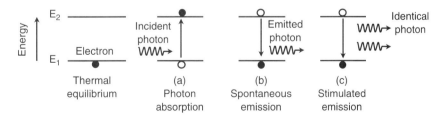

Figure 10.6   The three basic transition processes needed for laser operation.

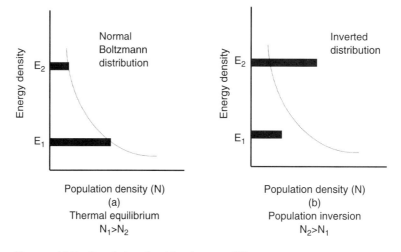

Figure 10.7   Population densities for two different systems: (a) Normal or Boltzmann distribution; (b) Inverted distribution.

Under equilibrium conditions, the rate of stimulated emission is extremely small when compared with the rate of spontaneous emission. This rate can be expressed by the following mathematical formula:

$$\frac{\text{Stimulated emission}}{\text{Spontaneous emission}} = \frac{1}{e^{h\nu/kT} - 1}$$

This result can be derived from Planck's blackbody radiation law discussed in Chapter 8. When the light source emits in the visible portion of the spectrum, T turns out to be on the order of 10^3 Kelvins. A quick calculation using Planck's law shows that, in this area of the spectrum, most of the radiation emitted will be due to spontaneous emission.

To increase the rate of stimulated emission, we must add energy to this system in a way that causes many more atoms to be in level E_2 than in level E_1. This condition is shown in Figure 10.7b. Let us consider an optical medium through which radiation in the visible portion of the spectrum can pass. As we know from our previous discussions, the electrons associated with the atoms of this medium can have various discrete energy levels as determined by Planck's law, $E = h\nu$. These levels can be E_1, E_2, E_3 and so forth. Let's consider two of these levels, say E_1 and E_2. Now, let E_2 be at a higher energy level than E_1. When this medium is in thermal equilibrium, then a Boltzmann's distribution of atoms per unit volume at the various energy levels will occur. For example, energy level E_1 will have a certain number of atoms specified by N_1 that will be greater than the number of atoms at energy level E_2 specified by N_2. Figure 10.7(a) shows a simplified graph of the energy density vs. the number of atoms per unit volume. This graph shows the population density for our two energy level system in thermal equilibrium. You can see that, as the energy level increases, the population density decreases. For this two-energy-level system, the distribution is given by the following expression:

$$\frac{N_2}{N_1} = \frac{e^{-E_2/kT}}{e^{-E_1/kT}}$$

If we can find a way to make N_2 greater than N_1, then the rate of downward transitions will exceed that of the upward transitions. This state, called a population inversion, is one of the main requirements for lasing action to occur within a laser. This condition represents a system not in equilibrium. Thus, to maintain this state there must be a constant input of energy. When a beam of light passes through this optical medium, amplification of this beam will occur because the amount of stimulated emissions will exceed the loss due to absorption. There are many ways to make N_2 greater than N_1. The method of adding energy to an amplifying medium of a laser is usually referred to as pumping. Some of the practical ways to accomplish this include adding energy by optical interaction, electrical excitation, or chemical reactions. Optical interaction was the method used to cause stimulated emission in the first optical laser. A flash lamp surrounding a ruby rod coupled its optical energy into the rod, thereby producing stimulated emission. The output from this laser was a brief optical pulse due to the very short flash duration. When a stimulated radiated photon is released by the process described above, it will have the same direction, energy, and phase as the original photon that induced the emission. When pumping a steady amount of energy into the gain medium, a continuous wave (cw) laser results. We will next consider the construction of the diode laser and its operation.

10.6 DIODE LASER CONSTRUCTION

In the structure of the diode laser, a double-heterojunction configuration confines both the carriers and optical energy. It achieves optical amplification by use of an optical feedback cavity known as a Fabry-Perot resonator. Typical dimensions of this cavity are approximately 250 to 500 micrometers long, 5 to 15 micrometers wide, and 0.1 to 0.2 micrometers thick. Figure 10.8 shows this structure in more detail.

The Fabry-Perot interferometer was discussed in Chapter 5. The context in which we considered its operation dealt with the wave nature of light only. To understand more fully how the diode laser operates, we must also consider the particle-like nature of light. The reader may want to review the concepts presented for the Fabry-Perot interferometer before proceeding. We will now consider this structure for use in a semiconductor diode laser by applying what we have already learned about the optical amplification process. The basic geometric requirement for this structure involves placing two reflecting mirrors facing each other separated by such a distance to provide a tuned cavity. This tuned cavity must be capable of supporting an optical standing wave between the mirrors. As we know, this distance must be an integral number of half wavelengths for resonance to occur. The mirrors are actually facets formed at the natural cleavage planes of the semi-

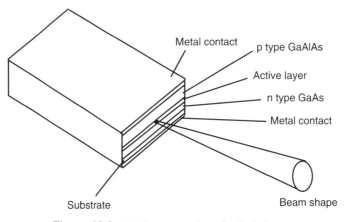

Figure 10.8   Basic construction of a diode laser.

conductor crystal. These partially reflective facets allow the amplified beam to exit the cavity. Greater reflectivities can be achieved by adding several layers of dielectric material at the rear of the laser. This increases the finesse of the optical cavity, resulting in a more efficient laser.

For lasing to occur within this Fabry-Perot structure, a population inversion must first be achieved. This can be done by injecting electrons into the semiconductor device, thus filling the lower energy states of the conduction band. The pumping action or electron injection provides the energy to maintain this population inversion. When the external pumping stops, the population inversion immediately ceases, and an equilibrium condition results. The amount of energy required to achieve lasing action depends upon many parameters. These parameters include the band gap energy of the semiconductor material, the gain coefficient of the material, the optical cavity geometry, and the losses associated with the interactions within this optical cavity.

Most of these losses are the result of material absorption, scattering, and mirror loss. The optical medium also has a gain associated with it when the population inversion occurs. Since the optical field intensity within the cavity is related to these parameters, we can express this relationship mathematically below:

$$I(z) = I(o)exp[(g-\alpha)z]$$

In this expression, I is the optical field intensity, z is the distance that the optical energy must travel, g is the gain coefficient of the optical cavity, and α is the absorption coefficient of the semiconductor cavity material. The intensity of the radiation will increase exponentially with the distance, z, that it travels in the optical cavity. Since the optical energy reflects between the two mirrors separated by a distance, L, the expression can be modified for the specific case of the diode laser in this discussion as follows:

$$I(2L) = I(o)R_1R_2exp[2L(g-\alpha)]$$

R_1 and R_2 are the reflectivities of the two mirror facets on each end of the laser diode cavity. At the threshold condition or just before lasing occurs, the gain must be equal to the total of the losses within the laser structure, or $I(2L) = I(o)$. The distance 2L equates to a round trip inside the cavity. Thus at this point, the original expression reduces to the equation below:

$$\exp[2(g-\alpha)L]R_1R_2 = 1$$

Lasing will continue if the conditions satisfy the above equation. Applying some typical numbers to this equation for a GaAs diode laser we get the following results. The facets for a GaAs device have reflectivities $R_1 = R_2 = 0.32$. This means that 32% of the optical energy reflects from each surface. The diode laser cavity length, L, is 500μm and the two coefficients are $g = 33\text{cm}^{-1}$ and $\alpha = 10\text{cm}^{-1}$.

The Fabry-Perot structure is basically a highly selective filter. An optical wave having an integral number of half-wavelengths will resonate within the cavity formed between the two mirrors. The optical medium in which spontaneous emission occurs must be selected such that this wavelength condition is satisfied. Figure 10.9 shows the basic steps needed for lasing to occur. In diagram A, the unexcited atoms of the optical medium are initially in the ground state. Adding energy to the optical medium as shown in diagram B causes the atoms to be in an excited state. The electrons of the optical medium are raised to a higher energy level by precisely the amount $E = h\nu_{12}$. Spontaneous emission immediately results in photons being emitted isotropically with the energy $E = h\nu_{12}$ as electrons in the excited medium drop back to the ground state. The photons produced in

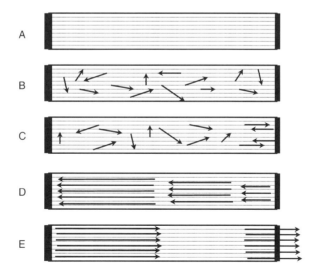

Figure 10.9   Basic steps needed for lasing to occur.

this way now have the correct wavelength to produce a standing wave within the structure. Since emission occurs isotropically, most are not aligned properly for this to occur. In diagram C, a few photons become started along the axis of the interferometer and experience reflection as they hit the mirrors at either end. The remaining photons still move isotropically and at random phases to each other. The initial few photons lucky enough to be emitted along the optical axis of the interferometer stimulate other transitions to occur having the same direction and phase. This stimulated emission will occur as long as the optical medium remains in the excited state. In diagram D, amplification occurs as a result of the multiple reflections between the mirrors as more and more transitions occur. When the population inversion becomes high enough to overcome the energy loss within the cavity, lasing starts. The point where the energy losses just balance the energy gain due to stimulated emissions is called the threshold condition. In diagram E, the optical energy continues to build as the threshold condition is exceeded. The optical beam will exit the mirror having the least reflectivity. In the case of a diode laser, the beam exits both facets. This light beam will be monochromatic and coherent as a result of the stimulated emission process.

The optical amplification process is shown graphically in Figure 10.10. This figure gives us the relationship between input laser diode drive current and optical output power. At relatively low currents, approximately 20 milliamperes or less, only spontaneous emission can be seen. The population inversion level at this point is not high enough to overcome the energy losses within the cavity. At the threshold current, I_{th}, the population inversion is high enough to overcome the energy losses. Lasing will occur with the addition of more energy to this system. In this case, the energy must be in the form of electrical current. The medium, at this point, consists of both spontaneous and stimulated emissions. From looking at the graph in Figure 10.10, we see that after reaching I_{th}, a small increase in current produces a much larger increase in optical power. In the lasing condi-

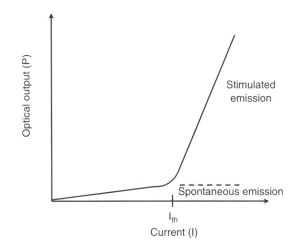

Figure 10.10   Optical output curve for a typical diode laser.

tion, a spectral width of about 1 or 2 nanometers occurs with a fully narrowed beam. Spontaneous and stimulated emissions are still present, but the amount of stimulated emissions far exceeds that of spontaneous emissions. The width of the output beam depends upon the geometry of the Fabry-Perot structure.

10.7   DIODE LASER CHARACTERISTICS

In this section, we will look at the important characteristics of a typical diode laser. You will find these characteristics given in most manufacturers' data books on the product. As with LEDs, some characteristics are given in graphical form, while others are given as a numerical value. To assist us in the explanation of these parameters, we will use an actual diode laser data sheet. This data sheet can be found at the end of the chapter with the Mitsubishi part number ML40123N. Please note that not all of the parameters to be discussed are found on this data sheet. The missing necessary information will be given as a graph or illustration in this section.

> *Relative Intensity vs. Wavelength*—The diode laser produces its light by the process of stimulated emission within a Fabry-Perot optical cavity. Light produced in this fashion is coherent, and thus has a more narrow spectral bandwidth, $\Delta\lambda$, than that of an LED. Figure 10.11 shows the spectral distribution in terms of relative intensity of a typical diode laser operating with an input current above I_{th}. The emitted wavelength range is on the order of 2 nanometers or less. During laser operation, the spectral emission pattern takes on a Lorentzian shape rather than a more spread-out Gaussian shape displayed when using LEDs. The peak wavelength also depends upon the case temperature of the diode laser and the forward current level.
>
> *Radiant Intensity vs. Forward Current*—When a diode laser operates at a current level of less than I_{th}, mostly spontaneous emission occurs. As it reaches threshold

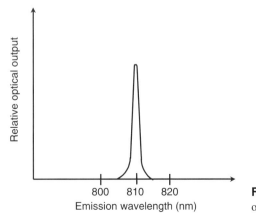

Figure 10.11   Output characteristic of a typical Fabry-Perot diode laser.

current level, stimulated emission begins to dominate the output of the diode laser. At current levels above I_{th}, the output power increases dramatically due to lasing action. Figure 1 of the ML40123N data sheet shows the relationship between optical output and drive current. The optical output is linear with input current, and increases greatly with a relatively small increase in drive current. Great care must be taken not to exceed the maximum input current, as the device could easily be destroyed. This fact makes the use of a complex drive circuit for diode lasers a requirement. Diode lasers are also more sensitive to temperature changes. To obtain a constant output at varying temperatures, the laser current must be monitored and the case temperature controlled. To assist in this task, the diode laser has a built-in monitoring photodiode. The output from this photodiode can be used to provide stable operation when used in a special feedback circuit. We will consider this feature of the diode laser later on in this section.

Characterization of Optical Power—The output emission or far-field pattern from a Fabry-Perot diode laser is displayed in Figure 10.12. The pattern takes on this characteristic because of the geometry of the active layer and the mirror facets. As we discussed earlier in this chapter, this Fabry-Perot structure contains cavity modes due to the resonance of the optical wave. Longitudinal cavity modes can be described by a sinusoidal power variation along the longitudinal axis. Other modes contained within this structure are related to the transverse and lateral distances. All of these modes determine the output beam's intensity distribution or pattern. The shape of this beam is not circular as might be expected. To characterize the forward beam, two specifications for beam divergence angle are required. The beam angle labeled θ// is the beam angle at half maximum for the divergence parallel to the active semiconductor layer. The angle labeled θ⊥ is the beam angle at half maximum for the divergence perpendicular to the semiconductor active layer. A closer look at the emission of the beam from the active area shows a lon-

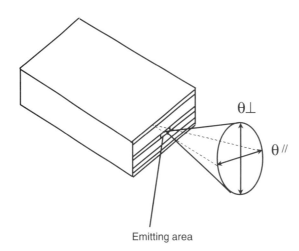

Emitting area

Figure 10.12 Forward beam from a diode laser showing the far-field emission pattern.

gitudinal separation between the emission points for the θ // and θ⊥ emissions. This astigmatism can be changed by using an external lensing arrangement to yield the desired beam shape.

As with LEDs, the radiant intensity is measured in milliwatts/steradian, and varies with beam angle and input current. On the ML4XX23 series data sheet at the end of this chapter, you can find two graphs of relative light output vs. angular off-axis measurement. Figures 2 and 3 show the graphs for the far-field patterns of the diode laser. Figure 2 gives the pattern in the plane parallel to the heterojunctions, θ //. Figure 3 gives the pattern in the plane perpendicular to the heterojunctions, θ ⊥. Notice that as the input current increases, the pattern also changes.

Although not displayed in Figure 10.12, a beam also exits the rear of the diode laser. This beam can be used for monitoring purposes, as mentioned above, to prevent the occurrence of an excess input current level. A monitoring photodiode placed in the path of the rear beam provides an electrical output for this purpose.

Polarization Ratio—Since the light output from a diode laser results from amplified stimulated emission, the beam will display a certain amount of polarization. The beam emission from the Fabry-Perot diode laser consists mostly of light polarized parallel to the active semiconductor layer. A small component of light polarized perpendicular to the semiconductor layer exists also. The ratio of these two intensities determines the polarization ratio. This ratio changes with light output. The appearance of these two components is the result of the Fabry-Perot structure since it supports longitudinal, lateral, and transverse modes. The strength of these modes determines the output beam characteristic.

Temperature Characteristics—Changes in temperature must be taken into account during the operation of a diode laser. A relatively small temperature change can easily destroy the device by allowing too much current to flow through the pn junction. As we know, an increase in current causes an increase in optical output. If this optical output increases beyond its maximum permitted value, the facets on the ends of the Fabry-Perot cavity will burn out causing the unit to be destroyed. Figure 10.13 shows the light output vs. forward current relationship at three different temperatures for a typical diode laser. This particular device has a maximum output optical power rating of 5 milliwatts when operated continuously (not pulsed). If, for example, this device had an output of 3.0 milliwatts at 25° C, and then the temperature decreased to 0°C, the input current would remain approximately the same but the current requirement for lasing would be much less. You can see from the curves that this condition would result in the device producing an optical output well in excess of 5 milliwatts thus causing permanent damage. We will see in the next chapter how the output from monitoring photodiode can be used in a feedback configuration to prevent this from occurring.

Monitoring Photodiode—As discussed previously, most diode lasers have a special built-in photodiode that monitors the output power level. Figure 4 on the data sheet displays the relationship between light output and monitor photodiode current. Since this relationship is linear, a circuit that limits the output power from the diode

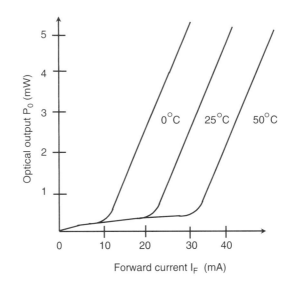

Figure 10.13 Diode laser light output characteristic for 3 different temperatures.

laser can easily be incorporated into the power supply for the device. This electronic function can be found in an automatic power control (APC) circuit.

In the previous several sections, we discussed the basic construction and characteristics of LEDs and diode lasers. With this information, we are ready to consider a very important quality that makes these devices so useful in electro-optical circuits. LEDs and diode lasers have the ability to be switched on and off very quickly. This process is known as modulation. We considered this process briefly in Section 10.4 when we discussed the pulse handling capability of the LED. In the next section, we will discuss in more detail the modulation of LEDs. In Section 10.9, we will then consider the modulation process of the diode laser.

10.8 MODULATION OF LEDS

Modulation of a light source is a basic requirement for transmitting optical information to a remote point. With the appropriate receiver at this point, the incoming light energy can be demodulated back to its original input signal. The modulation process can be done by directly switching the light source on and off or by varying the intensity of the light with time. The limiting factor for the modulation rate of the light source is the frequency response of the light source itself. During this modulation process, the transistor or driver switching speed also increases the time needed to completely respond to a step input condition. This frequency response or rise time directly affects the bandwidth of the transmitter in which the light source operates. In this section, we will consider the factors responsible for the frequency response of the light source itself. Techniques will also be given that demonstrate how to modulate an LED. In the next chapter, we will consider other limiting factors that may be encountered when interfacing a light source to electrical components.

The simplest way to modulate an LED is to apply current pulses in the forward bias mode as shown in Figure 10.14. The resistor in this circuit limits the amount of current that can flow through the pn junction and thus determines the intensity of the emitted light. The transistor acts as a switch to turn the LED on or off. We will neglect, for the time being, the response time of the transistor. In the next chapter, we will consider the response times for both devices. Thus, the maximum rate at which the LED can be modulated in this fashion depends primarily on the spontaneous recombination time of the carriers in the semiconductor (the carrier lifetime, τ_{sp}) and the capacitances associated with the semiconductor. As we know, the spontaneous recombination time depends upon the material composition and the carrier concentration. Response times can be as long as milliseconds for some devices to as short as fractions of a nanosecond for other devices. For GaAs based materials, this time period is on the order of a nanosecond at room temperature for a carrier concentration of 10^{19} cm^{-3}.

If the optical data modulation rate is less than about 50 Mbit/sec, the LED may be a good choice for a device to use. The drive circuit for an LED is much less complex than that of a typical diode laser. The useful lifetime of an LED can be substantially longer than that of a laser diode, and it is characterized by a gradual decrease in light output rather than an abrupt failure.

In most instances, the maximum speed at which an LED can operate is controlled mainly by the carrier lifetime, and partially by the semiconductor capacitances. This capacitance depends, in part, upon the structure of the semiconductor pn junction and the applied voltage. For simplicity, we will refer to these capacitances collectively as the semiconductor capacitance. A detailed study of these capacitances is beyond the scope of this book. The response time is measured by considering the rise and fall times associated

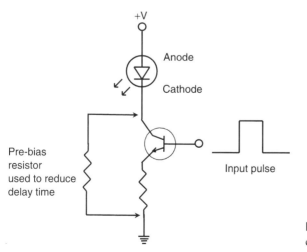

Figure 10.14 Basic modulation circuit for an LED.

with the optical signal. This response time has a direct relationship to the bandwidth of the device.

Using the simple circuit described in Figure 10.14, we apply an input forward voltage pulse to the LED instead of a constant forward voltage. In this case, it will stay on for the duration of this pulse period. We will assume that the input pulse is rectangular as shown in Figure 10.15a. When applying this step voltage to the pn junction, photons will be produced by the spontaneous emission process described earlier in this chapter. The response to this step increase is not instantaneous. At first, there is a short time period for the output pulse to achieve a 10% amplitude. Then, there are finite time intervals required for the output pulse to achieve its 90% and maximum amplitude. When the input voltage signal is removed, there are also finite time intervals required for the pulse to fall to the 90%, 10%, and zero output levels. Figure 10.15b shows the shape of a typical output pulse using the simple circuit in Figure 10.14. A typical rise time measurement, τ_R, is the time needed for the pulse to go from the 10% level of the maximum pulse amplitude to its 90% level. For the fall time, τ_F, the measurement is taken from the 90% level to the 10% level.

As stated before, the response time is dependent upon the semiconductor capacitance and the carrier lifetime of the LED. For a typical GaAlAs IR LED, the delay time has been found to be about 1 nanosecond for a pulse current of 80 mA. This time period can be reduced by applying a constant pre-bias current of 1 to 2 mA in the forward direction as shown in Figure 10.14. This small constant current flow reduces the semiconductor capacitance and the corresponding response time to about 1 nanosecond. The value of

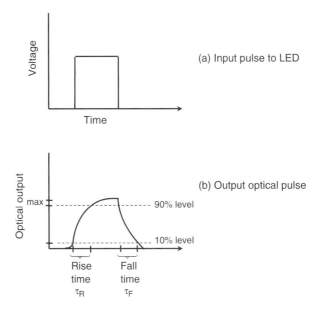

Figure 10.15   Response of a typical LED to an input square wave.

the pre-bias resistor depends upon parameters such as LED capacitance and how much the receiver circuit can tolerate a constant illumination level from the LED. With the correct amount of LED pre-biasing, the response time will depend mainly upon the spontaneous lifetime. In the next chapter, we will consider this relationship in more detail.

10.9 MODULATION OF DIODE LASERS

As with LEDs, diode lasers also have basic limitations on their modulation rate that relate to response time. This rate for diode lasers depends upon the stimulated and photon lifetimes associated with the processes of stimulated emission and optical amplification. Before stimulated emission can occur, spontaneous emission must be present within the laser cavity. When a population inversion state is achieved by pumping action, lasing can take place. In this population inversion state, photons are generated due to spontaneous and stimulated emissions. It takes a finite amount of time for a photon to stimulate an emission of another like photon. This time period is known as the stimulated lifetime, τ_{st}. For most diode lasers, this time period is on the order of 10^{-11} seconds. Next, the stimulated photon will encounter one of the end facets of the diode laser's optical cavity if it is in the proper mode. Reflection will then occur at this facet. The time period during which the photon remains in the Fabry-Perot optical cavity before it exits is known as the photon lifetime, τ_{ph}. A typical value for τ_{ph} is on the order of 10^{-12} seconds, which is about one order of magnitude smaller than τ_{st}. Thus, the photon lifetime establishes the upper limit to the response time of the diode laser.

The relatively short stimulated lifetime allows diode laser modulation at much higher speeds than LEDs. This can be accomplished by modulating the diode laser in its operating region above the threshold current, I_{th}. If the diode laser is completely turned off during pulse modulation, it will take a time period that depends upon the spontaneous lifetime, τ_{sp}, before stimulated emission occurs again. The device will effectively be no better than an LED in terms of response time. This time delay period can be eliminated if the diode laser is DC biased continuously at I_{th} or above. At this current level, lasing begins to occur and spontaneous emission is ongoing. The pulse can now be produced by increasing the current above I_{th} to some maximum value determined by the limiting resistor. In this way, the speed of response will be greatly increased. The speed of response in this case is limited by the stimulated lifetime associated with the diode laser.

A common method used for high speed digital or pulse modulation is shown graphically in Figure 10.16. This curve was initially introduced in Section 10.6 as the current-to-light output curve for a typical diode laser. To take full advantage of the diode laser's superior response time, the input current for the "off" or "0" state is adjusted to a level just above I_{th}. The "on" or "1" state is adjusted to a maximum value consistent with the operating range of the device that also meets the requirements for light output. At this current input level, the output is predominantly stimulated emission or coherent laser light. When the input signal calls for the "off" state, the diode laser is still biased slightly above I_{th}, thus taking advantage of the fast response time associated with stimulated emission. In this state, stimulated emission still occurs, but at a much lower output level. The diode

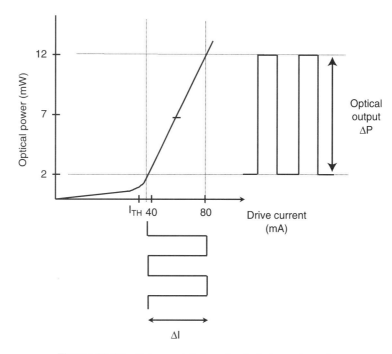

Figure 10.16   Pulse modulation of a diode laser above I_{TH}.

laser is pulse-modulated in this fashion with a specialized circuit that must correct for slow changes in temperature, and gradual aging of the diode laser itself.

Looking at Figure 10.16, we can see a typical modulation example using a diode laser designed for high speed optical communications. The threshold current for this device is typically 30 mA. The "off" level of current is chosen to be about 40 mA, which corresponds to an optical power output of 2 milliwatts at 25°C. The "on" level of current is chosen to be 80 mA, which corresponds to an optical power output of 12 milliwatts at 25°C. The variation in power output is 10 milliwatts, and the diode laser is still operating in the linear region of the curve. To provide this modulation of light, the current must be varied from 40 to 80 mA. In the next chapter, we will consider an actual circuit for a diode laser that can be used for optical communications.

10.10 COUPLING LIGHT SOURCES TO OPTICAL FIBER

In the previous sections, we considered two types of semiconductor light sources, the LED and the diode laser. Each source has its own unique electrical and optical parameters. When using either of these sources in a fiber optic transmitter, many of the parameters associated with both the device and the optical fiber must be considered. Typical device parameters include optical power distribution, size, and radiance. Typical fiber parameters include numerical aperture (NA), core size, and the difference between the core and cladding refractive indices. We will use simple geometrical optics in the treatment of this topic.

Typical semiconductor light sources emit optical power that can be measured in terms of radiance or brightness. Since we must measure in all three dimensions, we find it convenient to use a solid angle when specifying this optical emission. Radiance, L, is defined as the optical power radiated per unit surface area into a solid angle. Thus, L has the units of Watts/m^2·sr. When considering how much optical power can be coupled into an optical fiber, this spatial distribution pattern becomes just as important as the source's total output power. For example, a typical diode laser can emit a more intense beam of larger radiance compared to that of a typical IR LED. The spatial distribution of the optical energy for the diode laser is more compact than for the IR LED in the near-field case. The IR LED emits its optical energy within a larger beam. Thus, the diode laser will have a larger radiance or L value.

The active surface of an IR LED emits diffuse light in a Lambertian or hemispherical output pattern. When making an off-axis measurement, the radiance will follow a cosine function where the greatest optical power occurs normal to the surface of the emitter. Thus, at normal incidence, the optical power will have the greatest radiance. At 90° from normal incidence, it will be zero. The diode laser, on the other hand, emits light defined by a much narrower beam. This beam has a more complex pattern. Since the beam is highly directional, more light can be coupled into an optical fiber.

When coupling light into a multimode fiber, the primary losses are due to area mismatch, numerical aperture, and Fresnel reflection at the fiber's end face. We will consider each of these loss mechanisms separately.

An area mismatch occurs when the source emitting area is larger than the core diameter. A large amount of optical power could be lost in this instance. This problem can be solved by using a lens to collect the emitted light. The lens can be used to reduce the size of the source to allow for better coupling. Unfortunately, this reduced source image size will have a larger solid angle. The beam divergence on the other side of the lens will be greater, thus creating a problem associated with numerical aperture. Since most optical fibers have a relatively small numerical aperture, this method may not yield any improvement.

Consider the case when the source emitting area is smaller than the diameter of the fiber core. In this case, we will find another problem to solve. A Lambertian source such as an IR LED will usually emit light into a larger cone than that formed by the fiber's acceptance angle. In Figure 10.17, we show a small Lambertian source placed as close as possible to a step-index multimode fiber. For simplicity, the source can be considered circular with a radius r_s less than the radius of the fiber's core. Any light emitted outside the fiber's acceptance angle will not enter into the fiber. Neglecting any losses due to Fresnel reflection, the amount of power coupled into the fiber can be calculated by using the following mathematical relationship:

$$P_{LED} = \pi^2 r_s^2 L_0 (NA)^2$$

where $r_s <$ the fiber's core radius, and NA is the fiber's numerical aperture.

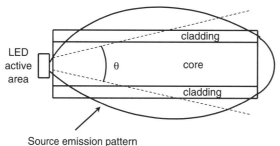

LED active area

θ

cladding

core

cladding

Source emission pattern

θ is the acceptance angle of the fiber

Figure 10.17 Coupling an optical source to optical fiber. Light emitted outside of the acceptance angle is lost.

As discussed earlier in this chapter, the emission pattern of a typical diode laser has a full-width half-maximum (FWHM) of about 30° in the plane perpendicular to the junction of the active layer and about 11° in the plane parallel to the junction. This angular distribution being greater than the fiber's acceptance angle presents a problem for efficient coupling. To improve this coupling efficiency, spherical lenses are sometimes used. Figure 10.18 shows an example of a micro ball lens coupling light from a diode laser into a glass fiber. The micro ball lens is placed very close to the chip in the diode laser. Micro ball lenses can also be used with IR LEDs or in coupling two fiber ends together. The ball lens has a MgF_2 anti-reflection coating to eliminate back reflections into the diode laser. Figure 10.18 also shows a lens in the form of a cylinder (LENSLET) that directs the light from the diode laser into the single mode fiber.

The third loss mechanism concerns light loss due to Fresnel reflection. If the indices of refraction of the fiber core and the medium separating the light source are the same, then no loss will occur due to Fresnel reflection. When a difference occurs, such as at the air-glass interface, there will be reflection losses. The following mathematical expression can be used to calculate this loss:

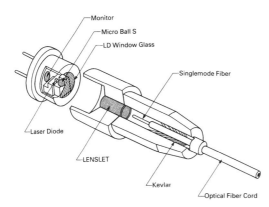

Monitor

Micro Ball S

LD Window Glass

Singlemode Fiber

Laser Diode

LENSLET

Kevlar

Optical Fiber Cord

Figure 10.18 A micro ball lens is used to couple light from a diode laser into an optical fiber. Reprinted with permission of Nippon Electric Glass Co., Ltd.

$$R = [(n_2 - n_1/n_2 + n_1)]^2$$

In the above equation, n_2 is the fiber core index and n_1 is the index of the separating medium.

Example 10.2

An IR LED has a circular emitting area of radius 25 μm. Assuming that the optical emission pattern is Lambertian with a radiance of 75 W/(cm²·sr), how much optical power can be coupled into a step-index fiber having a numerical aperture of 0.20 and a core diameter of 100 μm?

Using the mathematical expression for power coupling, we get:

$$P_{LED} = \pi^2 r_s^2 L_o(NA)^2 = (3.14)^2 (2.5 \times 10^{-3} cm)^2 (75\ W/cm^2)(.20)^2$$
$$P_{LED} = 185\ \mu W$$

SUMMARY

In this chapter, we have seen how semiconductor emission processes can be controlled to produce a desired optical output. Spontaneous emission is the basic process by which the LED gives off its light. The emission wavelength can be controlled, to some extent, by the addition of different amounts of dopants. In the operation of the diode laser, spontaneous emission also occurs, but this emission process only initiates the amplification process of light known as stimulated emission. This amplification process can occur when we control both the geometry of the device structure and the semiconductor construction.

The optoelectronic emitters considered in this chapter are used in typical transmitter circuits. In the next chapter, we will consider how to integrate these optoelectronic emitters into typical transmitter circuits. Since the LED is the simpler of the two basic emitters discussed in Chapter 10, it will be considered first in some basic types of applications. We will later consider the circuit parameters required to safely operate a diode laser.

SIEMENS

SFH 484
SFH 485
GaAlAs INFRARED EMITTER

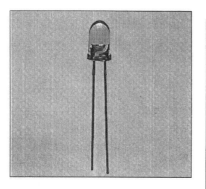

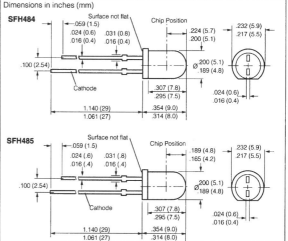

Dimensions in inches (mm)

SFH484

Surface not flat
.059 (1.5)
.024 (0.6) .031 (0.8)
.016 (0.4) .016 (0.4)
.100 (2.54)
Chip Position
.224 (5.7)
.200 (5.1)
.232 (5.9)
.217 (5.5)
Ø 200 (5.1)
.189 (4.8)
Cathode
.307 (7.8)
.295 (7.5)
.024 (0.6)
.016 (0.4)
1.140 (29)
1.061 (27)
.354 (9.0)
.314 (8.0)

SFH485

Surface not flat
.059 (1.5)
.024 (.6) .031 (.8)
.016 (.4) .016 (.4)
.100 (2.54)
Chip Position
.189 (4.8)
.165 (4.2)
.232 (5.9)
.217 (5.5)
Ø 200 (5.1)
.189 (4.8)
Cathode
.307 (7.8)
.295 (7.5)
.024 (0.6)
.016 (0.4)
1.140 (29)
1.061 (27)
.354 (9.0)
.314 (8.0)

FEATURES

- **T1$\frac{3}{4}$ Package**
- **Blue Tinted Plastic Lens**
- **Long Term Stability**
- **Very High Power, 25 mW Typical at 100 mA**
- **Good Spectral Match with Silicon Photo Detector**
- **Gallium Aluminum Arsenide Material**
- **SFH 484–16° Narrow Beam,
  SFH 485–40° Medium Beam**
- **Smoke Detection Application: SFH484-E7517 (UL Recognized)**

DESCRIPTION

SFH 484, an infrared emitting diode, emits radiation in the near infrared range (880 nm peak). The device comes in a T1$\frac{3}{4}$ (5 mm) plastic package. Uses for SFH 484 include IR remote control, smoke detectors, and other applications requiring high power, such as IR touch screens.

The SFH 485 contains the same IR emitter chip as the SFH 484 but features a wider beam.

Maximum Ratings

Operating and Storage Temperature
  Range (T_{OP}, T_{STG}) .... −55° to +100°C
Junction Temperature (T_J) .... 100°C
Reverse Voltage (V_R) .... 5 V
Forward Current (I_F) .... 100 mA
Surge Current (I_{FSM}) t=10 µs .... 2.5 A
Power Dissipation (P_{TOT}) .... 200 mW
Thermal Resistance (R_{thJA}) .... 375 K/W

Characteristics (T_A=25°C)

Parameter	Symbol	Value	Unit	Condition
Peak Wavelength	λ_{PEAK}	880±20	nm	I_F=100 mA, t_p=20 ms
Spectral Bandwidth	$\Delta\lambda$	80	nm	I_F=100 mA
Half Angle SFH 484 SFH 485	φ φ	±8 ±20	Deg. Deg.	
Active Chip Area	A	0.16	mm²	
Active Chip Area Dimensions	L x W	0.4 x 0.4	mm	
Switching Times, I_E, 10% to 90% and 90% to 10%	t_R, t_F	0.6/0.5	µs	I_F=100 mA
Capacitance	C_0	25	pF	V_R=0 V, f=1 MHz
Forward Voltage	V_F V_F	1.5 (≤1.8) 3.0 (≤3.8)	V V	I_F=100 mA, t_p=20 µs I_F=1 A, t_p=100 µs
Reverse Current	I_R	0.01 (≤1)	µA	V_R=5 V
Temperature Coefficient, I_E or Φ_E	TC_I	−0.5	%/K	
Temperature Coefficient, V_F	TC_V	−2	mV/K	
Temperature Coefficient, λ	TC_λ	0.25	nm/K	

Radiant Intensity I_E in Axial Direction	Sym	solid angle of Ω=0.01 sr		solid angle of Ω=0.001 sr		Unit	Condition
		SFH 484-1	SFH 484-2	SFH 485-1	SFH 485-2		
	I_{Emin}	50	80	16	25	mW/sr	I_F=100 mA, t_p=20 ms
	I_{Emax}	100	—	32	—	mW/sr	
	I_{Etyp}	700	900	220	340	mW/sr	I_F=1 A, t_p=100 µs

Reprinted with permission from Siemens Components, Inc. Optoelectronic Division.

Figure 1. Radiation characteristic—SFH484 $I_{REL}=f(\varphi)$

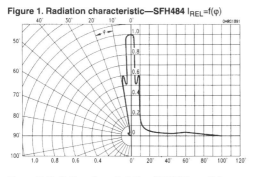

Figure 2. Radiation characteristic—SFH485 $I_{REL}=f(\varphi)$

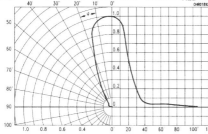

Figure 3. Relative spectral emisson $I_{REL}=f(\lambda)$

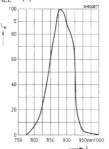

Figure 4. Radiant intensity I_E/I_E 100mA$=f(I_F)$, Single pulse, $\tau=20\ \mu s$

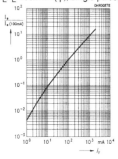

Figure 5. Maximum permissible forward current $I_F=f(T_A)$

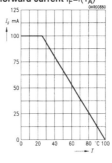

Figure 6. Forward current $I_F=f(V_F)$ Single pulse, $\tau=20\ \mu s$

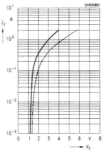

Figure 7. Permissible pulse handling capability $I_F=f(\tau)$, $T_A=25°C$, duty cycle D=Parameter

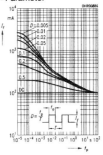

Figure 8. Maximum forward current vs lead length, package bottom and PC board $I_F=f(l)$, $T_A=25°C$

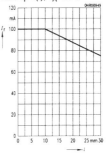

MITSUBISHI LASER DIODES
ML4XX23 SERIES

AlGaAs LASER DIODES

TYPE NAME **ML40123N**

DESCRIPTION

ML4XX23 is a AlGaAs laser diodes which provides a stable, single transverse mode oscillation with emission wavelength of 785nm and standard continuous light output of 3mW.

ML4XX23 is hermetically sealed devices having the photodiode for optical output monitoring.

ML4XX23 is produced by the MOCVD crystal growth method which is excellent in mass production and characteristics uniformity.

FEATURES

● Output 5mW (CW)
● Built-in monitor photodiode
● MQW* active layer
● Low droop
   * : Multiple Quantum Well

APPLICATION

Laser beam printing, digital copy

ABSOLUTE MAXIMUM RATINGS

Symbol	Parameter	Conditions	Ratings	Unit
Po	Light output power	CW	5	mW
V_{RL}	Reverse voltage (Laser diode)	–	2	V
V_{RD}	Reverse voltage (Photodiode)	–	15	V
I_{FD}	Forward current (Photodiode)	–	10	mA
Tc	Case temperature	–	– 40~ + 60	℃
T_{stg}	Storage temperature	–	– 40~ + 100	℃

ELECTRICAL/OPTICAL CHARACTERISTICS

Symbol	Parameter	Test conditions	Min.	Typ.	Max.	Unit
I_{th}	Threshold current	CW	–	18	40	mA
I_{OP}	Operating current	CW, Po = 3mW	–	40	70	mA
V_{OP}	Operating voltage	CW, Po = 3mW	–	2.0	2.5	V
η	Slop efficiency	CW, Po = 3mW	–	0.15	–	mW/mA
λ c	Center wavelength	CW, Po = 3mW	770	785	800	nm
$\theta_{//}$	Beam divergence angle (parallel)	CW, Po = 3mW	9	11	15	deg.
$\theta_{\perp}$	Beam divergence angle (perpendicular)	CW, Po = 3mW	22	29	36	deg.
I_m	Monitoring output current (Photodiode)	CW, Po = 3mW, V_{RD} = 1V, R_L = 10 Ω (Note 1)	–	0.6	–	mA
I_D	Dark current (Photodiode)	V_{RD} = 10V	–	–	0.5	μ A
C_t	Capacitance (Photodiode)	V_{RD} = 5V, f = 1MHz	–	7	–	pF
D	Droop	CW, Po = 3mW	–	3	–	%

Note 1 : R_L = the load resistance of photodiode

MITSUBISHI ELECTRIC

MITSUBISHI LASER DIODES
ML4XX23 SERIES

AlGaAs LASER DIODES

OUTLINE DRAWINGS

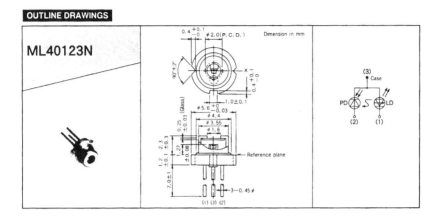

ML40123N

 MITSUBISHI
ELECTRIC

AlGaAs LASER DIODES

TYPICAL CHARACTERISTICS

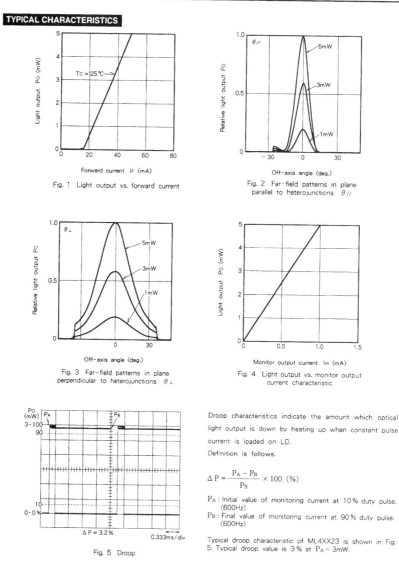

Fig. 1 Light output vs. forward current

Fig. 2 Far-field patterns in plane
parallel to heterojunctions $\theta_{//}$

Fig. 3 Far-field patterns in plane
perpendicular to heterojunctions $\theta_\perp$

Fig. 4 Light output vs. monitor output
current characteristic

Fig. 5 Droop

Droop characteristics indicate the amount which optical
light output is down by heating up when constant pulse
current is loaded on LD.
Definition is follows.

$$\Delta P = \frac{P_A - P_B}{P_B} \times 100 \ (\%)$$

P_A : Initial value of monitoring current at 10 % duty pulse.
(600Hz)
P_B : Final value of monitoring current at 90 % duty pulse.
(600Hz)

Typical droop characteristic of ML4XX23 is shown in Fig.
5. Typical droop value is 3 % at $P_A = 3$mW.

MITSUBISHI ELECTRIC

11

Optical Transmitters

In a typical optical transmitter design, the first consideration will most likely be the choice of a particular light source. Usually, this choice can only be made after careful consideration of receiver requirements such as responsivity and bandwidth. Optical receiver components and their typical applications will be covered in Chapters 12 and 13. In this chapter, we will be concerned with the applications involving two basic semiconductor light sources, the LED and the diode laser.

When designing an optical transmitter, you obviously want the best possible performance. So, your first choice may naturally be the diode laser, since this source has many of the fine optical characteristics such as narrow line width and fast response. But let's not forget the LED. The LED should be considered for use in applications where source line width and speed of response are not so critical. It requires relatively simple electronic circuitry, and can deliver a relatively large optical output. By contrast, a diode laser requires complex circuitry to prevent a catastrophic failure due to even the most brief overcurrent situation. If the total light output intensity is an important consideration, then the IR LED may be the better choice. IR LEDs find applications in television remote control units and in smoke detection devices. For these devices, the total optical output power, P_o, is more important than a spectrally pure, collimated light beam. As mentioned before, the diode laser can easily fail catastrophically with the application of just a relatively short transient. By contrast, IR LEDs are much less susceptible to catastrophic failure due to the same level of transients. The reason for this difference can be found in the device construction. When using a diode laser, a nanosecond transient above its maximum allowable level can easily destroy the reflecting facets on both ends of the optical cavity. The construction of a typical IR LED allows it to withstand these same transients without being completely destroyed.

The diode laser may be the correct choice for an optical source if spectral width, speed of response, and polarization are important considerations. A typical CD player makes good use of all three of these diode laser qualities. In this application, the optical beam must be collimated and reduced to a small size by a special lens. This small spot of laser light "reads" the stored information on the underside of the disk by processing the reflected laser light. This information is stored in pits of different lengths that reflect the laser light to the receiver section of the player. A relatively small optical power level can be used due to the close spacing of the optical read head and the CD itself. A somewhat higher optical power level may be required for the purpose of recording information on the compact disk media.

A diode laser usually requires a much more complex circuitry than that used with an LED. Parameters such as temperature dependence on light output require that the diode laser's optical output be carefully monitored and controlled. As discussed in the last chapter, the optical output from a diode laser can vary significantly with a change in temperature of just $10°$ F. By contrast, LEDs usually have a much smaller temperature dependence on light output than diode lasers. In this chapter, we will consider some of the basic design parameters involved with both of these semiconductor light sources.

11.1 CIRCUITS USING LIGHT EMITTING DIODES

The LED converts a forward current flow into a light output. From the discussions presented in the last chapter, we saw that this light output is proportional to input current over the acceptable operating range of the device. Figure 11.1 shows a simple circuit used

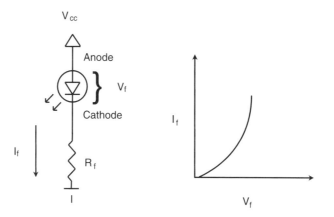

Figure 11.1   In this circuit, the forward current of the LED is limited by R_f. The graph of V_f vs. I_f is shown for a typical LED.

to bias an LED for proper operation. In this circuit, resistor R_f limits the forward current through the LED. V_{cc} is the applied circuit voltage, and V_f is the voltage across the LED when forward biased. As an example, let us find the value of R_f for a forward current I_f, that will produce a desired luminous intensity. This process of determining just how much current to allow through the LED is called DC biasing.

Example 11.1

From a device data sheet, we find that I_f must be 20 milliamperes to produce a desired optical output. What value should we select for R_f if V_{cc} is 10 volts?

The data sheet tells us that the typical forward voltage, $V_{f\,(typ)}$, is 1.4 volts @ 20 milliamperes. We must remember that this forward LED voltage will change as the current changes. The graph of V_f vs. I_f must then be carefully considered for any design. You can see that this relationship is not a linear one. We have the following solution for R_f:

$$R_f = V_{cc} - V_f /I_f$$
$$R_f = (10 - 1.4) \text{ V} / 20 \text{ mA} = 430 \text{ ohms.}$$

The LED in this example will operate with a continuous light output. If the LED must be modulated, or pulsed on and off, we will need to use additional circuitry to accomplish the switching action. This case will be considered in our next example. In our present example, we must also select the correct resistor in terms of power dissipation. Since the resistor will also experience 20 mA of current, it must be capable of handling this amount. To calculate the power dissipation, we use the following formula:

$$P = I^2 R = (.02)^2 430 = 0.172 \text{ Watts}$$

We see from our results, that the resistor in Example 11.1 can have a 1/4 watt rating. It must also be mounted on the circuit board such that air can circulate freely around it to allow any generated heat to dissipate.

As we learned in the last chapter, a very important relationship exists between current input and radiant intensity. Figure 4 of the data sheet for the Siemens device SFH 484 at the end of Chapter 10 shows this relationship. As the current is increased in the simple circuit of Example 11.1 by either increasing V_{cc} or decreasing R_f, the radiant intensity will also increase in a linear fashion. This parameter has limits because increasing the power also increases the heat generated by the LED. The resultant rise in junction temperature results in less efficiency. In the extreme case, the LED will be destroyed. It is very important not to exceed the maximum forward continuous current, I_f, for the particular device.

It is possible to exceed this forward current specification substantially by pulsing the LED on and off. The average dissipation can be kept low with a small duty cycle. For example, the current level can be increased by a factor of ten for some infrared LEDs if the pulse "on" time period is a few hundred microseconds or less. The "off" time period must be on the order of a second or two. This pulse handling capability was discussed in the last chapter. We learned that a current level of about 1 ampere can safely flow through a typical device for a period of 10^{-5} seconds. This condition produces a much larger optical output intensity during the "on" time period.

A circuit capable of switching the LED on and off for specified time intervals is shown in Figure 11.2. Modulation of the LED can be accomplished by using the switching ability provided by the transistor. In this circuit, the base of the transistor receives a pulse from an electrical source that places the collector-emitter junction into saturation. At this point, current will flow from V_{cc} through R_1, the LED, and the transistor. The calculation of the forward current flow, I_f, is given below:

$$I_f = V_{cc} - V_f - V_{CE} / R_1$$

First, we must consider all voltage drops. In above equation, V_{cc} is the supply voltage, V_f is the forward voltage across the LED, and V_{CE} is the saturation voltage of the transistor. The forward current, I_f, must then be carefully calculated using the duty cycle relationship to find the maximum allowable current. When using this circuit, the power dissipation values of all electrical components must also be considered. For example, the transistor used to switch an LED must be capable of operation at the current level chosen. For faster circuit operation, resistor R_2 can be added to the circuit as shown. This resistor should have a value such that a residual or pre-bias current of about 1 mA flows continuously through the LED. This will reduce the delay time period for an input pulse. This resistor is used in applications where infrared LEDs transmit in high speed communications systems. A darlington transistor may be used instead of the previous transistor for the switching device when the input base current is low. We will next consider a practical circuit that can be used to modulate an IR LED.

Example 11.2

A particular IR LED must be pulsed on for a time period of 100 μsec with a forward current of 1.0 ampere in order to achieve a high enough radiant intensity. If the circuit in Figure 11.2 is used with $V_{cc} = 10$ volts, find the resistance required for R_1. The saturated voltage, V_{CE}, for the transistor is 0.4 volts at this current level. R_2 is not used in this case.

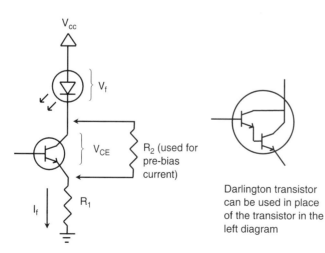

Darlington transistor can be used in place of the transistor in the left diagram

Figure 11.2   A transistor can be used to modulate an LED. The darlington transistor to the right is sometimes used.

Looking in the data book for this device, we find that for pulsed operation at 1.0 ampere, the forward voltage of the device is 3.0 volts. Since the transistor used at this current level will have a saturated voltage of 0.4 volts, we can use the above equation to solve for R_1. If we substitute in the values, we get the expression below:

$$R_1 = 10v - 3.0v - 0.4v/1.0\ a = 6.6\ \text{ohms}$$

You can see from this example that a relatively large amount of current can be used by the IR LED for a very short time period. If this device were allowed to operate continuously at this current level, the device would soon experience a catastrophic failure. This device can continue to operate by emitting these short pulses, provided that the time period between each short pulse meets the permissible pulse load requirement. A typical device specification for the permissible pulse load requirement can be found in Figure 7 of the SFH 484 data sheet at the end of Chapter 10. Some LED manufacturers also refer to this as the duty cycle specification. Going back to our last example, we can see that at 1.0 ampere, the duty cycle ratio D cannot exceed 0.05. Using a pulse length of 100 μseconds, we get for the period T between pulses (the "off" time):

$$T = \tau/D = 10^{-4}/.05 = 2\ \text{milliseconds}$$

This is the minimum time required between pulses at this current level. We can also see from the data sheet that the maximum current level allowed for a 50% duty cycle is 200 milliamperes. A digital transmission of random sequences of high and low data bits uses this duty cycle in its operation. If we use this duty cycle and current level, we find that R1 must be increased to allow the forward current to be 200 milliamperes. We do this simple calculation below using the circuit in Figure 11.2:

$$R_1 = 10v - 1.4v - 0.3v/.20 = 41.5\ \text{ohms}$$

We must note here that the values of V_F and V_{CE} are somewhat lower due to the reduction of current used. These values are given only for the purpose of illustration, and may vary substantially depending upon the device used.

Using the correct transistor in a transmitter circuit is another important consideration. Parameters such as the maximum collector current, I_C, and transistor dissipation must be considered, as these place an upper limit to the current carrying capability of the device. The forward current transfer ratio determines the amount of base current required to drive the transistor. If switching speed is an important consideration, then the transistor's switching characteristics must be considered.

Transmitter circuits must usually be interfaced to TTL electronics. Discrete transistors can be used to perform this function, but in some cases, an integrated circuit containing the required transistors for this interfacing can be used. An integrated circuit, called an LED driver, makes the task of interfacing digital signals much easier. This driver can switch high current devices such as IR LEDs within a few nanoseconds. As with LED devices, the current rating must be observed to prevent catastrophic failure. For digital operation, a typical device is usually operated at a 50% or less duty cycle. These devices also

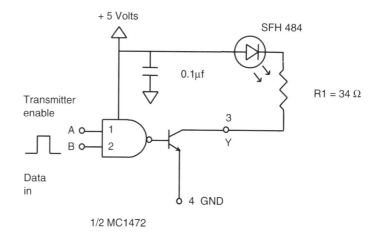

Figure 11.3   Using a peripheral driver, input TTL digital data can modulate an IR LED.

have a high enough input resistance to allow compatibility with CMOS, MOS, and TTL electronic devices.

Figure 11.3 shows a simple transmitter circuit using a peripheral driver MC 1472 to drive an IR LED. This 8-pin device contains two independent driver circuits as shown in the figure. This allows two IR LEDs to operate separately. An alternate peripheral driver, such as the SN75452, may be used also. For our purposes, we will consider only one of the two driver circuits available in this IC.

Digital data is transferred by a sequence using high (1) and low (0) bits for binary codes. A 1 bit ideally is a 5-volt pulse with some predetermined pulse width. A 0 bit ideally is a 0-volt condition for the same time period. For a 50% duty cycle, we can operate the IR LED at 100 mA peak current or 50 mA average current. It will be turned on, producing a 1 bit when the transmitter enable input is driven to a logic high. Conversely, it will then turn off, producing a 0 bit when the transmitter enable input is driven to a logic low. In this fashion, the device can modulate the IR LED from an input TTL data

Table 11.1   Truth Table

A	B	Y
L	L	
L	H	H (off state)
H	L	
H	H	L (on state)

H = Logic one
L = Logic zero

stream. The MC 1472 can handle a maximum continuous current of 300 mA and the supply voltage should not exceed 7.0 volts. Let us now calculate the value of the load resistor, R_L, that will allow the circuit to operate as described above. We perform this calculation by using Ohm's law.

$$R_L = V_{cc} - V_F - V_{sat}/I_F$$
$$R_L = 5 - 1.4 - 0.2/.100$$
$$R_L = 34 \text{ ohms}$$

Thus, by using a 34 ohm resistor, and keeping the other circuit parameters constant, we will be assured of not exceeding the current specification of the IR LED. A truth table describing circuit operation is in Table 11.1.

A very practical application of the digital switching of light can be found in a remote control transmitter. The information to be transmitted is encoded in the form of bits, or ones and zeros as described previously. When using IR LEDs for this purpose, the duty cycle must be taken into consideration. The average current must not exceed the maximum forward current rating, I_f, of the device. Figure 11.4 shows the electrical schematic diagram for a relatively simple transmitter circuit having the feature of adjustable duty cycle. This circuit can be used for a single instruction, and has the advantage of low power consumption, a requirement for battery operation. We will next provide a description of how this circuit operates.

The circuit as shown in Figure 11.4 uses two CMOS NAND gates to produce a stable multivibrator condition. This condition provides the oscillation frequency needed for

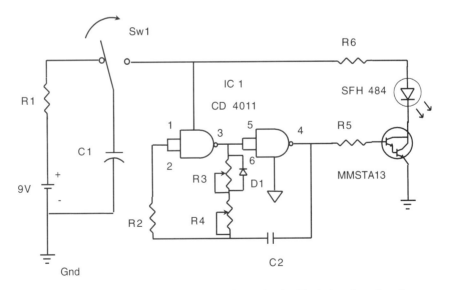

Figure 11.4 Remote control transmitter circuit. Typical values for the above components are R1 = 270Ω, R2 = 100KΩ, R3 = 30KΩ, R4 = 10KΩ, R5 = 4.7KΩ, R6 = 40Ω, C1 = 1000μf, C2 = 1nf.

digital transmission. Other NAND gates may work, but close attention must be paid to the low current and voltage requirements. Switch 1 charges C1 through resistor R1 using the available electrical energy from the 9-volt battery. When the switch changes state, the charged capacitor delivers this electrical energy to the remainder of the circuit. The multivibrator configuration consisting of IC 1, R2, R3, R4, D1, and C2 sets up an oscillation of square wave pulses. These pulses drive the darlington transistor at this frequency. Current flows through R6 and the IR LED, producing modulated light pulses at the same frequency. Great care must be taken to specify resistor R6 and the darlington transistor such that the current and voltage specifications are not exceeded. Since the circuit operates from the stored charge in capacitor C1, there will be only a single pulse with duration of approximately 5 milliseconds for the values shown. This relatively wide pulse is the envelope for the modulated pulses produced by the multivibrator shown in Figure 11.5.

The duty cycle determination is also shown in Figure 11.5. By varying the resistances of R3 and R4, the times t_1 and t_2 can be adjusted to obtain the desired duty cycle. For convenience in adjusting this duty cycle, R3 and R4 can be potentiometers as shown in Figure 11.4. Due to the tolerances involved with the CMOS IC, the oscillation frequency of the multivibrator can be calculated approximately as shown below.

$$f = 1/T \approx 1 / 1.1(R3 + 2R4)C2$$

Using the suggested values for components in Figure 11.4, we get an oscillation frequency of approximately 30 KHz.

In the last chapter, we discussed the parameters responsible for the finite rise time of an LED. When given a step input signal, the LED will take a certain amount of time to go from no light output to some specified light output level. We specified this rise time as the period of time required for the pulse to go from its 10% amplitude level to its 90% level. This time period determines the frequency response or bandwidth of the LED. When adding other electrical components to the LED circuit, such as a load resistor, transistor, or a driver, the frequency response of the combination must be taken into account.

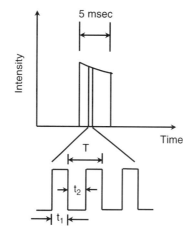

Figure 11.5 Duty cycle determination for the circuit of Figure 11.4.

This total rise time period results from the various delays associated with each component. As we know, delay times in an LED are caused by the internal capacitances and the carrier lifetime characteristic. Switching delays also exist when using transistors and drivers. To determine the total rise time in this case, we must take the root-sum-square of the rise times for the LED and the driver device. Thus, when specifying an LED driver or transistor in a modulation circuit, the frequency response for both circuit elements must be considered. The mathematical expression for the total rise time is given below:

$$\tau_R \text{ (optical output)} = [(\tau_{R(LED)})^2 + \tau_{R(Driver)})^2]^{1/2}$$

Next, we will consider a typical example to illustrate how to determine this total rise time.

Example 11.3

The measured rise time for an IR LED is 10 nsec when used with a particular load resistor. When using this combination with an LED driver, find the resultant rise time. The data sheet rating for the rise time of the driver is 5 nsec.

We substitute the values into the above equation to find the total rise time:

$$\tau_R = [(10 \times 10^{-9})^2 + (5 \times 10^{-9})^2]^{1/2}$$
$$\tau_R = 11 \text{ nsec}$$

When considering the design of a digital optical data link, the rise time must be small enough not to cause degradation of data. A complete communication system contains a transmitter, transmitting medium, and receiver. All three of these devices must be taken into account to determine the system rise time. In general, the total rise time should not exceed 70% of the bit period when using an NRZ code (Section 11.5). In this chapter, we will consider only the transmitter portion of the complete system. The receiver portion will be considered in the chapters that follow.

11.2 LED DISPLAYS

A very common application of visible LEDs can be found in everyday devices as electronic instruments, digital clocks, and kitchen appliances. These LED displays are available in various sizes, shapes and colors. The sizes range from those used in surface mount (SMT-LED) applications to font segments used to produce numerical digits. The brightness of each LED or segment depends upon the amount of current being used, and this means that the value of the load resistor must be carefully considered.

We will first consider an application that uses a seven-segment LED display such as found in electronic instruments or alarm clocks. These displays are available in colors such as standard red, high efficiency red, green, and yellow. We will consider the simple drive circuits discussed in the last section for this application. To keep the LED on continuously, only a voltage source, V_{cc}, and a load resistor, R_L, will be required. The value of R_L can be calculated as follows for this simple circuit:

$$R_L = V_{cc} - V_f/I_f$$

In the above relationship, V_{cc} is the supply voltage, and V_f is the specified forward voltage of the LED at the forward current level I_f. We must consult the data book for the LED device to be sure that V_f and I_f are not exceeded. The LED can be switched on or off easily by using the transistor circuit discussed in the last section. This may be necessary to achieve a higher brightness level. When using a transistor, the value of the load resistor must also take into account the saturation voltage, V_{CE}. Thus, this resistor value now becomes:

$$R_L = V_{cc} - V_{CE} - V_f/I_f$$

When driving several LEDs or segments, the driver element can usually be replaced with an integrated circuit (IC). This makes the design process much easier and more efficient. Many ICs exist that can be used in place of several transistors. But before we can select a particular IC driver, we must determine how the LEDs will be configured in the display circuit. To start with, we will consider the case of LEDs arranged in a parallel display. To access each individual LED, we can either have all the anodes tied together (common anode) or all the cathodes tied together (common cathode). Figure 11.6 shows the schematic diagrams for both of these cases. The LED configuration selected will determine which IC to use as a driver. Another factor that must be considered when selecting a driver is the high-low state of the driver's output with a particular input.

For a design example, we will use a common anode LED display with a seven-segment driver. The driver consists of NAND gates, input buffers and seven AND-OR-INVERT gates. Four bit binary-coded-decimal (BCD) can be accomplished to drive a seven-segment LED display. The inputs in this case correspond to A, B, C, and D. The circuit schematic shown in Figure 11.7 is part of a digital display panel. For our next task,

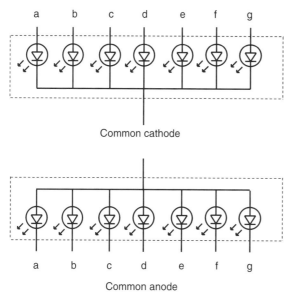

Figure 11.6 Common cathode and common anode displays.

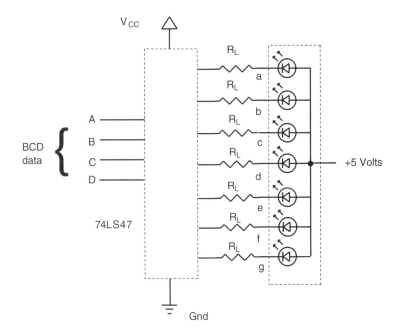

Figure 11.7   Circuit schematic for a seven-segment display.

we must determine the value of each load resistor, R_L, to be used with each segment. We want to operate somewhere in the mid-range of the specified current for this display. If the current level is too low, there may be uneven illumination between each LED in the display. Operating at too high a current may cause heating of the LED chip. This shows up as a non-linear output with input current. For the circuit shown in Figure 11.7, we will use V_{cc} as 5.0 volts, V_f for each LED as 2.1 volts, V_{on} as 0.35 volt, and I_f as 15 mA. V_{on} is the output voltage of the display driver. We can now calculate the value of the load resistor as follows by using Ohm's law:

$$R_L = V_{cc} - V_f - V_{on}/I_f = 5.0 - 2.1 - 0.35/.015$$
$$R_L = 170 \text{ ohms}$$

Once a particular drive current has been established, the luminous intensity, I_V, for dc operation can be established. Some data books give this intensity in a graph form directly. The luminous intensity can then be determined by selecting a given current level. Other data books present a graph of relative luminous intensity vs. dc current. In this case, the luminous intensity, I_V, can be calculated by multiplying the value on the graph by an output factor for some specified current level given by the manufacturer. For example, let's assume that each LED segment in a display has a luminous intensity of 205 μcd (min) at 5 mA. Using the graph of relative luminous intensity, we find that for a current level of 15 mA, the multiplication factor is 4. Thus, the minimum luminous intensity for each LED segment will be 820 μcd.

11.3 PRECAUTIONS TO OBSERVE FOR DIODE LASERS

Before discussing the operation of diode lasers in actual circuits, we must consider some very important safety issues involved with their use. We will also present some device considerations that must be observed to maintain reliability. When operating a diode laser, close attention must be paid to operating temperature and current consumption.

1. Permanent damage to the diode laser can occur due to static electricity. Before handling these devices, static precautions must be observed. A typical grounding strap worn around the wrist and connected to earth ground limited by a nominal one megohm resistor will protect against static electricity.

2. Be careful not to drop or cause mechanical shock to the device, as this action can result in permanent damage.

3. The diode laser emits very intense infrared light energy that the human eye cannot detect. Never look directly into the device during operation. Eye protection should be worn to avoid accidental exposure.

4. The diode laser will be operated using a relatively large amount of current. This means that there will be a corresponding power dissipation to be considered. Mount the device to a thermal radiator or heat sink depending upon the operating time and output power.

5. Large surge currents must not enter the diode laser. Surge currents, in excess of the rated value, can damage the reflecting facets within the optical cavity, even if they exist for a very short period of time. These large surge currents can easily occur during powering up the device from a power supply. The selection of a power supply with a slow starter circuit is required for safe diode laser operation. The power supply itself must contain the necessary filters and noise suppression to protect against surge currents from the AC source.

6. Looking at the diode laser construction, you can see that the package includes the diode laser chip plus a monitoring photodiode. A small portion of the diode laser's output beam shines onto the surface of the monitoring photodiode for purposes of stabilization. The electrical output from the photodiode goes into a comparator circuit that helps to maintain a constant output power. This assures that the diode laser will not receive too much current resulting in a catastrophic failure.

We will next consider a simple power supply that can be used to operate a diode laser continuous wave (no intensity modulation). This circuit diagram, shown in Figure 11.8, should be used mainly for laboratory testing purposes, since no control exists for the output beam intensity with temperature variations. Thus, one must pay particular attention to current and voltage by using an ammeter and voltmeter. To achieve stable operation, the use of an automatic power control (APC) can be used. An APC circuit monitors the light output from the laser to achieve a constant output with temperature variations. We will consider such a circuit in the next section.

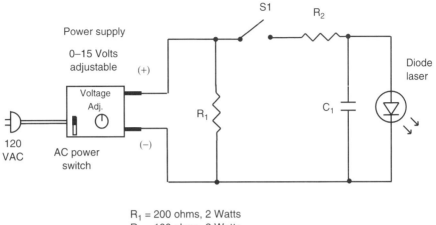

R₁ = 200 ohms, 2 Watts
R₂ = 100 ohms, 2 Watts
C₁ = 0.05 μf

Figure 11.8 Laboratory power supply for a diode laser.

To build the circuit in Figure 11.8, you must use a well-filtered and regulated variable power supply. The value for R_2 must be calculated to yield the desired current for a particular operating voltage. The maximum current specified for the diode laser must not be exceeded. When the circuit is completed, set the power supply to the minimum voltage level. With switch S1 closed, slowly increase the power supply voltage to the desired level while monitoring the ammeter for over current. The diode laser must not experience an over-current condition even for a brief moment. As we discussed before, when the input current increases, so does the optical output. If this output becomes great enough, the mirror facets within the diode laser's structure could experience too much optical energy. This will result in a catastrophic failure. During continuous operation, the diode laser must be monitored carefully since changes in temperature will affect the optical output. The use of a larger heat sink may be required to stabilize device operation. When you want to conclude diode laser operation, reverse the procedure described above.

11.4 INTENSITY MODULATION OF DIODE LASERS

In Chapter 7, we discussed how the amplitude of a continuously operating laser can be varied for signaling purposes by using an external electro-optic modulator. This method of modulation results in a stable output wavelength because the light beam intensity varies by controlling its polarization. The polarization was controlled after the beam was emitted from the laser. One disadvantage with this method is the relatively high voltage level required to control the polarization of such a light beam. Circuits using high voltages are difficult to design for use in a compact size. In this section, we will consider how to modulate the diode laser using current pulses to the laser itself. This results in voltage levels that are much lower than those used in external electro-optic modulation. This means that these designs can be easily integrated into small compact spaces. This direct modulation method requires very close attention to such variables as temperature changes and electrical transients. As we discussed in Chapter 10, the diode laser is much more susceptible to temperature changes and electrical transients than the LED. This means that, generally, the electronic circuitry involved with diode laser operation will be more complicated than that used with LEDs.

Before considering an actual electronic circuit using a diode laser, we must know about its temperature dependence. In Chapter 10, we showed the light output vs. current curve for a typical device. As the diode laser's current level increases, more photons are produced. At a current level defined as I_{TH}, lasing begins to occur because a population inversion condition exists within the Fabry-Perot oscillating cavity. From this point on, a relatively small increase in current will produce a large increase in light output by the process of light amplification. This nice feature comes with a penalty. Unfortunately, a relatively small temperature change will also produce a large variation in light output for a given current level. Complex circuitry must be used with the diode laser to prevent a catastrophic failure with excessive optical output.

For convenience, we show again a typical diode laser's light output variation with temperature in Figure 11.9. This device is chosen for illustration purposes only. The output curve for a particular device may be quite different. We will now consider the laser output curves resulting at three different temperatures. As you can see from each curve, as the

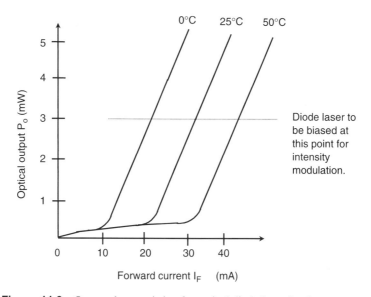

Figure 11.9   Output characteristic of a typical diode laser for three temperatures. This temperature dependence on output requires stabilization circuitry to be used with diode laser.

temperature increases, I_{TH} also increases. For every 10°C increase, I_{TH} increases by about 4 mA. In the lasing mode, an increase of 4 mA of current represents a significant amount of light production. You will also notice from this graph that all three curves have the same shape. They are displaced along the x axis according to the temperature variation experienced by the device. A light output of 3 mW occurs approximately in the middle of the linear lasing portion for all three curves. This important characteristic can be used in the design of a stable circuit to operate a diode laser at these different temperatures.

As with the external modulation example given in Chapter 7, we will want to bias the diode laser's operating point. We do this electronically by keeping it biased at the 3 mW output level. This will ensure that when we modulate the device by changing the input current, it will be operating on the linear portion of the lasing curve. This will allow for coherent light output and a larger bandwidth. The ultimate factor limiting bandwidth for the laser is the photon production time, τ_p. This holds true when the laser operates above I_{TH}. When the laser operates at a current level below I_{TH}, the ultimate factor limiting bandwidth is the time needed for spontaneous emission or τ_{sp}. As you remember, τ_p is a much smaller time period than τ_{sp}.

The diode laser has a built-in monitoring photodiode for the purpose of stabilizing the light output at a desired operating point over a specified temperature range. Since the diode laser emits light through the front and rear surfaces of the Fabry-Perot structure, the monitoring photodiode can be placed in the rear output beam path. The photodiode's output will be linear with the light output of the front beam. We can see this relationship by reviewing the data sheet at the end of Chapter 10 for the diode laser device ML40123N. Figure 4 on this data sheet shows that at a light output of 3 mW, the monitoring photodiode produces a current of 0.60 mA. As the temperature changes, we can keep the light output at 3 mW by simply adjusting the current into the diode laser so that the monitoring photodiode produces 0.60 mA of current. We will next consider an electrical circuit that performs this function, and at the same time, protects against damaging electrical transients.

The circuit described in Figure 11.10 utilizes the optical feedback from the diode laser into the monitoring photodiode. This circuit also uses conventional electronic components, and requires only one power supply. The component values are typical, and may require adjustment to operate a particular diode laser. We will start our discussion of the operation of this circuit by considering the monitoring photodiode. In this schematic diagram, the monitoring photodiode is called the photodetector. First, we need to determine the output current from this photodiode at the desired bias level by using the graph included in the data sheet. For the purposes of this example, we will choose an operating point of 3 mW. Once this has been determined, the value for R_1 feedback resistor can be calculated by using Ohm's law. Resistor R_1 is used in transimpedance amplifier A_1 to establish an output voltage that keeps the laser at 3 mW, the midpoint bias level. Amplifier A_1 can then produce voltage excursions equally above and below this bias point. The simple calculation of R_1 is given below:

$$R_1 = V_{CC}/2\, I_{PD}$$

I_{PD} is the output from the monitoring photodiode at the midpoint bias condition. This condition is analogous to the half-wave voltage condition for the electro-optic modulator de-

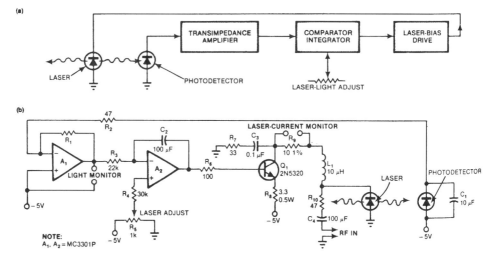

Figure 11.10 Electrical schematic for a circuit using a diode laser. Reprinted from EDN Magazine (March 5, 1980) © CAHNERS PUBLISHING COMPANY 1996 A Division of Reed Elsevier, Inc.

scribed in Chapter 7. As you remember, for efficient modulation, we want the output of the laser to swing above and below the midpoint bias level.

Once we have established the correct midpoint operating voltage set up by R_1, we must next determine experimentally a reference voltage for integrator/comparator A_2. This IC compares the output from amplifier A_1 to the reference voltage value from variable resistor R_5. This reference voltage can be determined by adjusting resistor R_5 to produce the desired light output. The voltage across R_9 must be monitored to determine the current level corresponding to the 3 mW bias operating point, I_B. Any deviation in light output causes A_2 to produce a proportional error voltage from this set point.

The output from A_2 controls the base of transistor Q_1. This transistor acts as a current amplifier for dc bias to the diode laser. The diode laser's current path consists of inductor L_1, and resistors R_9, Q_1, and R_8. The condition for operating the diode laser at the midpoint of its linear region must be obtained by adjusting the values of the components associated with A_1 and A_2. Once this operating condition has been achieved, a modulating current with changes about I_B will swing the light output about the 3 mW bias point.

We will next consider how to intensity modulate the diode laser. Figure 11.11 shows the light output vs. current input curve for a typical diode laser. One condition for modulation requires that the diode laser be biased to a certain current level. This condition should result in laser operation approximately in the middle of the lasing curve. This bias current level, I_B, must be determined from the device specifications. You can see that I_B has a resultant light output level of P_B. In this example, we vary the modulation current, ΔI, to produce a square wave with a 50% duty cycle. An analog input signal, such as a sine wave, can also be used. As long as the input current variations stay within the two vertical dotted lines, the diode laser will operate on the linear portion of its lasing curve.

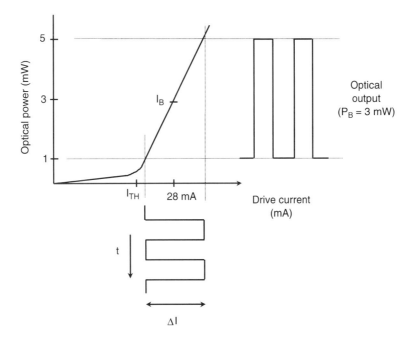

Figure 11.11   Modulated light output vs. modulated input current.

To maintain this operating condition, P_B must remain constant. This is one function of the circuit described in Figure 11.10. If the current excursions drift into the knee of the curve, the light output will become distorted. Operation below the knee produces only sponta-neous emission with a slower response time.

The circuit must then maintain the I_B level while allowing for the fast changes in current of ΔI. To assure that this happens, the value of C_2 must be sufficiently large such that changes caused by the input modulation current do not affect the operating bias point, I_B. This capacitor slows down the response time of A_2, and protects the diode laser from transient surges on power up due to this delay time.

Applying modulation voltage to the laser (RF IN) through C_4 and R_{10} impresses changes in current of ΔI about the bias point, I_B. This arrangement offers a simple way to convert this input voltage into a modulation current while providing good linearity. One drawback to this technique is that most of the input power appears across R_{10}. Try to keep the value of C_4 as small as possible for good high frequency response. During the modula-tion process, R_7 and C_3 serve the function of dampening the resonant effect caused by in-ductor L_1.

As mentioned in Chapter 7, the method used for direct modulation of the diode laser has a disadvantage over the process of external modulation. When current increases quickly to the diode laser, the carrier density within the structure also increases. This re-sults in an increase in optical power. Unfortunately, this changing carrier density also causes changes in the refractive index of the active region leading to variations in the emitted optical wavelength. This change in wavelength during the modulation process is known as chirp. Chirp can introduce a dispersion penalty in optical fibers since the optical energy within a pulse contains more wavelength components than when it runs continuous wave. As discussed in Chapter 4, these wavelength components travel at different speeds within the optical fiber. Given enough traveling distance, a relatively narrow pulse will spread out into a wider pulse due to the dispersion of light energy. Of course, this dispersion also depends upon factors such as wavelength and fiber type. In some cases, this effect may cause a severe limitation on the transmission distance before a repeater is required to amplify the signal to a useful level.

When modulating a diode laser, an important parameter known as the modulation index gives the circuit designer an indication about the extent of this modulation. This index, m, can be defined as:

$$m = (0.5)\ \Delta I/I'_B$$

where ΔI is the total current variation, and $I'_B = I_B - I_{TH}$. A modulation index of 0.9 to 1.0 will provide a signal of good transmission quality. An example using the light output curve in Figure 11.11 will help to illustrate how this works.

Example 11.4(a)

A 780 nm diode laser with a threshold current of 20 mA is used to modulate a digital signal. If I_B is 41 mA, find the peak-to-peak current swing needed to produce a modulation index of 0.9.

First, we substitute the values into the above equation:

$$0.9 = (0.5)\ \Delta I/(41 - 20)$$

After rearranging terms and solving for ΔI, we get our answer:

$$\Delta I = 37.8\ \text{mA}$$

Half of this current is used to swing the signal below I_B, and the other half is used to swing the signal above I_B. Thus, the current swings from a minimum of 22.1 mA to a maximum of 59.9mA.

Example 11.4(b)

What minimum and maximum voltage values should be measured across resistor R_9 during this modulation?

We use an oscilloscope to measure the pulses, and then use Ohm's law to verify our measurements.

$$V_{MIN} = IR = (.0221)(10) = 0.221V$$
$$V_{MAX} = IR = (.599)(10) = 0.599V$$

In addition to the precautions mentioned in the last section, there are others that must be observed when using diode lasers with optical components. The diode laser is very sensitive to its own back-reflected light. Examples of situations where light reflection could cause a problem include coupling laser light into a glass fiber having an opened connector, or using optics having no anti-reflection coating. If light travels back into the laser's resonant cavity, increased noise may result. This could especially happen when light reflects back very close to the device. The reflected light varies with amplitude and phase, thus disturbing the lasing process within the Fabry-Perot structure. In other cases, the reflected light can actually affect the monitor photodiode. If reflected light strikes this photodiode, the compensation circuitry will be affected due to the excess signal current produced there. To prevent these situations from occurring, use lenses with anti-reflection coatings or place an optical isolator after the diode laser. Tilting the optical elements slightly so that light does not reflect directly back may be another alternative.

11.5 LINE CODING

LEDs and diode lasers have the ability to be pulsed or intensity modulated as described in the previous sections. These pulses of light must be set up in a particular pattern using a set of rules so that the received signal can be decoded. This process, called line coding, will be discussed in this section. We will only be concerned with binary codes since these are most commonly used. A binary code uses two possible states, 0 or 1 to transmit an optical signal. Thus, a transmitted optical message using a binary code would consist of bit sequences of 0's and 1's. Specifically, the 0's would be transmitted when the light source was turned off. The 1's would be transmitted when the light source was turned on. There are a large variety of modulation codes in use today. Each code has a set of rules that must be understood so that efficient information transfer can occur. The selection of a particular code can optimize data rate and system performance. We will look at some of the most popular codes in current usage.

A typical coded sequence is shown in Figure 11.12. The binary signaling technique uses symbols as a high or low level (1 or 0) held in that position for a given length of time. This length of time can be considered as the symbol width or bit period. You can also think of this as a time slot. There are some codes requiring more than one symbol to define a bit. This technique is commonly used in error detection. Each bit has an assigned time slot, Δt, where the pulse can either be present or not. The simplest of these codes is the NRZ code (Non-Return to Zero). For an NRZ code, the entire bit slot, Δt, is filled with either a 0 or a 1. A low level with this code is a 0, while a high level is a 1. If there are two successive 1's, the level does not return to zero. (You can see how this code got its name.) One problem with the NRZ code occurs when a long string of 1's or 0's becomes transmitted. In this case, the timing information can be easily lost, since no transition levels exist during this time period. Thus, when using the NRZ code, there is a system requirement for a highly stable clock signal for both the transmitter and the receiver.

Another code, known as the Return to Zero or RZ format, uses two symbols per bit slot. Thus, twice as many symbols are used. The main advantage to using this format is reduced power consumption. For this format, a transition occurs for every bit transmitted. The pulse width is less than the bit interval so that a return to zero excursion can occur. This format will require more bandwidth than the NRZ code.

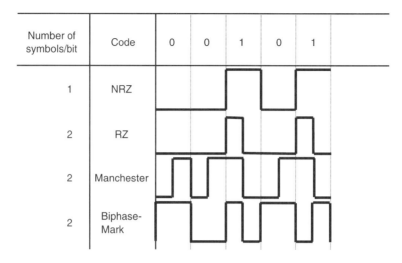

Figure 11.12   Description of modulation codes.

The Manchester code requires two symbols per bit cell period. With this code, a 0 starts in a low position, then in the middle of the bit cell period a low to high transition occurs. The 1 starts in a high position, then in the middle of the bit cell period a high to low transition occurs. This mid-cell transition that occurs for each bit allows for clock recovery of the signal.

The Biphase-Mark code uses the edge of each symbol to determine a 1 from a 0. This differs from the NRZ and RZ codes that are impulse codes. With the Biphase-Mark code, you can see that each bit cell period begins with an edge. Two symbols per bit are also needed for this code. For a 0, no transition or edge occurs during the bit cell period. A 1 has an additional transitional edge during the bit cell period. This type of code allows for clock recovery of the signal since the beginning of each bit cell has an edge. Clock recovery methods such as phase-lock or one-shot can be used.

Next, we will consider the use of these codes in the modulation of a light source. Though we have seen many examples of how light can be modulated when using an LED or diode laser, the specific types of modulation have yet to be discussed. There are two basic types of modulation, analog and digital. Analog modulation was considered in Chapter 7 with the example of external modulation of a laser using a birefringent crystal. This design could also be used to modulate the laser in a digital fashion. With analog modulation, a linear reproduction of the input signal is required, otherwise distortion can easily degrade the output. The signal example used in Chapter 7 was a continuously varying sine wave. Linear reproduction of this signal occurred as long as the variations stayed in the linear portion of the operating curve. Digital modulation, which we considered in the diode laser example in Section 11.4, involves the presence or absence of pulses. These pulses can tolerate slight distortions because the pulse height, width, and location are most important to faithful reproduction of the transmitted signal. The pulse parameters can vary slightly and still be interpreted as a 1 or a 0.

The speed at which these pulses are generated is called the bit rate, B, or pulse transmission rate. Each pulse will have a time duration T_b called the bit period. These two fundamental parameters are related by the mathematical relationship shown below:

$$B = 1/T_b$$

We will consider only digital modulation in this chapter.

The transmitter circuit converts the input electrical digital signal into output pulses of light. The simplest digital transmission scheme requires the light source to be fully on, for a 1 to become transmitted. Conversely, when the light source is fully off, a 0 becomes transmitted. These symbols occur during time slots of width T_b.

In practice, there can be may factors that distort the initial optical signal pulses. These factors include attenuation within the transmission medium, electronic noise, and other noise sources. Transmission errors may result if these factors are not brought under control. A measure of system performance is the bit error rate or BER. This can be calculated by dividing the number of errors during a given time interval by the number of time slots or symbols transmitted during this same time interval. Typical bit error rates for an optical fiber communications system range from 10^{-6} to 10^{-10}. The receiver circuit relies on the integrity of the transmitted signal. Generally speaking, the better the transmitted signal quality, the fewer errors will occur. A lower limit exists to the received power of the optical signal. If this signal power level is not observed, the bit error rate will increase.

In Example 11.3 of this chapter, we considered the rise time for a simple transmitter composed of an IR LED, resistor, and a driver. As we know, this finite rise time will restrict the bandwidth of the device, since the output pulse does not rise instantaneously. The 3 dB electrical bandwidth of the transmitter can be determined from the pulse rise time displayed by the output pulse. The standard empirical formula is given below:

$$B = 0.35/\tau_R$$

In this formula, B is the bandwidth, and τ_R is the rise time measured on the pulse from the 10 to the 90 percent points of the rising edge. We will next consider a typical example of a 3 dB bandwidth calculation.

Example 11.5

The transmitter circuit discussed in Example 11.3 displays a rise time of 11 nanoseconds to an input square pulse. Find the 3 dB bandwidth of this transmitter.

Substituting this rise time into the above formula, we calculate the bandwidth as shown below:

$$B = 0.35/11 \times 10^{-9} = 31.8 \text{ MHz}$$

This result means that the circuit described in Example 11.3 provides an electrical 3 dB bandwidth of 31.8 MHz. For input signals faster than this, the 3 dB limit will be exceeded due to the limitation on rise time.

Optical modulation codes are also used in the operation of many consumer devices. Typical devices include the IR wireless remote control unit. In the next section, we will consider a relatively new modulation code or signaling scheme that is presently used in consumer devices such as calculators, printers, and personal computers.

11.6   THE INFRARED DATA ASSOCIATION (IrDA)

For many years, the manufacturers of infrared wireless transmission systems have used their own modulation signaling schemes. These different schemes had nothing in common, resulting in relatively high production costs in most cases. This situation changed with the formation of the Infrared Data Association (IrDA) in 1993. A serial infrared (SIR) standard was adopted by companies like HP, IBM, and Sharp that used a 115.2 Kbit/sec. transmission rate. The transmission range for this first revision, IrDA-1.0, was a distance of up to 1 meter. Today, this standard is at revision IrDA-1.1, specifying a transmission distance of up to 3 meters between compatible devices. The transmission data rate was increased to 4 Mbit/sec. The membership over the last few years has increased dramatically to include many international companies. Since this is a standard protocol, IrDA compliant equipment from different companies will be compatible. The advantages to using IrDA compliant equipment include simple setup, low power consumption, and efficient data transfer. With the adoption of this standard, a large reduction in the number of protocols was experienced. This means that IC manufacturers can offer IR controllers, encode/decode ICs, transceivers, and many other related ICs at lower costs due to the larger volumes involved. This has caused the infrared market to grow at a tremendous rate.

Important applications of infrared communications can be found in the electrical appliances we use every day. These include notebook computers, desktop PCs, printers, and calculators. Optical transmission of data can be implemented, for example, between a desktop PC and a laser printer without setting up cumbersome cables and using the same manufacturer for both devices. This infrared data transmission protocol also uses a simple point and shoot approach.

The IrDA standard covers at least three areas. These areas are the Physical Layer Specifications, the Link Access Protocol (IrLAP), and the Link Management Protocol (IrLMP). Our discussions in this section will emphasize the Physical Layer Specifications since the components necessary to implement an IrDA data link are covered here. We will next consider each of these areas separately.

IrDA Physical Layer

The physical layer of the IrDA standard concerns the specifications for the modulation code, pulse width, and other physical parameters. To start with, an IrDA-compliant transmitter must emit optical radiation at a peak wavelength between 850 and 900 nm. The radiant intensity of this emission in the axial direction must be between 40 to 500 mW/sr within a half-angle range of ± 15 to ± 30 degrees. The modulation code used is a return to zero inverted (RZI). With this code, a 0 transmits as a single pulse, and a 1 transmits as the absence of a pulse. The pulse width can be as short as 1.6 μsec to a maximum of 3/16 of the bit period. Short duration pulses help to reduce the power dissipation. This will extend the operational time period for battery operated devices.

To achieve the correct specification pulse, an efficient high-speed IR LED must be used in a high-speed drive circuit. Peak drive currents of about 250 mA or more are usually required. Higher current can be used depending upon the duty cycle. For example, if an IR LED has a maximum specified average forward current of 100mA, a peak current level of 500 mA can be used for a 20% duty cycle. In the case of an IrDA minimum pulse width of 1.6 μsec. at a transmission speed of 9600 bits/sec, a peak current of 1 ampere can be used. This will not exceed the average allowable forward current of 100 mA since the duty cycle here calculates to be 1.54%. Table 11.2 lists a few of the data transfer speeds and applications for both IrDA versions.

Emitter and receiver components must be carefully chosen for the application. Since the transmitter's parameters have been specified, the job of designing a compatible receiver has just become a little easier. Some of the specifications for the receiver include a 30° viewing angle. This aids in the reduction of outside optical interference. When using a photodiode having an acceptance angle of 135°, lenses must be used to restrict this angle to meet the specification. Many photodiode manufacturers now produce devices that conform to this specification. The amount of optical energy at the receiver must be between 4 μWatts/cm^2 to 500 mWatts/cm^2. This may present a design problem when using discrete components. For example, a receiver optimized to receive optical energy at the 4 μWatts/cm^2 level may easily saturate when it receives a much higher level of 500 mWatts/cm^2. In this case, an automatic gain control circuit (AGC) may be the solution to provide a wider dynamic range. By recessing the photodiode a certain distance inside the aperture, saturation may also be prevented since the light source cannot physically get to a zero distance. Also, many optical receiver ICs are now available that meet this specification for wide dynamic range.

There are many devices available to assist in the design of an IrDA data link. A device such as a transceiver module contains the IR LED, photodiode, and the transmitter/receiver IC all in one package. An advantage gained by using this device over discrete components is EMI immunity due to device integration. A better signal-to-noise ratio may be realized by using such an integrated package. Transceiver ICs or ASICs that lack the

Table 11.2   IrDA Applications

Data Transfer Rate (Kbits/sec)	IrDA Protocol	Application
9.6	1.0, 1.1	Cellular-phone, Modem
28.8	1.0, 1.1	High Speed Modem
115.2	1.0, 1.1	Serial port
4000	1.1	Enhanced parallel port

optoelectronic components are also available to the designer. These devices are helpful in a design where particular optical characteristics need to be incorporated. In this case, the designer will be required to specify the IR LED and receiver element using the appropriate lenses to meet the IrDA standard. EMI immunity may also require attention due to the use of more external components. The transmitter and receiver need to be interfaced to other electronic hardware that provide the function of encoding and decoding of the serial bit stream. ICs are available that perform this function. In Chapter 14, a typical IrDA transceiver design will be discussed.

Link Access Protocol (IrLAP)

The next protocol stack in the IrDA standard covers the software required to connect to other machines, and resolve addressing conflicts, connection start up, information exchange, and disconnection. IrLAP was derived from an existing asynchronous data communication standard known as HDLC. The frame and byte structure of the IR packets are described by this protocol as well as the error detection methodology. An important function of this protocol concerns the relationship between a primary and one or more secondary communicating stations. All transmissions over a link go to or from the primary station.

Link Management Protocol (IrLMP)

The IrLMP software manages the functions and applications provided by the IrLAP. It determines what equipment and services are available on each machine, and then negotiates parameters such as bit rate and link turn around time. For example, it can provide for multiple data link connections over the single connection provided by the IrLAP.

SUMMARY

The two major types of semiconductor light sources used in electro-optical systems are the LED and the diode laser. Both light sources use the same basic semiconductor materials but have vastly different constructions. LEDs use a relatively simple circuit to control their emission. They also deliver a relatively wide near field output optical beam. In contrast, the diode laser requires more complex circuitry to control its emission. It also delivers a fairly narrow near field output optical beam. The complex circuitry used with the diode laser helps to prevent catastrophic failure which can easily occur during even a brief period of excess current.

In the next chapter, we will begin our study of another major electro-optical system component, the receiver. The receiver must be designed to sense the optical radiation emitted from the transmitter. This sensing is accomplished by a detector component. The process by which a detector senses emitted photons is very much the reverse of the light emitting process discussed in the last few chapters. In this process, photons are absorbed by the semiconductor material to produce an electrical current flow. We will devote the next chapter to the discussion of the physics involved with these detector components.

12

Photodetectors

A photodetector device converts incident electromagnetic energy from the optical portion of the spectrum into a useful electrical current. The physical processes involved with this conversion are the reverse of the stimulated emission processes discussed in Chapter 10. Being the first element in the receiver circuit, the photodetector must meet some very rigid performance requirements. It is essentially an open eye on the environment, and as such, can be subjected to a host of changing conditions. Among these requirements are high sensitivity to the particular emission wavelength of the transmitter, fast response, and low noise. It must meet these requirements while being insensitive to changing environmental conditions. These requirements will be studied in this chapter.

There are several types of photodetectors with vacuum tube configurations or based upon a semiconductor structure. A very useful vacuum tube type device, known as the photomultiplier, consists of a photocathode and an electron cascaded multiplier. The capabilities of this device include very high gain with low noise, a practical combination. It finds applications primarily in astronomical work where the influx of photons is extremely small. Unfortunately, its large size and high voltage requirements make it impractical for the types of applications described in this book. This leaves the semiconductor-based devices such as photodiodes, phototransistors, opto-couplers, and photoconductors. The operation of these semiconductor devices will be covered in this chapter. The photodiode will be considered first since it is the basic building block for many of the other optoelectronic detector components.

Since photodetector operation involves the conversion of photons to electrons, the wave and particle natures of light must be considered. The particle-like nature can be seen when considering the energy and momentum exchange during current production within the device. The wave-like nature is displayed when we consider the wavelength dependence upon the device's responsivity. Specifically, the energy level of the incident electromagnetic radiation must be of a certain amount before it can interact with the detector material. This energy level is inversely proportional to the wavelength of the incident radiation, or stated mathematically below:

$$E = hc/\lambda$$

One of the main objectives of this chapter is to provide the reader with a working knowledge of the physical principles involved in the operation of these devices. With sufficient knowledge of these physical principles, the correct design parameters can then be chosen to produce a circuit having the desired output characteristics. The concepts presented in this chapter will rely heavily upon the discussions presented in the last three chapters.

12.1 PHYSICAL PRINCIPLES OF THE PHOTODIODE

The photodiode is the light receiver of choice for many applications. It offers excellent linearity in photocurrent over 7 to 9 decades of light intensity while providing fast response. Its low cost and low noise characteristic make it a versatile device to be used in many applications involving the conversion of light into an electrical current or voltage. The above characteristics, plus the fact that there are photodiodes with peak sensitivities in the near infrared spectral region, make it an excellent choice for use in fiber optic communication systems. The near infrared region at about 1330 nanometers happens to coincide with the spectral region for which glass fiber experiences very low losses due to absorption. This results in low signal attenuation when using IR LEDs or diode lasers having peak emissions in this wavelength band. Luckily, a type of photodiode is available that also has a peak spectral sensitivity in this same wavelength region. There are several types of photodiodes. They are classified in terms of wavelength sensitivity, function, and construction. We will consider many of the most common types in this chapter.

As with the light emitters discussed in Chapter 10, these light receivers also involve the interaction of radiation and matter. This means that we will use some of the basic physical principles of semiconductor device physics in this chapter to explain their operation. Energy band diagrams will be helpful in explaining this interaction within the light sensitive receiver. A review of these principles can be found in Chapter 10.

The first type of photodiode to consider is a planar diffusion pn type. The planar diffusion photodiode is a semiconductor device consisting of a light sensitive p-layer material and a substrate n-layer. A depletion or charged space region forms at the pn junction due to the voltage difference that exists there. In this depletion region, electron-hole pairs are generated by photons possessing the required amount of energy. These electron-hole pairs then drift toward the n and p layers respectively due to an internal electric field set up by the voltage difference between the p and n layers. The addition of an external circuit with a reverse-bias voltage across the device will assist in the current flow. Figure 12.1 shows a cross section of this device. Unfortunately, this type of photodiode has a relatively low conversion of photons to electron-hole pairs due to the shallow depletion width. The response times involved are also relatively slow due to the random diffusion process. As you can see from the figure, not all of the incident radiation becomes converted into electron-hole pairs. Radiation composed of shorter wavelengths becomes absorbed near the surface of the device, while longer wavelengths make it further though the device before interaction occurs. We will see how some of the difficulties listed above are eased in the next type of photodiode to be discussed.

A distinct improvement can be made to the pn junction photodiode by adding an intrinsic layer between the p and n layers. This intrinsic (i) layer is made from a very lightly n-doped compound, and only valence and conduction states are involved. This device has a similar cross section structure to that shown in Figure 12.1 with the addition of an intrinsic layer. Figure 12.2 also shows the relative positioning of the three layers, and an energy band diagram. The high resistance intrinsic layer added between the p and n layers assists greatly in the production of photocurrent from incident photons. When configured in this way, the photodiode is known as a pin (p − intrinsic − n) photodiode due to its physical structure. This type of photodiode is commonly used due to its simplicity, high sensitivity, stability, and faster response. It also exhibits a higher conversion of photons to electron-hole pairs due to the wider intrinsic region. This wider region results in a greater probability for electron-hole production. This is a distinct improvement over the planar diffusion type photodiode. Thus, in the construction of the pin photodiode, the widths of all three

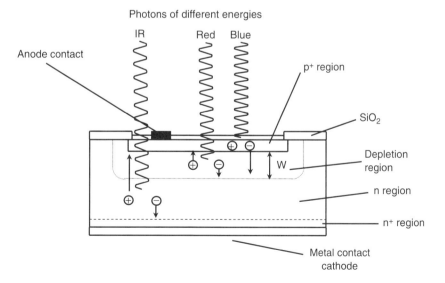

Figure 12.1   Cross sectional view of a silicon photodiode.

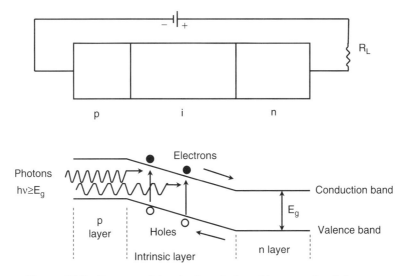

Figure 12.2   Structure of the pin photodiode and its energy band diagram.

regions must be optimized to produce a device with good photosensitivity and fast response. For example, as we just stated above, a wider intrinsic layer will provide for a higher conversion of photons to electron-hole pairs. Unfortunately, a somewhat slower response time will result. The width of this region must be designed by considering the appropriate compromises. When photons possessing an energy greater than or equal to the band gap energy of the semiconductor material strike the detector, electrons will be excited from the valence band into the conduction band. During this time, electron-hole pairs or photocarriers will be generated in the intrinsic region. Figure 12.2 provides a simplified electrical illustration showing this occurrence within the photodiode structure.

An ideal photodiode will generate one electron-hole pair per incident photon. As shown in Figure 12.2, a reverse-bias voltage exists across the structure. The resultant electric field will cause the photocarriers to separate, and then collect across the reverse-biased junction. This results in a current flow in the external circuit that can be converted to a voltage. Subsequently, this voltage can then be amplified by conventional electronic amplifiers. The energy band diagram below the pin structure in Figure 12.2 shows this interaction. If the incident light consists of photons having an energy greater than the bandgap level, E_g, the electrons of the material will be pulled up into the conduction band. This action leaves vacancies or holes in their place. In this way, electron-hole pairs are generated by the interaction of photons with the photosensitive material.

Silicon tops the list of materials used to construct photodiodes. With a bandgap energy of 1.1 eV, the peak spectral sensitivity for silicon occurs for optical radiation at about 900 nanometers. From Chapter 10, we learned that GaAs LEDs also have a peak spectral emission in the same wavelength region. This fact allows silicon to be a good match for a receiver material to use with these LEDs. The silicon pin photodiode and IR LED are used in applications such as television remote control and camera focus control. A data sheet for a typical pin silicon photodiode can be found at the end of this chapter. This data sheet is for the Siemens device SFH 205F. It has a daylight filter that rejects the shorter visible wavelengths. The device's relative spectral sensitivity can be found on the second page of this data sheet. You can see that the peak spectral sensitivity for the SFH 205FA occurs at 900 nanometers.

There are still other applications requiring a relatively high sensitivity and fast response in the visible region of the spectrum. Blue-enhanced silicon photodiodes are manufactured to provide a greater spectral sensitivity and fast response to these shorter wavelengths. As you can see from Figure 12.3, silicon has its peak spectral sensitivity at about 900 nm. This sensitivity drops dramatically for shorter wavelengths. At wavelengths shorter than about 700 nm, the light energy becomes absorbed mostly in the p-layer (diffusion layer) of the device. The sensitivity to these shorter wavelengths can be enhanced by reducing the thickness of this p-layer. This action will bring the pn junction closer to the device's surface resulting in the absorption of more shorter wavelength photons in the intrinsic region. Figure 12.1 shows the basic structure of the photodiode and how this occurs within the device.

To further enhance the photodiode's sensitivity to shorter wavelengths, the material used for the device window must also be considered. The typical glass window used in the construction of these devices absorbs optical energy at wavelengths less than 300 nm. If ultraviolet wavelengths are important, then fused silica windows can be used to allow optical energy below 300 nm to pass.

There are at least two other materials that can be used in the construction of near IR pin photodiodes. These materials are Ge and InGaAs. The peak spectral sensitivities of these two materials are at about 1300 to 1600 nanometers, respectively. These two types

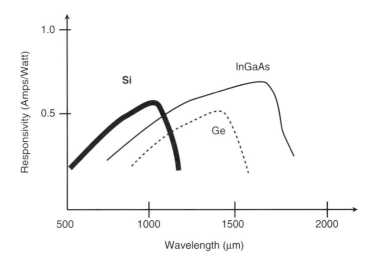

Figure 12.3   Responsivity vs. wavelength for three popular photodiode types.

of pin photodiodes find useful applications in fiber optic receivers since glass fiber displays the least amount of attenuation due to absorption in this wavelength region. More discussion on the various types of photodiode materials will be given later in this chapter.

A photodiode's efficiency can be increased with the use of a wider intrinsic region. This increases the probability for electron-hole pair production and recombination at the n and p regions. The quantum efficiency of a photodiode can be defined as the number of electron-hole pairs generated per incident photon having the required bandgap energy. As discussed before, the generation of electron-hole pairs produces a photocurrent, I_p, when these carriers migrate across the junction. If we ignore the photons reflected at the entrance of the photodiode, this photocurrent can be expressed by the equation below:

$$I_p = \frac{e \eta P_0}{h \nu}$$

In this equation, e is the electron charge (1.6×10^{-19} Coulomb), P_0 is the incident optical power, $h\nu$ is the photon energy, and η is a conversion factor known as the quantum efficiency of the device. Note that the quantum efficiency is not constant at all wavelengths and varies according to the photon energy. The mathematical expression for quantum efficiency can be found by rearranging the above equation:

$$\eta = \frac{I_p / e}{P_0 / h\nu}$$

$$\eta = \frac{\text{Number of electron} - \text{hole pairs generated}}{\text{Number of incident photons}}$$

On the first page of the data sheet for SFH 205F, the quantum efficiency or yield is given as 0.77. This means that typically 100 incident photons will create about 77 electron-hole pairs. This ratio could be increased by increasing the thickness of the intrinsic region to allow for more absorption of photons. However, the thicker this region becomes, the longer it takes the photocarriers to travel across the junction. This increased thickness results in more photocurrent and a longer response time for the device. Thus, a compromise must be made between these two parameters. On the second page of the data sheet, you will find the graph of photocurrent as a function of input optical energy for the SFH 205F. More discussion will be given on the subjects of photocurrent and response time later in this chapter.

Another factor that affects the number of incident photons, and thus the quantum efficiency, is the package construction housing the photosensitive chip. Most photodiodes have a glass window as a mechanical protection for the fragile semiconductor material. Losses due to reflections at this window must be considered. To reduce such reflection losses, the glass can have an anti-reflection coating. Other losses due to the electrical contacts at the front of the device can be minimized by making these contacts as thin as possible. Thus, all of these factors contribute to the net quantum efficiency of a photodiode.

Example 12.1

A silicon pin photodiode receives an optical power of 5 μwatts at a wavelength of 880 nm. If this photodiode has a quantum efficiency of 0.8 at this wavelength, find the photocurrent produced.

Using the equation for photocurrent, we substitute the values into the equation:

$$I_p = (1.6 \times 10^{-19} \text{ coul})(0.8)(5 \times 10^{-6})/(6.63 \times 10^{-34})(3.4 \times 10^{14})$$
$$I_p = 2.84 \text{ μamp}$$

For each semiconductor material, there exists a limited wavelength range of operation. The upper range or cut-off wavelength can be determined by the bandgap energy of the material. A relatively simple expression to determine this wavelength is given below. When E_g is expressed in units of electron volts (eV), then λ_c (cut-off) can be calculated in micrometers (μm):

$$\lambda_c(\text{μm}) = \frac{hc}{E_g} = \frac{1.24}{E_g(\text{eV})}$$

In the above expression, h is Planck's constant and c is the speed of light. For wavelengths longer than the cut-off wavelength, the incident photon does not have enough energy to excite an electron from the valence to the conduction band. This can be seen by the relatively abrupt drop in responsivity curve for each material displayed in Figure 12.3. At the shorter wavelength range, photoresponse decreases due to photons being absorbed very close to the material's surface rather than in the intrinsic region. In the extreme case, the electrons and holes generated close to the surface recombine before they can be collected by the external circuit. This overall effect causes a gradual decrease in photoresponse as the wavelength decreases from the cut-off wavelength. Now, if we return to Figure 12.3, we can see how the two mechanisms described above help to produce the shape of the photoresponse curve. The SFH 205F spectral sensitivity curve does not look like this because its built-in daylight filter blocks out the visible wavelengths. Detailed knowledge of photoresponse vs. wavelength is required to determine the maximum efficiency of an electro-optical receiver. In some systems, a combination of silicon and germanium photodiodes can be used to increase the spectral response range, since their spectral sensitivities peak in two different wavelength regions. These two photodiodes are also available in a one-package device. We will now consider an example using the above relationship.

Example 12.2

A particular photodiode has a long wavelength cut off at an energy level of $E_g = 0.89$ eV. To what wavelength does this correspond? Using Figure 12.3, determine the type of photodiode this may be.

Using the above formula to calculate the cut-off wavelength we get:

$$\lambda_c = 1.24/E_g = 1.24/0.89 = 1.39 \text{ μm}$$

Looking at Figure 12.3, we can see that this cut-off wavelength corresponds to that of a germanium photodiode.

Instead of using the term photoresponse, we will use, from this time forward, the more common term known as responsivity. A mathematical expression for responsivity is given below:

$$\Re = \frac{I_p}{P_0} = \frac{\eta e}{h\nu}$$

Responsivity is the amount of photocurrent generated per unit of optical power, usually expressed in amperes per watt. The above equation can be obtained by rearranging the terms in the first equation of this chapter. Thus, you can see that responsivity is a linear function of the optical power while quantum efficiency is independent of the power level. The following example will help to demonstrate this relationship. Looking at the data sheet for the SFH 205F, we find that this device has a responsivity of 0.59 at a wavelength of 950 nanometers.

Example 12.3

Find the responsivity of the photodiode used in the situation of Example 12.1 above.
Since we already have the input optical power and the output current, we can easily calculate the responsivity from the above equation:

$$\Re = I_p/P_o = 2.84/5.00 = 0.57$$

This result seems to be consistent with the information given in Figure 12.3. You can also see from this figure that responsivity decreases with wavelength. As an example, if the incident radiation was changed to red light (650 nm), the responsivity would drop to about 0.39. Thus, you can see the wavelength dependence on responsivity from Figure 12.3.

12.2 RESPONSE TIME

When placing a photodiode in a circuit, there will be at least three factors affecting its response time. As you remember, the response time directly affects the bandwidth of the device. In Chapter 10, we discussed the response time associated with the production of a light pulse from a step input electrical pulse. Here, we will discuss the response time required for this light pulse to form an electrical pulse. Important factors to consider during this process are listed below:

1. The time period required for the photocarriers to travel across the intrinsic layer. This parameter is known as the transit or charge collection time.
2. The time required for the photocarriers to diffuse out of the p and n layers.
3. The RC time constant of the photodiode and its associated circuitry.

There are many photodiode parameters involved with the three factors listed above. These parameters include the width, W, of the intrinsic layer, the capacitances associated with the photodiode, and the absorption coefficient, $\propto$. Other parameters contributing to response time effects include amplifier capacitance, load resistance, and the amplifier's input resistance. All of these factors must be considered in the final design of an electrical circuit using a photodiode. In this chapter, we will be primarily interested in the first two factors, since these deal with the construction of the photodiode itself. In the next chapter, we will be more concerned with factor number 3. This factor must be considered when you integrate the photodiode into useful circuits.

The transit time required for both electrons and holes to travel across the intrinsic layer places an upper limit on the photodiode response time. The transit time, t, depends upon the carrier velocity, V, and the width of the depletion layer, W. It can easily be determined by using the following expression: $t = W/V$. Figure 12.4 shows a schematic of the pin structure when placed in the reverse-bias mode. Below this figure you will find the electric field profile of this pin structure for two bias voltage conditions. These conditions are for 0 and 5 volts applied across the structure. When applying a reverse-bias voltage, the electric field intensity will be greatest across the intrinsic region due to its high resistivity. This electric field assists in the movement of the photocarriers within the intrinsic region. Now, let's consider the case when photons enter the pin structure through the p-layer. When they reach the intrinsic region, electron-hole pairs will be produced. Next, these photocarriers or charge carriers will be separated by the electric field resulting in a current flow (drift current). Thus, the velocity of the charge carriers is a function of the electric field strength. With little or no electric field present, electrons and holes will move at lower velocities. A maximum or saturated velocity occurs with the application of only a few volts. The specific voltage needed to achieve this velocity depends upon the device construction. For a typical silicon pin photodiode, maximum velocities for electrons and holes are typically 8.4×10^6 and 4.8×10^6 cm/sec, respectively. This occurs for a field strength of about 20 Kvolts/cm. The electric field strength can be calculated from Gauss' law by using the expression below:

$$E = \rho W/\varepsilon$$

In the above expression, ρ is the charge density of electrons or holes, W is the width of the intrinsic layer, and ε is the permittivity of the material. The electric field strength must be great enough so that both electrons and holes travel at saturated velocities during the transit time. With the application of a large enough reverse-bias voltage, the resulting carrier velocities will be quite constant. Figure 12.4 shows the electric field distribution across the device for bias voltages of 0 and 5. For the 0 volt bias condition, the velocities across the intrinsic layer will not be constant. This results in a large difference between electron and hole velocities. The speed of response will most likely be too low for use in receiving high speed pulses of light. To provide a faster response to an input optical signal, reverse-biasing the device with a 5 volt level will be sufficient in this case. With this 5 volt bias, the photocarrier velocities will remain essentially constant across the intrinsic layer. Thus, for high speed operation, the photodiode should be reverse-biased by a predetermined voltage. The specific amount of voltage depends upon the device specifications

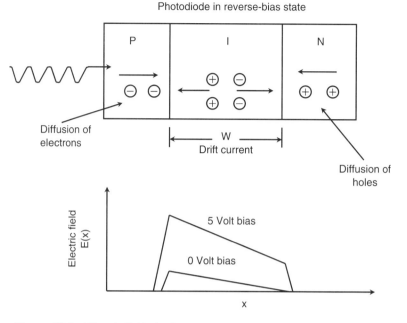

Figure 12.4 Electric field distribution across a pin structure for different bias voltages.

and the circuit application. The 0 volt or unbiased condition has useful applications. It can be used when the requirement for frequency response is below approximately 100 KHz. This unbiased operation usually provides a better signal-to-noise performance due to the lack of dark current produced when reverse-biasing the device.

Looking at Figure 12.4, one can easily see that a thinner intrinsic layer should yield faster response times for the device. But, as previously discussed, the quantum efficiency will decrease when decreasing the thickness of the intrinsic layer. This happens because there are less interactions between photons and the semiconductor material. Also, the capacitance of the device will increase with decreasing thickness. Increased capacitance will cause an increase in the response time. Increasing the active area of the device also tends to increase the capacitance. Thus, there are design trade-offs that must be made. If we are interested in the fastest response time or maximum bandwidth, we must have the thinnest possible intrinsic layer without increasing the capacitance appreciably. The active area of the device must also be taken into account. Unfortunately, decreasing the intrinsic layer's thickness also decreases its quantum efficiency. For any given active area, there exists an optimum intrinsic layer thickness that yields the maximum bandwidth. For example, if a 3 dB bandwidth of 1 GHz is needed in a particular InGaAs photodiode design with an active area of 10^5 μm^2, an intrinsic layer thickness of 0.5 μm can be used. This combination will yield a quantum efficiency of about 0.45. By increasing the thickness to 1.0 μm, and the active area by a factor of 10, the quantum efficiency now increases to 0.68.

The diffusion of electrons and holes out of the p and n regions becomes the second factor affecting response time. This is a slower process than the transit velocity of carriers discussed previously due to the lack of a relatively high electric field to assist in charge carrier movement. To reduce this diffusion time, the thickness of the p and n layers must be carefully considered. As you remember, the absorption depth of incident photons in the intrinsic region determines how long this diffusion time will be. Ideally, the photodiode should be designed such that both the charge collection and diffusion times are equal. To do this, the widths of all three regions must be carefully manufactured such that most of the incident photons are absorbed in the intrinsic region. The p and n regions must obviously be made much thinner to allow for the slower diffusion current. Figure 12.4 shows the relationship between diffusion and transit times.

Long diffusion times can result if carriers are generated outside of the depletion or intrinsic region of a particular photodiode type. To see how this can happen, we must consider the basic structure of a photodiode as shown in Figure 12.1. The structure of the pin photodiode will have a thicker intrinsic region in place of the depletion region shown in the diagram. This structural feature will result in most of the photons interacting in the intrinsic region for a pin type photodiode.

We can see that the average depth that a photon penetrates into the structure depends upon the wavelength, and thus the photon's energy. For example, when photons possessing energy near the cut-off wavelength strike the active surface of the device, a maximum amount of charge carriers will be produced in the depletion region. With the correct amount of electric field there, these charge carriers will move at saturated velocities across the pn junction to contributing to the photocurrent. When light of an even longer wavelength (IR) strikes the active surface, some of these photons will penetrate even further into the structure. As shown in Figure 12.1, this depth can correspond to the region well beyond the depletion layer for planar types. The carriers generated here will travel at a slower speed as they diffuse toward the pn junction. Though not as efficient a process as that involved with charge collection, a certain percentage of these carriers will reach the pn junction to contribute to the total photocurrent. Since these carriers will take a longer time to reach this junction, the overall response to an input square pulse will be a rounded pulse shape with a slow tail. This results from a longer rise and fall time. Planar diffusion photodiodes tend to have this characteristic to a larger extent than pin types due to their relatively thin depletion layer.

We can see the combined affect of transit and diffusion times on the response to a square optical pulse. Figure 12.5 shows the response of a typical photodiode for the conditions of 0 and 5 volt reverse-bias. The input square optical pulse is shown below the response curves. In both instances, the charge carriers generated in the intrinsic region will contribute significantly to the initial stage of the rise time period. Since the diffusion process results in slower velocity carriers, the later stages of the rise time period will be the result of these carriers. These slow carriers were produced in the p and n layers.

A typical response to an input optical pulse can be seen in Figure 12.6. It takes a finite amount of time before reaching maximum voltage and a corresponding amount of time to reach minimum voltage. The rise time can be measured by considering the time it takes the pulse to go from 10 to 90 percent of its maximum point. This parameter is

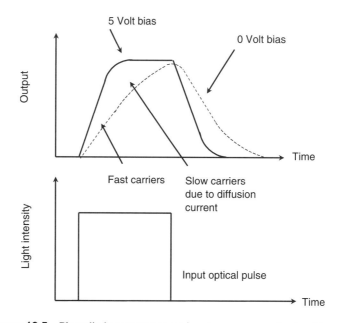

Figure 12.5 Photodiode response to an input square wave for two bias voltages. Notice that diffusion current contributes to a slower response for the 0 volt bias condition.

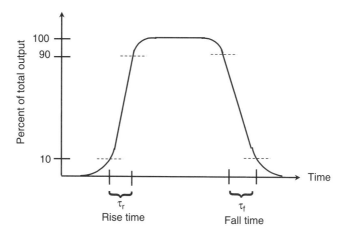

Figure 12.6 Response to an input optical pulse where rise and fall times are equal.

known as τ_r. The fall time can be determined by measuring the time it takes the pulse to go from 90 percent to 10 percent of its minimum point. This parameter is known as τ_f. For a properly designed and biased device, these two parameters will usually be the same as shown in the figure.

The third factor affecting response time is the RC time constant of the photodiode and its associated circuitry. R, in this case, is electrical resistance in ohms. The parameter C must not be confused with the value for the speed of light. C, in this case, is capacitance. The units of capacitance are usually given in microfarads, or µf. This RC combination results in a delay time involved in discharging the capacitor. In most cases, this time will be the limiting factor for determining the photodiode response time. This assumes that charge collection and diffusion times are greatly reduced by virtue of the device construction. The photodiode has a junction capacitance, C_J, associated with it due to its construction and biasing. Generally, its capacitance will increase with increasing active area. The semiconductor material also has a resistivity that causes a series resistance in the photodiode. These factors, combined with connecting the photodiode to a load resistance, produce an RC time constant that determines the maximum speed or bandwidth of the device. In effect, this RC time constant acts as a low-pass filter limiting the response time. The following expression can be used to determine the bandwidth:

$$B = 1/2\pi R_t C_t$$

In the above expression, R_t and C_t are the sum total of the resistances and capacitances, taking into account the external load resistance. An approximation can be determined by measuring the output response pulse as shown in Figure 12.6. Here, we must measure the rise time of the pulse from the 10 to 90 percent points. This rise time results from the three basic factors previously discussed. A good approximation of this rise time can be calculated by considering all three components. The total rise time can be expressed as shown in the mathematical expression below:

$$\tau_r = [(\tau_{cc})^2 + (\tau_{dif})^2 + (\tau_{rc})^2]^{1/2}$$

For simplicity, we will consider the limiting factor to the rise time to be τ_{rc}. The first two factors can be reduced considerably by correct device construction and matching the input optical wavelength so that the maximum production of charge carriers occurs within the intrinsic region. With these considerations, the approximate bandwidth can then be determined by the expression below for rise time. This method will be used later in this chapter for a typical application example:

$$\tau_r \approx 0.35/f_c$$

In the above expression, f_c is the cutoff frequency or bandwidth B. This cutoff frequency corresponds to the point where a 3 dB decrease in photocurrent occurs relative to the low frequency response. The data sheet at the end of this chapter gives the value of junction capacitance as a function of reverse-bias voltage for the SFH 205F. This can be found on the second page of the data sheet with the graph labeled "Capacitance." Notice that this junction capacitance varies with reverse-bias voltage. This fact allows increased bandwidth when using a reverse-bias configuration. The first page of this data sheet gives typical rise and fall times for circuit conditions of $R_L = 50\ \Omega$, $V_R = 5$ volts, $\lambda = 850$ nm, and $I_P = 800\ \mu A$. We will next consider an example using the above relationship.

Example 12.4

A photodiode specification gives a 10% to 90% rise time of 50 nanoseconds to an optical pulse. To produce this output, it has a reverse-bias voltage of 10 volts when used with a 1000 ohm series load resistor. Find the 3 dB bandwidth of this device for the above conditions.

Rearranging the above formula to yield an expression for f_c, we determine the 3 dB bandwidth as shown below:

$$f_c = 0.35/\tau_r = 0.35/50 \times 10^{-9} = 7\ \text{MHz}$$

12.3 NOISE SOURCES IN PHOTODIODES

Noise consists of random fluctuations that distort a received signal. Understanding the sources of noise and how to minimize their effects are important in the design of any optical receiver. The noise present in the photodiode provides the lower limit of signal detection. Since photodiodes have no internal gain, the photocurrent signal must be converted to a voltage, and then amplified electronically. This fact makes noise a critical factor. Noise shows up as a current generated with the signal during this amplification process. The amount of noise generated during this process limits the gain that can be applied to the receiver circuit. One source of noise in a photodiode results from the randomness associated with the arrival of photons from the signal source. Other sources of noise exist that are not signal related. The noise sources that we will consider in this section are (1) dark noise, (2) shot noise, and (3) thermal noise. When operating a photodiode in the reverse-bias mode, the noise current generated will be some combination of all three noise sources.

Dark Noise—This noise results from the leakage current produced when operating a photodiode with a reverse-bias voltage. This current, I_d, flows through the photodiode even when no light strikes the photodiode's active area. The dark current is measured at specified reverse-bias voltages under the condition of no illumination. Its specification can usually be found in the manufacturer's data book for the particular device. The amount of leakage current depends upon the device construction and the level of voltage applied across the diode's junction. The spectrum of the noise source is described by a Poisson distribution. The mathematical expression for this noise is given below:

$$i_n = (2eBI_d)^{1/2}$$

In the above expression, I_d is the dark current in the photodiode, B is the bandwidth, and e is the electron charge (1.6×10^{-19} Coulomb). Operation of the photodiode in the photovoltaic mode results in this noise approaching zero since I_d does not exist. This reduction in noise allows for greater sensitivity in this case due to a larger signal-to-noise ratio. The trade-off here is a somewhat slower response. The dark current specification for the SFH 205F is given as a graph on the second page of the data sheet. You can see that, for this device, a linear relationship exists between dark current and reverse-bias voltage. This fact must be taken into account when determining just how much reverse-bias voltage to use.

Shot noise—This noise is signal dependent and proportional to the amount of light incident upon the photodiode. The shot noise current results when photons generate electron-hole pairs in the detector material. Since photons arrive at the photodiode in a random fashion, this condition results in a statistical nature to the production of electron-hole pairs. This shot noise shows up as a fluctuation in the photocurrent. Thus, shot noise current exists in the photodiode for both biased and unbiased cases. The spectrum of this noise source is also described by a Poisson distribution. The shot noise current is given by the following expression:

$$i_{sh} = (2eI_pB)^{1/2}$$

As you remember from Section 12.1, the photocurrent, I_p, can be expressed mathematically as shown below:

$$I_p = e\eta P_o/h\nu$$

In this expression, I_p is the photocurrent and B is the bandwidth of the photodiode as discussed in the previous section.

Thermal noise—This noise source is also referred to as Johnson noise. It results from the random thermal motion of electrons in the circuit material. Thus, thermal noise does not depend upon the input signal current, and can be found in any linear passive resistor. The shunt resistance in the photodiode is a common source of this noise. In a typical circuit, the photodiode uses a load resistor for the purpose of producing a signal voltage. This load resistor usually has a much smaller value of resistance than the input impedance of the amplifier. Thus, the load resistor itself must be considered as another source of noise. This thermal noise can be expressed as a noise current in the following equation:

$$i_J = (4KTB/R_{SH})^{1/2}$$

In the above equation, K is Boltzmann's constant, T is the absolute temperature in Kelvins, B is the bandwidth, and R_{SH} is the shunt resistance of the photodiode. When considering the noise caused by the load resistance, replace R_{SH} with R_L. As you can see, this noise can be reduced by using a larger value load resistor. The trade-off involved with doing this is a reduction in detector bandwidth. We will consider this case in a circuit example in the next chapter.

The total noise current in the device can be determined by finding the root mean square sum of the individual noise currents. Using the three noise currents described above, we get the following mathematical expression:

$$i_T = (i_n^2 + i_{SH}^2 + i_J^2)^{1/2}$$

We will now present an example using the concepts described above.

Example 12.5

A silicon pin photodiode receives an input optical power of 5 µwatts from a diode laser emitting at a wavelength of 880 nm. The quantum efficiency of the photodiode is 0.8 at this wavelength.

(a) Find the dark noise present when a reverse-bias voltage produces a dark current of 2 nA. Let B = 1 Hz.

Using the equation for dark noise current, we get:

$$i_n = (2eI_dB)^{1/2} = [(2)(1.6 \times 10^{-19} \text{ C})(2 \times 10^{-9} \text{ A})(1 \text{ Hz})]^{1/2} = 2.5 \times 10^{-14} \text{ Amp}$$

This amount of noise current flows through the photodiode in the reverse-biased mode under the condition of no input optical power.

(b) Find the shot noise produced as a result of the input optical power. Again, use B = 1 Hz.

Before we can make the substitution into the equation for shot noise current, we must first find the photocurrent produced by the photodiode. We show this step below:

$$I_p = e\eta P_0/h\nu = (0.8)(1.6 \times 10^{-19} \text{ C})(5 \times 10^{-6})/(6.63 \times 10^{-34})(3.4 \times 10^{14})$$
$$I_p = 2.84 \times 10^{-6} \text{ Amp (This is the same result obtained in Example 12.1)}$$

Using the result above, we can now find the shot noise current:

$$i_{sh} = (2eI_pB)^{1/2} = [(2)(1.6 \times 10^{-19} \text{ C})(2.84 \times 10^{-6})]^{1/2}$$
$$i_{sh} = 9.53 \times 10^{-13} \text{ Amp}$$

(c) Find the thermal noise produced in the photodiode if the shunt resistance value is 7×10^8 ohms with a temperature of 300 K. Use B = 1 Hz.

Using the equation for thermal noise current in a resistor, we get thermal noise produced by the photodiode's shunt resistance.

$$i_J = (4KTB/R_{SH})^{1/2} = [(4)(1.38 \times 10^{-23})(300)/(7 \times 10^8)]^{1/2}$$
$$i_J = 4.87 \times 10^{-15} \text{ Amp}$$

(d) Find the total noise current present in the above case.

To find the total noise, we must find the root mean square sum of the individual noise currents. This mathematical step is shown below:

$$i_T = [(2.5 \times 10^{-14})^2 + (9.53 \times 10^{-13})^2 + (4.87 \times 10^{-15})^2]^{1/2}$$
$$i_T = 9.53 \times 10^{-13} \text{ Amp}$$

This example shows that the shot noise due to the input optical signal is the dominant component in the total noise present. The next largest noise source in this example is the dark noise. This dark noise can be neglected when the reverse-bias voltage equals zero. This subsequent reduction in total noise will allow for greater sensitivity. Of course, the trade-off in this case becomes a somewhat slower response.

(e) If a 50 ohm load resistor is used with the photodiode in the above example, find the increase in thermal noise.

We use the thermal noise current equation to find this thermal noise:

$$i_J = [(4)(1.38 \times 10^{-23})(300)/(50)]^{1/2} = 1.82 \times 10^{-11} \text{ Amp}$$

You can clearly see that the thermal noise now becomes the dominant noise source in this circuit. In order to use this relatively small value load resistor, a much larger optical signal may be necessary. Of course, this will also result in a larger shot noise current. In the next chapter, we will consider the trade-offs involved when using photodiodes in practical circuits.

A very useful detector parameter, known as the Noise Equivalent Power (NEP), gives circuit designers some knowledge of the photodiode's lower detection limit. This parameter gives the optical power needed to produce a signal-to-noise ratio of one. Factors affecting the noise level in a photodiode, such as reverse-bias voltage and active area, will help determine this value. Typical NEP values range from about 10^{-12} Watts/(Hz)$^{1/2}$ for large area photodiodes, to about 10^{-15} Watts/(Hz)$^{1/2}$ for small area photodiodes. According to the data sheet for the SFH 205F photodiode, this device has a NEP of 4.3×10^{-14} W/(Hz)$^{1/2}$ for a reverse-bias voltage of 10 volts. Its active area (radiant sensitive area) is given as 7.00 mm^2.

If we take the inverse of the NEP, we get still another very useful parameter known as detectivity or D. Detectivity is a measure of the detection sensitivity of a photodiode. We can determine the relative sensitivity of a detector by multiplying the detectivity by the square root of its effective sensitive area. The resultant parameter is known as D* (D-Star). This parameter has the units of cm·Hz$^{1/2}$/Watts. The higher the value for D*, the better the detector. According to the data sheet for the SFH 205F photodiode, this device has a D* of 6.2×10^{12} cm·Hz$^{1/2}$/Watts. The relationship between NEP and D* is given below:

$$D* = \frac{[\text{Effective Sensitive Area (cm}^2)]^{\frac{1}{2}}}{\text{NEP}}$$

12.4 PHOTODIODE CIRCUIT OPERATING MODES

There are two basic modes of operation in which the photodiode may be placed in a circuit, the photovoltaic and the photoconductive modes. A photodiode operating in the photovoltaic mode uses no reverse-bias voltage. In this mode, it generates photocurrent in a fashion similar to a photocell when illuminated. When applying a specified voltage in a reverse-bias configuration to the device, it operates in the photoconductive mode.

The photovoltaic mode finds its application primarily in low frequency and low noise circuits. The decision to use a particular mode of operation depends mainly upon the frequency response requirement for the given application. By eliminating the need for a reverse-bias voltage, receivers using the photovoltaic mode have simpler circuitry with no dark current noise. For a frequency response requirement of approximately 100 KHz, photovoltaic detectors offer a better signal-to-noise performance than do photoconductive detectors. Above this requirement, the slower response of photovoltaic detectors places them at a disadvantage.

The photoconductive mode finds applications primarily in fast response systems, such as in fiber optic communications. In this mode, the bandwidth requirement is usually

100 KHz or greater. To meet this requirement, there must be an application of an external bias voltage in the reverse-bias direction of the pn junction. The current generated, in this case, comes from two sources: the photoinduced current and the reverse leakage (dark) current. This reverse leakage current will remain constant for a fixed bias voltage level and fixed temperature. The reverse-bias voltage also tends to decrease photocarrier transit times by setting up an electric field within the device. This reverse-bias voltage causes a subsequent reduction in the junction capacitance thereby minimizing capacitive loading effects on the frequency response. Photodiodes in this mode can operate from D.C. to above 1 GHz. The noise generated, in this case, results from a combination of shot noise and noise due to the dark current. Thus, the dark current associated with the reverse-bias voltage level across the photodiode is responsible for the increased amount of noise when configuring a photodiode in the photoconductive mode.

It is also important to point out that pin photodiodes are designed to be operated at specified reverse-bias voltages. They can also be operated in the photovoltaic mode with good results. The reverse-bias voltage range can be found by reviewing the electrical parameters for the device. The maximum reverse-bias voltage must not be exceeded as the electric field created will destroy the device. If the device is not operated at the recommended bias voltage, many of the characteristics listed on the data sheet will be different. As an example, the following parameters will increase as bias voltage increases from zero

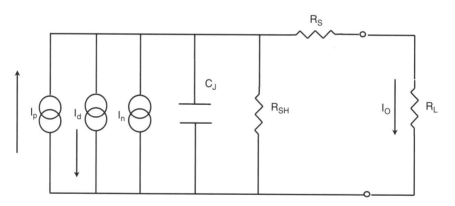

I_p = Current generated by the incident light.

I_d = Dark current that flows in the absence of light. Reverse biasing increases this current. This is a noise source.

I_n = Noise current.

C_J = Junction capacitance.

R_{SH} = Shunt resistance. Resistance of the photodiode with zero bias.

R_s = Series resistance. This is the resistance of the photodiode itself. It is usually several ohms or less. This becomes more important in high frequency applications.

I_O = Output current that flows through load resistor R_L.

Figure 12.7   Equivalent circuit for a photodiode.

to the recommended amount: dark current, noise current, depletion depth, and bandwidth. The following parameters will decrease: junction capacitance, series resistance, rise and fall time.

As you can see, the photodiode acts as a current generator. The equivalent circuit for the photodiode is shown in Figure 12.7. This equivalent circuit shows the relationships between the parameters previously discussed. Most data books use these parameters to specify each device.

12.5   CHARACTERIZATION OF DETECTOR RESPONSE

Two relatively simple experiments can be performed to demonstrate some of the physical principles involved in the operation of a photodiode. Figure 12.8 gives the electrical schematic for these two simple experiments. The transmitter portion of the circuit employs a GaAlAs infrared emitting diode with a peak emission wavelength at 880 nanometers. The period of time to pulse it on is 120 microseconds. In the first experiment, we use three photodiodes with different active areas that find uses in common electro-optical circuits. In this experiment, they are used separately in the simple receiver circuit shown in Figure 12.8. Each photodiode must also be separated by the same distance from the emitter in each case. To minimize the affects of ambient light, we must place a light-tight covering over both devices without interfering with the transmission of infrared light from the transmitter. The photodiodes are identified as A, B, and C. We use for photodiode A, Motorola part MRD 721 (pin type). Its data sheet can be found in Appendix A. For photodiode B, we use the Siemens part SFH 205F (pin type). Its data sheet can be found at the end of this chapter. For device C, we use a Vactec part VTP 8350 (planar diffused type). Its data sheet can be found in Appendix A. To start with, we reverse-bias each photodiode

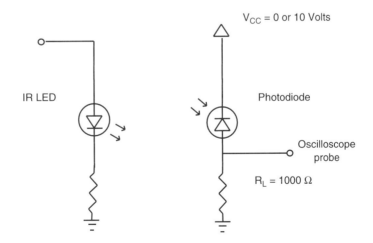

Figure 12.8   Experimental design to determine detector response.

by applying 10 volts to its cathode. A load resistor with a value of 1000 ohms acts as a current-to-voltage converter allowing voltage measurements using an oscilloscope. This simple circuit places the photodiode in the photoconductive mode of operation. The measurements of pulse width and rise time must be taken for each device. In the second experiment, the same photodiodes are used in the design described above. We change only one variable to this design. Now we remove the reverse-bias voltage for each device. The same oscilloscope measurements as performed in the first part of this experiment can now be taken. For convenience, a comparison of the output pulses for the 0 and 10 volt bias conditions is given in Figure 12.11 using only device C.

The measurements of rise time and amplitude for each device during the first experiment can be found in Figure 12.9. These same measurements were taken during the second experiment after removing the 10 volt bias voltage from the cathode of each photodiode. Figure 12.10 gives the output pulses for the second experiment. For device C, the complete pulse is shown for the purpose of detailing the unequal rise and fall times. Table 12.1 summarizes the results obtained from these signals, and also lists the active surface areas for the three devices. It is important to keep in mind when reviewing these figures that the rise time for the transmitted pulse was 5 nanoseconds in every case.

As discussed previously, the response time is the result of three basic factors. These factors are: (1) the carrier collection time, (2) the diffusion time, and (3) the RC time, or the time required to discharge the junction capacitance. Thus, the bandwidth can be calculated by taking the rise time measurement of the pulse, and then using the rise time equation discussed in Section 12.2. You must keep in mind that this result includes all of the accumulated response times from the transmitter through to the oscilloscope. Even a typical oscilloscope probe causes a slight delay due to its capacitance of about 2 pf. The measured rise times and bandwidths can be found in Table 12.1. The main purpose of these experiments is to provide information about the bandwidth of the devices. You must realize that it will usually become necessary to add some level of electrical amplification to the photodiode. The addition of amplifying circuitry will reduce the system bandwidth. The amount of reduction depends upon the amplifier used and the application. This situation will be considered in the next chapter.

As stated previously, one purpose of these simple experiments was to show relative response times for planar diffuse and pin silicon photodiodes of various active areas. To

Table 12.1   Rise Time Measurements for Devices A, B, and C

Device	Active Area (mm)^2	Rise Time 0 Volt	10 Volts	Bandwidth (KHz) 0 Volt	10 Volts
A	0.6	791 n	661 n	440	530
B	7.0	2.02 μ	952 n	170	368
C	7.45	23.7 μ	1.53 μ	14.8	229

Note: In the rise time columns, n = nanoseconds and μ = microseconds.

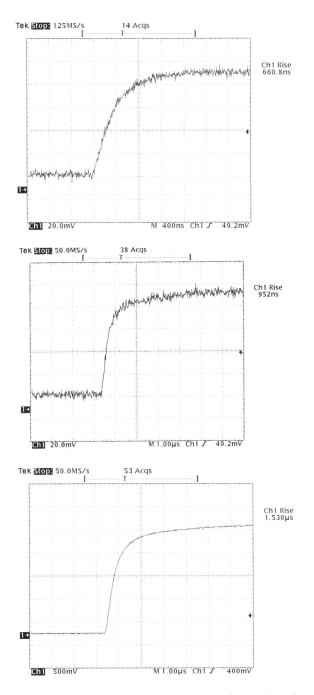

Figure 12.9   The first experiment shows rise times for the three photodiodes using a 10 volt reverse-bias.

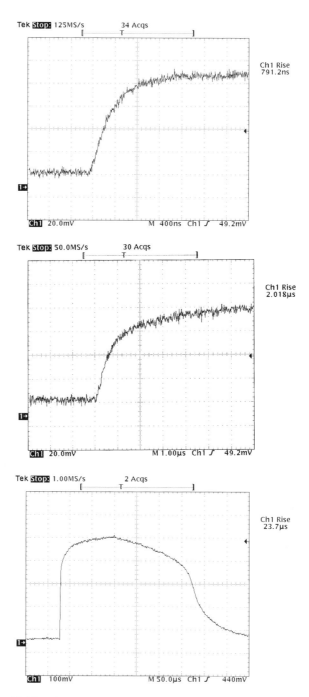

Figure 12.10   The second experiment shows rise times for the three photodiodes when the 10 volt reverse-bias was removed.

do this easily, the same circuit configuration was used in each case. You must be careful when reviewing the results obtained from these two experiments. The rise times obtained here do *not* represent the optimum response for each device. The experiments described above were carried out by using a typical application for which any of the three devices could be used. There are other applications for which the response times may deviate substantially from what is shown in Table 12.1. Each application must be evaluated separately for bandwidth requirements. We will consider this subject in more detail in the next chapter.

Figures 12.11a and 12.11b clearly show why a photodiode must be reverse-biased for high speed operation. This result compares the response characteristic of device C when using a 10 volt bias (Figure 12.11a) to the response characteristic when using a zero volt bias (Figure 12.11b). In the case of the reverse-biased photodiode, the waveform shows that a maximum level is reached after a finite time, then there is a corresponding fall time. When removing the bias voltage, the waveform becomes distorted. A flat maximum voltage level is never achieved. This results from the unequal transit and diffusion times associated with electrons and holes in the depletion region. A pulse shape such as that from device C, without external bias voltage, could not be used in digital communications circuits. In this application, the pulse must be distinctly square.

Finally, we list two general procedures that may help the circuit designer when selecting a photodiode device on the merits of response time.

1. Test the device using its intended light source. The frequency response can differ significantly for two different input optical wavelengths due to the wavelength dependence on the absorption coefficient. Figure 12.1 shows why this occurs. When photons of different energies strike the active surface of the device, they interact at different levels within the photodiode's structure as shown. This can result in different transit and diffusion times.

2. Try to characterize the photodiode using the conditions of the particular application. This means that the photodiode should be tested using the intended optical illumination level(s), pulse characteristics, and operating speed.

What can we conclude from the results of the previous experiments? One important thing to keep in mind is that for high speed operation, the pin photodiode must be reverse-biased. As discussed previously, this bias voltage produces an electric field within the device structure that allows the electrons and holes to move with decreased transit times at essentially the same velocity. The amount of bias voltage needed for this to occur depends upon the device construction. If high speed operation is of primary importance, then a pin photodiode, such as device A in the experiment above, could be used. Device A was designed for ultra high speed response, and finds its application in position encoders and high speed logic circuits. Device B has a somewhat longer response time and a much larger active area. This device was designed to be used in digital IR sound transmission and remote control circuits. Its characteristics were discussed in detail in the last few sections of this chapter. The larger active area allows it to detect a transmitted signal from a relatively large distance. Most devices used for this application come with a visible light

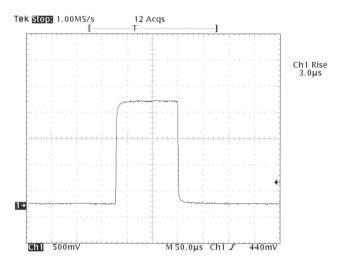

Figure 12.11a   Response for device C with a 10 volt reverse-bias.

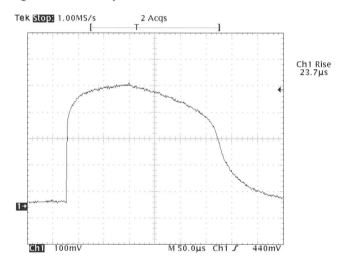

Figure 12.11b   Response for device C with no bias voltage applied.

filter integrated into the device construction. This filter restricts the passage of background optical radiation in the visible wavelengths while allowing near IR radiation to pass to the active area. It helps to increase the signal-to-noise ratio substantially, allowing for a more efficient detector. Device C obviously was not designed for high speed applications, but may be used for relatively low data rates involving low optical radiance levels. Typical applications for device C can be found in exposure meters.

12.6 AVALANCHE PHOTODIODES

Some applications require more sensitivity than can be provided by the typical pin photo-diode. Such applications include airborne particle counters and fiber optic receivers. These applications basically require high gain under low light conditions. An improved device, the avalanche photodiode, has a construction that allows for much higher reverse-bias voltages. The stronger internal electric field produced by this high bias voltage causes a multiplication of photogenerated carriers. The advantage gained with this configuration shows up as increased photocurrent due to this process.

As with the pin photodiode, there are different types of avalanche photodiodes, hereafter termed APD. The particular type depends upon the semiconductor material used in construction. Silicon APDs have a sensitivity in spectral range from 400 to 1100 nm. These devices find useful applications in low light photometry and as laser monitors. For longer wavelengths, the germanium APD has a spectral sensitivity from 800 to 1550 nm. The semiconductor material indium gallium arsenide (InGaAs) has a spectral sensitivity that overlaps that of germanium. APDs made from InGaAs tend to have lower noise and better frequency response. Unfortunately, these devices are also more expensive.

The APD uses an electric field in excess of 10^5 V/cm caused by increasing the reverse-bias voltage to a point just below the breakdown voltage of the semiconductor. This voltage level can be as high as 200 volts or more. When electrons and holes produced by the incident photons encounter this high electric field, they will obtain enough kinetic energy to produce additional electron-hole pairs through inelastic collisions within the semiconductor structure. The net effect shows up as additional electron-hole pairs due to this impact ionization. The generation of a greater number of electron-hole pairs results in a higher photocurrent within the device. The APD also requires a much more complex electronic circuitry than that of the typical pin photodiode. This includes a careful monitoring of device temperature and bias voltage. Feedback control circuits are used to accomplish stability in APD operation.

Figure 12.12 shows a schematic of the APD structure. In this structure, depletion or intrinsic regions occur in two different locations within the device. When photons strike the device, they enter through the P^+ layer. They then travel to the first intrinsic region labeled i or π to produce photocarriers. These initial photocarriers are produced by the same process described previously, the exchange of energy between photons and the silicon material. After the generation of electron and hole pairs, they become separated by the influence of the electric field in the π region. You can see that the electrons will continue to drift through the π region, and then to the pn^+ junction where a higher electric field exists. This higher electric field results from the use of a high resistivity p-type material. When electrons drift into the high electric field of the second intrinsic region, the narrow p-region, the multiplication process takes place due to impact ionization. As a result, there will be an increase of photocurrent within the device.

This internal current multiplication can be as much as 100 or more. The specific amount of gain in a particular structure varies with photon wavelength, bias voltage, and temperature. The mathematical expression below describes the multiplication, M, of carriers within the photodiode:

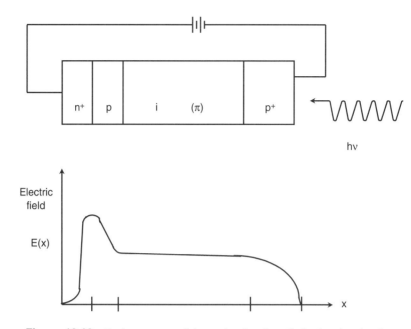

Figure 12.12 Basic structure of the avalanche photodiode showing the electric field profile.

$$M = I_m/I_P$$

I_m = the average value of total output current
I_P = the initial photocurrent in the π region before the multiplication process

Now, we can describe mathematically how the responsivity of the device changes. For convenience, we give the mathematical expression for the responsivity for a typical pin photodiode below:

$$\Re = I_P/P_O = \eta e/h\nu$$

This expression shows that when the photocurrent, I_P, increases, there will be a corresponding increase in responsivity. When using an APD, we can see that the responsivity will increase by the factor M. The mathematical expression below gives the responsivity of the APD.

$$\Re_{APD} = (\eta e/h\nu)M$$

The advantage of high responsivity in the APD structure comes with the disadvantage of increased noise. This excess noise results from the statistical nature of the avalanche process. Specifically, this noise depends upon many factors such as device structure, gain, and illumination. The avalanche gain is an average, since not all photocarriers undergo the same multiplication factor. Thus, the noise current increases by the square of the multiplication factor, M, and shows up as increased shot noise. Much smaller noise currents exist, but these will not be discussed here. The expression for the shot noise of the APD then becomes:

$$i_{SH} = (2eM^{(2+x)}IB)^{1/2}$$

It has been found that as M increases to higher levels, this factor must be increased accordingly as M^{2+x} due to a faster increase in noise. Values for x vary between 0 and 1.0.

12.7 PHOTOTRANSISTORS

Now that we understand how the photodiode converts light energy into electrical energy, we can use this basic building block to construct other useful light sensing devices. The phototransistor will be the next such device to consider. It uses in its construction a photodiode and another semiconductor layer to form two junctions instead of just one. This extra junction gives the phototransistor an advantage over the photodiode—internal gain. As you remember, the photodiode has unity gain, and requires an external amplifier to produce a useable signal. It is possible to use a phototransistor to amplify an input optical signal without using an external amplifier such as an op-amp. This can offer a cost savings, since very few external electrical components are needed in most cases. If the objective of the circuit design is to switch a device on and off at a predetermined light level, then the phototransistor may be the best choice for a light sensing device. We will see that there are some limitations to the operation of the phototransistor, but it may be the better choice for a particular application.

The basic construction of a phototransistor is shown in Figure 12.13. Here, we see a three terminal device having emitter, base, and collector connections (NPN phototransistor). The light sensitive pn junction (collector-base) acts as the photodiode that can be reverse-biased. The output of this junction is fed into the base of a small signal transistor. When light produces a photocurrent, I_p, in the collector-base portion, base current is generated and amplified by the current gain of the transistor. Some phototransistors don't have a separate base connection, so the base cannot be externally biased. When this base terminal is available, external biasing is possible, and the resultant emitter current can be expressed by the following formula:

$$I_C = I_P(h_{FE} + 1)$$

In this equation, I_P is the photocurrent produced by the input light signal, and I_C is the collector-emitter current that results from the amplification process. The term h_{FE} is a number that specifies the transistor DC current gain. Values of h_{FE} can range from 100 to over 1500. A data sheet for a typical NPN phototransistor can be found in Appendix A with the part number Vactec VTT1015. Included with this data sheet is a device description with various electro-optical parameters. A detailed specification for the phototransistor chip is also included in this appendix with the Vactec part number VTT-C50. We will discuss these parameters in this section.

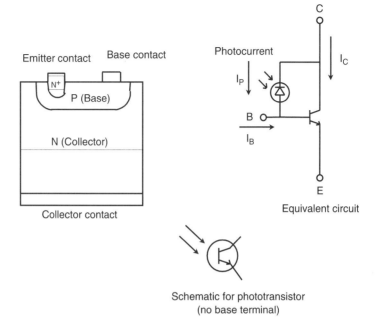

Figure 12.13 Basic construction of the phototransistor. Its equivalent circuit and schematic representation are also shown.

The current amplification process is accomplished by introducing a bias current, I_P or I_B, at the base terminal. Light having a wavelength within the device's spectral range of photosensitivity will produce this photocurrent. It is this photocurrent that drives the base of the transistor. With the exception of this photocurrent, the operation is the same as would be expected from an ordinary small signal NPN transistor. This bias current affects the collector-emitter current, I_C. It must be noted that the parameter h_{FE} is not a constant. This parameter can vary with bias voltage, collector-emitter voltage, irradiance, wavelength, and temperature. The current-voltage characteristic curves for a typical phototransistor are shown in Figure 12.14.

This graph shows the output of a typical phototransistor for four different values of I_B. The base current, I_B, is the photocurrent produced from the light sensitive section of the device. Each value of I_B represents a corresponding amount of optical power converted into a photocurrent. The optical power required to produce a certain amount of photocurrent can be calculated by using the responsivity equation discussed in the first section of this chapter. The gain can only be specified for a particular irradiance, wavelength, and collector-emitter voltage. For example, a particular device has a current gain of 550 when irradiated by a 950 nanometer source at 0.5 mW/cm^2 for the condition of V_{CE} equal to 5 volts. In this example, a photocurrent of 1μA to the base of the device will produce a collector current of 550 μA.

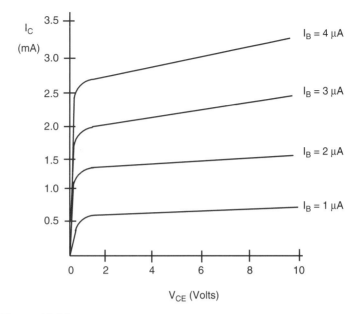

Figure 12.14 V_{CE} versus I_C for four different input base current levels, I_B.

There can be a distinct disadvantage when using a phototransistor in an application that requires a very fast response. The phototransistor's response time is affected by the usual RC time constant of the collector-base junction and the current gain. In some devices, the base terminal is not available for a circuit connection. This results in a high input impedance. In general, this condition can be stated simply as follows: the higher the gain, the slower the response time. Later, we will consider this problem in a typical application using a phototransistor.

Another disadvantage that must be considered when using a phototransistor in place of a photodiode is the fact that many different factors affect the current gain. The current gain varies with irradiance, collector-emitter voltage, wavelength, base drive current, and temperature. Variations in the current gain can be seen in the current-voltage curves in Figure 12.14. Take for example, the curve for I_B of 4.0 μA. When V_{CE} is 2 volts, the output, I_C, becomes about 2.8 mA. When V_{CE} increases to 8 volts, the output increases to 3.1 mA. All other parameters are constant in this example. Variations in the current gain also occur with variations in the light intensity. This nonlinear effect must be considered when the design involves large variations in light input. At relatively low light intensities, the gain will be small but will increase with increasing amounts of light. This will occur until a peak gain is achieved. After this peak gain is achieved, any increased amount of light will result in a gain decrease.

The phototransistor finds several applications as a light sensitive switch. Figure 12.15 provides two basic electrical schematics for this application.

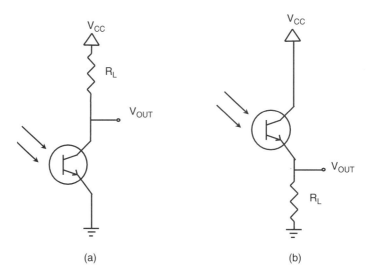

Figure 12.15   For case (a), the output will go from a high to a low value. For case (b), the output will go from a low to a high value.

In Figure 12.15a, V_{out} goes from a high value to a low value when light strikes the base of the phototransistor. In this case, the phototransistor will become saturated and a current will flow through R_L. The output will initially be high before it switches, and then it will go low after the switching action occurs. In Figure 12.15(b), the inverse of this operation occurs. When light strikes the base of the phototransistor, current will flow through R_L also. The voltage across R_L will initially be low, then when switching occurs, the voltage will be high.

In these circuits, the phototransistor acts like any general purpose small signal transistor. The only difference here is how the base of this transistor is driven. When using a phototransistor, incident light drives the base. For a typical small signal transistor, an applied base current drives the device to produce the same results obtained when using the circuits in Figure 12.15a and Figure 12.15b.

When using a phototransistor in a switching application, the time delay must be considered. A fast rise in the photocurrent, I_P, will not result in an equivalent rise in the collector-emitter current. When light strikes the device's sensitive area, a photocurrent will be produced. This current must first charge the capacitances associated with the base circuit. Once this occurs, photocurrent will flow within the device as previously described. The period of time required for this to occur depends upon many factors such as the photodiode capacitance, amplifier gain, etc. This time period is substantially longer than if just a photodiode of the same active area was used. When the incident light level drops suddenly, a similar situation occurs. The capacitances must be discharged through the device before the current level drops to zero or a low turn off value. All of this assumes that the leakage current is negligible. Thus, this characteristic delay time depends only upon the device construction. The rise time associated with this device will increase even more with the addition of an external load resistor. This increase in resistance changes the RC time constant.

Careful attention must also be paid to the amount of optical radiation used to switch the phototransistor. If too much light strikes the sensitive base region, a saturation condition can occur. The switch on time will occur as previously described but the switch off

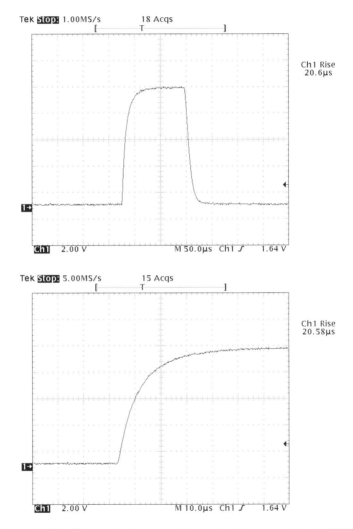

Figure 12.16   The speed of response for a typical phototransistor, VTT1015. (a) Output pulse with amplification. (b) An enlarged view of the rising edge. Notice rise time value.

time may be substantially longer. This longer switch off time occurs because the excess stored charge must be discharged before it can turn off. To minimize this effect, the value of R_L must be reduced, or less optical radiation should be allowed to strike the device's active area.

To demonstrate the phototransistor's speed of response, a Vactec VTT1015 phototransistor can be used in the same design detailed in Figure 12.8. We use a voltage to the collector terminal of 10 Volts. For simplicity, we use the same input optical pulse of 120 microseconds at a wavelength of 880 nanometers. Figure 12.16a shows a typical shape that may be obtained for the full pulse. Figure 12.16b shows a magnified rising edge of this pulse to be used for the purpose of calculating the bandwidth. The active area of the device is substantially less than that of photodiode device B (SFH 205F photodiode). The results show a rise time of 20.58 microseconds with a voltage height of almost 9 volts. This voltage level is the result of the internal amplification of the device. Finally, if we do the calculation for bandwidth, we get 17 KHz. You can see from these simple experiments that the phototransistor could not be used in applications requiring very fast switching speeds.

12.8 OPTOCOUPLERS

The optocoupler contains both an emitter and a receiver separated by a transparent dielectric medium. This transparent medium transfers the optical power. Most optocouplers use an IR LED for the transmitter and a phototransistor or photodiode for the receiver. The transmitter and receiver are totally enclosed in an IC-type package. The input and output can only be modified electronically to provide the desired circuit performance. This makes the optocoupler a simple electro-optic system, and thus cannot be considered an optoelectronic component. When using optocouplers in a circuit, you do not need to worry about the optics involved since this has already been integrated into the structure. The concerns to be addressed are the electrical characteristics, speed of response, voltage drops, and limitations of the emitter and detector elements.

The main advantage gained by using an optocoupler is electrical isolation between the two internal elements. The degree of isolation depends upon the dielectric material and the physical separation between the transmitter and receiver elements. In Appendix A, you will find a data sheet for a commonly used optocoupler with the Motorola part number 4N35. This data sheet shows what a typical optocoupler looks like and also lists its electro-optical specifications. Optical glass has proven to be an efficient dielectric material to use in the construction of these devices. It allows for high voltage isolation with good transfer of optical energy, and protects the receiver circuit from any damaging overvoltage produced from switching or lightning induced surges. The maximum voltage that the dielectric can withstand is known as the isolation voltage. To utilize these devices in an electrical circuit, the characteristics of both internal elements must be understood. These optoelectronic devices have been previously studied, and the reader can turn to the appropriate chapters for review. A simple electrical schematic for the optocoupler is given in Figure 12.17.

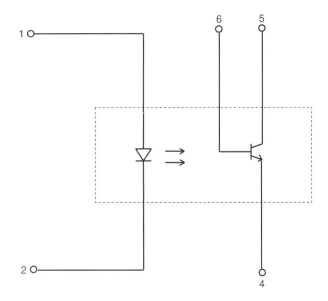

Figure 12.17   Schematic diagram for the 4N35 optocoupler.

The basic operation of the optocoupler involves passing a predetermined forward current through the IR emitter section to produce an optical signal. The receiver element then converts the optical energy back into electrical energy for use in the isolated electrical circuit. While we have just described one of the simplest optocouplers, there are many other types available. The receiver element could be a photodiode, phototransistor, photodarlington, or light sensitive SCR (silicon controlled rectifier). Each of these types provides a special electrical application based upon speed of response, sensitivity, or other desired switching characteristics.

Figure 12.18 shows the electrical schematic and setup using a the 4N35 optocoupler. This device contains an IR LED and a phototransistor in a 6 pin DIP package. Pins 1 and 2 provide electrical access to the IR LED using R1 to limit the current flow, I_F. Pins 4, 5, and 6 provide electrical access to the emitter, collector, and base respectively of the NPN phototransistor. When emitted light from the IR LED strikes the base of the phototransistor, current will flow from the collector to the emitter. A load resistor converts this current into a signal voltage. For pulsed operation, the output can be taken at either pins 4 or 5. The output from pin 4 will give a positive going pulse with a positive input current into the IR LED, while pin 5 will provide an inverted pulse. A manufacturer's data sheet for the 4N35 can be found in Appendix A. This data sheet contains the device parameters, and has a section for typical circuit applications.

An important optocoupler parameter, known as the coupling efficiency (η) or dc current transfer ratio (CTR), tells the circuit designer how much collector current, I_C, will

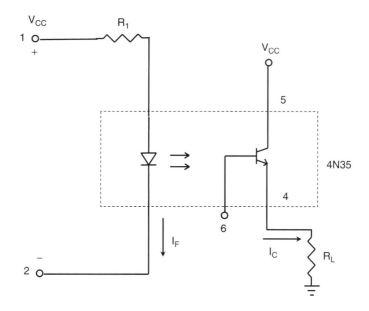

Figure 12.18   Typical electrical components used with the optocoupler.

be produced from a given forward LED current, I_F. This number gives a measure for the effectiveness of the electrical signal transfer from the transmitter to the receiver. A simple formula used to calculate the coupling efficiency is given below:

$$\eta = I_C/I_F$$

The CTR will vary with V_{CE} and I_F, so the graphs for a particular device must be consulted before using the device in a circuit. For example, a typical CTR value for the 4N35 is 1.0 when $I_F = 10$ mA and $V_{CE} = 10$ Volts. CTR is usually expressed as a percentage. For the above example, a CTR of 1.0 is 100%. This means that when I_F is 10 mA, I_C will also be 10 mA.

The switching speed associated with the optocoupler tends to be fairly slow when compared to that of switching transistors. This is partially due to the large photosensitive base-collector area. As we discussed in the last section on phototransitors, this large area causes a relatively high capacitance that contributes to slower response. For the highest speed possible, you can operate the phototransistor as a photodiode. To do this, the bias voltage must be supplied between the collector and base terminals, resulting in an unused emitter (see Figure 12.13). Use a load resistor having the lowest possible resistance that provides good circuit operation. One disadvantage to the phototransistor used in this fashion is that the output signal will require amplification.

When using a phototransistor type optocoupler, it is possible to increase the switching speed by adding a resistor, R_{BE}, between the base and emitter terminals as shown in Figure 12.19. Resistor R_{BE} removes the stored charge from the base region faster than

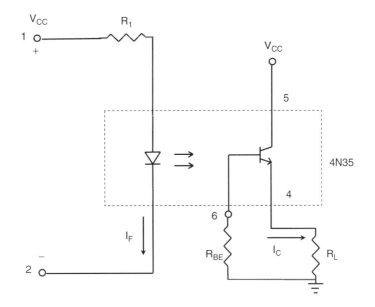

Figure 12.19   The switching speed can be increased by using resistor, R_{BE}, as shown.

when using no base terminal resistor. The recombination time for the carriers within the structure determines the response characteristic. When using this method, trade-offs must be made in terms of performance. The use of resistor, R_{BE}, will require a larger light output from the IR LED since this resistor now provides another path to ground for the photocurrent. The device will now experience faster turn-on and turn-off times. The value of R_{BE} must determined by the trade-off allowed with increasing I_F and decreasing R_{BE}. Another alternative may be to decrease the pulse width. This will reduce the amount of stored charge in the base region, resulting in faster response. Thus, parameters such as R_L, R_{BE}, I_F, and pulse width must be studied to optimize the operation of the device for a particular application.

Example 12.6

A 4N35 optocoupler is used in a circuit as described in Figure 12.18. During operation, it is found that when V_{CC} is 10 Volts and R_L is 1Kohms, I_C becomes 7.0 mA. If the CTR for this device is 1.4 or 140%, find the value of R_1.

The first parameter that we must find is the IR LED forward current, I_F. This can be found by using the given values for CTR and I_C. The forward IR LED current can now be calculated as shown below:

$$I_C/I_F = 1.4 = 7.0/I_F$$
$$I_F = 5 \text{ mA}$$

Using Ohm's law, we can easily find the value for R_1:

$$R_1 = V_{CC} - V_F/I_F = 10 - 1.3/0.005$$
$$R_1 = 1740 \text{ Ohms}$$

12.9 DETECTORS—NEAR AND MID-IR SPECTRAL REGIONS

The near and mid-infrared regions of the spectrum offer very useful information for analysis by specialized detectors. Typical applications include, but are not limited to, terrestrial imaging, industrial gas analyzers, automotive exhaust analyzers, and flame detection. We will consider only the intrinsic types of IR detectors in this section. Intrinsic detectors operate on the band gap principle described earlier in this chapter. Some detectors operating in this spectral region require cooling below room temperature to be effective. We will briefly consider a few detector types and then discuss some typical applications.

We can readily see why the IR spectral band becomes so important by considering Planck's radiation law. In Chapter 8, we discussed this law in detail, and then considered some practical examples. For a blackbody source, a relationship exists between the temperature of the source and its peak spectral emission wavelength. For convenience, we will restate this relationship below:

$$\lambda_{max} = (2899)1/T \ (\mu m)$$

In the above equation, T must be in Kelvins for the wavelength λ_{max} to have units of micrometers (μm). When considering objects that we all experience everyday, we see from the above relationship that they emit radiation strongly in the IR spectral band. A quick calculation using this relationship shows that the human body, with a temperature of approximately 308 K, has a peak spectral emission wavelength of 9.4 μm. Thus, a detector having a spectral sensitivity in this area of the spectrum will be useful in sensing these objects.

PbS and PbSe Types

The first detector of this type to consider, lead sulfide (PbS), has a spectral sensitivity from 1.3 to 3.0 μm at room temperature. Cooling this detector to 77°C extends its sensitivity out to about 4.2 μm. An alternate detector that can be used in this wavelength region is lead selenide (PbSe). Its long-wavelength spectral response and bandwidth exceeds that of PbS. Figure 12.20 compares the detectivities of PbS and PbSe at various temperatures. The dotted lines give spectral sensitivity data for PbSe (BH and BX types). The solid curves in the upper left are for PdS (A type), while the solid curves in the lower right are for PbSe (B type). You can see that while PbSe has a broader wavelength range, PbS has a greater detectivity to radiation at 2.0 μm than PbSe at the same temperature. The choice of detector to use depends upon the particular application.

A very useful application of PbSe detectors can be found in flame detectors. Flame detectors are used extensively in aircraft hangers, fuel loading stations, and other places where flash fires may occur. Research with hydrocarbon fires has shown that they all have a fire "signature." The emission spectrum of every hydrocarbon fire is widely dis-

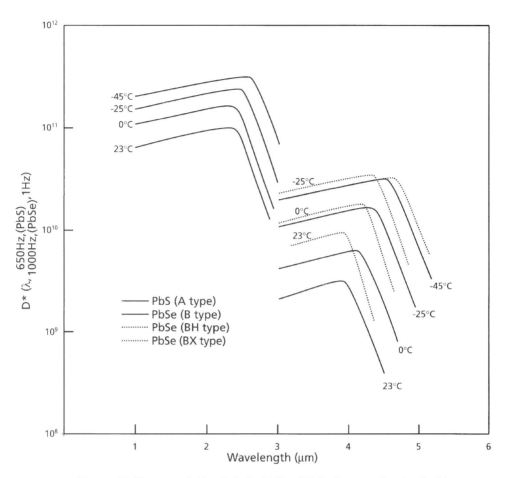

Figure 12.20 Detectivities (D*) for PbS and PbSe detectors. Reprinted with permission of Cal Sensors, Inc.

tributed over the UV to IR regions with a strong emission component occurring at 4.3 μm. This emission spectrum results from the interaction of matter and radiation. Specifically, when CO_2 molecules are excited, photons are emitted in the 4.3 μm region. In the world of fire detection, this is known as the CO_2 "spike." Detectors sensitive to this wavelength are used in many types of flame detectors. To make flame detectors less prone to false alarms, they usually employ two sensing elements, one at 4.3 μm and another at UV wavelengths. In this case, both sensing elements must detect sufficient radiation before the decision for an alarm is made. For example, an arc welding source emits strongly in the UV but does not emit sufficient radiation in the IR. With a UV/IR flame detector, false alarms due to arc welding and many other sources can be eliminated.

HgCdTe Type

Another very useful detector material is mercury cadmium telluride (HgCdTe). By adjusting the ratio of HgTe to CdTe, peak spectral response can be adjusted from 1 to about 30 μm. Detectors using this compound can be used in both the photovoltaic and photoconductive modes. As with the PbS and PbSe compounds, cooling increases sensitivity. It can also be used at room temperature for certain applications. The frequency response depends upon material composition and operating temperature. As you remember, the frequency response is related to the lifetime, τ, of the electrons in the semiconductor material. This relationship is given by the relationship below:

$$f_c = 1/(2\pi\tau)$$

Thus, tradeoffs exist in performance that must be evaluated for any particular application. Typical applications for HgCdTe detectors can be found in thermal imaging, missile guidance, night vision, FTIR (Fourier Transform Infrared) spectroscopy, and gas detection. Figure 12.21 compares the detectivities of various alloys of HgCdTe at a temperature of 77 K.

The HgCdTe photoconductive detectors have a relatively low impedance. Actual impedance depends upon alloy composition, temperature, device construction, and carrier concentration. Typical detectors used for FTIR spectroscopy have impedances ranging from 18 to 120 Ω/square. They can be placed in an electrical circuit in a fashion similar to silicon photodiodes. A load resistor converts the photocurrent generated in the detector into a signal voltage which can be amplified by conventional amplifiers. Figure 12.22a gives a sketch of typical HgCdTe detector, and Figure 12.22b gives an electrical schematic used in its operation.

These detectors can be used in quick response systems such as video imaging in the IR spectral region. This area of the spectrum provides for good detection of many types of hazardous gases that absorb strongly in the near to mid-IR. For example, radiation from a

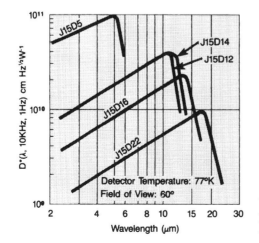

Figure 12.21 Detectivities (D*) of various HgCdTe alloys at 77 K. Reprinted with permission of EG & G Judson

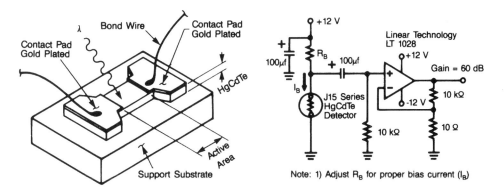

Figure 12.22   (a) A typical HgCdTe detector. (b) Electrical schematic showing its integration into a circuit.

tunable laser in the 9 to 11 μm region will experience a substantial absorption of photons at specifically tuned wavelengths. This absorption is due to the interaction of radiation and matter. The matter in this case is the gas. To detect a gas such as Freon 11, the laser can be tuned to provide an optical output at 9.22 μm. When photons emitted from this laser encounter Freon 11 molecules, a large percentage will be absorbed showing up as a signal loss in the HgCdTe detector. Of course, the amount of absorption depends upon factors such as gas density and the separation distance between the transmitter and the receiver. Another gas that has a strong absorption characteristic in this area of the spectrum is sulfur hexafluoride (SF_6). This gas absorbs photons with a wavelength of 10.55 μm.

12.10 PHOTOCONDUCTIVE CELLS

Photoconductive cells or Light Dependent Resistors are very useful in both digital and analog applications. Their low cost and wide dynamic range make them the best solution for many applications. In the field of photography, they provide accurate control of film exposure and camera focusing. They are also used in automatic headlight dimmers, copy machines, TVs, position sensors, and card readers. Other useful characteristics include operating voltages of 50 to 400 volts, and operation in the visible and near IR wavelength region.

    This type of light receiver has no p-n junction as in a photodiode. In its operation, the entire layer of material changes its resistance to current flow as light strikes its sensitive surface. This is called the bulk effect. Thus, with these devices, resistance changes with the level of illumination. Some materials used in the construction of photoconductors include Cadmium Sulfide (CdS) and Cadmium Selenide (CdSe). As the level of illumination increases, the resistance of the cell decreases. The actual value of the resistance at a particular light level depends upon factors such as the material used, cell size, light history, and the wavelength of the light. This change in resistance with light is analogous to the change in resistance with temperature when using a thermistor. A data sheet for a typical photoconductive cell can be found in Appendix A with the part number VT300.

Each light sensitive material used in photoconductive cells has a different spectral response. That is, the material will be sensitive to some wavelength regions and insensitive to others. A plot of relative sensitivity vs. wavelength yields a spectral response curve for each photoconductor type. These different photoconductor types also have different peak spectral response characteristics. For example, a type 2 material has a peak spectral response at about 515 nm while a type 1 material has its peak spectral response at 730 nm. The curve for each type should be consulted for a particular application. Specific data for type 0 material can be found at the end of this section. Table 12.2 lists some common material types and their peak spectral sensitivity.

The sensitivity of the photoconductive cell is expressed by the relationship between incident light and the corresponding resistance of the device (Resistance versus Illumination). Instead of using a monochromatic light source for this measurement, the manufacturers of photoconductive cells agreed upon an industry standard. This standard specifies the use of a tungsten lamp operating at a color temperature of 2850 K. This lamp displays a blackbody curve having its peak output at a wavelength of about 1000 nm. As you remember from Chapter 8, a blackbody radiation source at this temperature emits a considerable amount of optical energy in the visible and near IR regions. This makes such a light source very useful for calibration of photoconductive cells. Most importantly, this spectral distribution can be repeated at other locations by using a tungsten filament lamp adjusted to the correct color temperature. A resistance vs. illumination curve for material 0 can be found at the end of this section.

The speed of response is a very important characteristic for any receiver. This is usually defined as the time required for the photoconductive material to respond to a step increase or decrease in light illumination level. When given a step increase in light, the material's rise time is the time period from initial illumination to 63% of its final output value. Conversely, when given a step decrease in light, the material's fall time is the time period from the initial light decrease to 37% of its final output value. Of course, these

Table 12.2   Photocell Materials

Material Type	Composition	Peak Spectral Sensitivity (nm)
0	CdS	565
1	CdSe	735
2	CdS	515
3	CdS	550
4	CdSe	675
7	CdS	615

times depend upon the level of illumination used. Typical response times for these photo-conductive cells are in the range of 5 to 100 msec. when illuminated at a level of 1 foot-candle. Response time versus illumination curves for material 0 can be found at the end of this chapter.

SUMMARY

In this chapter, we considered the physics of photodetectors. Photodetectors convert the energy of the photon into a useful electrical current by the process of the interaction of radiation and matter. In the pin photodiode, the absorption of a photon results in the production of an electron-hole pair that contributes to this current flow. Today, we find a wide variety of photodiodes. The wavelength sensitivity of these devices depends upon the semiconductor used in its construction.

The current produced in a photodiode must usually be converted to a voltage before it can be used in an external circuit. This conversion involves the use of a load resistor. As we know from Ohm's law, when a current flows in a resistor, the voltage drop across the resistor will be the product of the current times the resistor or $V = IR$. In the next chapter, we will discuss how to integrate the photodiode into useful circuits. The very first electrical component added to the photodiode is usually a resistor. This combination of photodiode and resistor can now be used in an external circuit to produce a useful electrical output. We will also discuss many of the important circuit parameters involved with photodiodes to produce a stable circuit.

Typical Characteristic Curves @ 25°C - Type Ø Material

TYPE Ø MATERIAL

This is a general purpose Cadmium Sulfoselenide material. Its characteristics include a good temperature coefficient and fast response time, especially at very low light levels. Cells of this type have relatively low dark history. Type Ø material is often used in lighting controls such as nightlights, and security lighting.

The resistance for any standard catalog cell is controlled at only one light level. If the resistance at other light levels is of concern, please contact the factory.

RESISTANCE VS. ILLUMINATION

To obtain the typical resistance versus illumination characteristic for a specific part number:

1. Look up 2 footcandle resistance in table.

2. Insert resistance given and draw a curve through that point and parallel to the closest member of the family of curves shown for the appropriate type of photo-sensitive material.

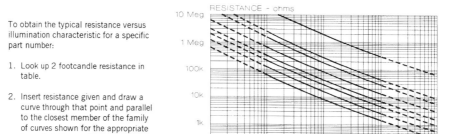

RESPONSE TIME VS. ILLUMINATION
(RISE TIME)

RESPONSE TIME VS. ILLUMINATION
(DECAY TIME)

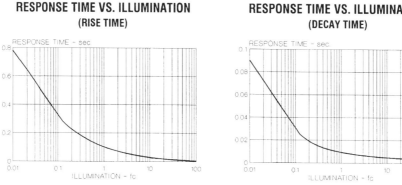

Data curves reproduced with permission of EG & G Vactec.

Typical Characteristic Curves @ 25°C - Type Ø Material

RELATIVE SPECTRAL RESPONSE

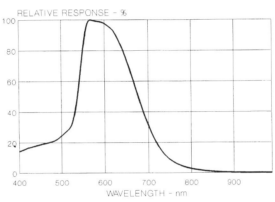

RELATIVE RESISTANCE VS. TEMPERATURE

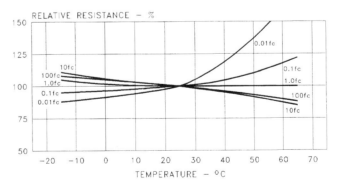

SIEMENS

SFH205F
SFH205FA

SILICON PIN PHOTODIODE
DAYLIGHT FILTER

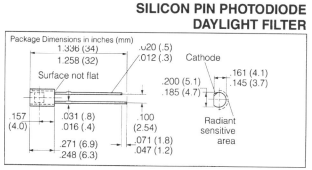

Package Dimensions in inches (mm)

FEATURES

- **Built-in Daylight Filter**
- **High Reliability**
- **0.1" (2.54 mm) Lead Spacing**
- **Fast Switching Time**
- **Black Plastic Encapsulated Package**
- **Suitable for IR Sound Transmission**

APPLICATIONS

- **IR Remote Control of Hi-Fi, TVs, VCRs, Dimmers, etc.**
- **Light Reflecting Switches for Steady and Varying Intensity**

DESCRIPTION

The SFH 205F/205FA silicon planar PIN photo-diode is housed in a plastic package that serves as both a filter and a window for infrared emission. Its terminals are solder tabs at 0.1" (2.54 mm) lead spacing. The cathode marking is stamped at the package edge.

Key features include low junction capacitance, high cut-off frequency, short switching times.

This versatile photodetector can be used as either a diode or as a voltaic cell.

Maximum Ratings

Operating and Storage Temperature
Range (T_{OP}, T_{STG}) .... −55 to +80°C
Soldering Temperature
(≥2 mm from case bottom) (T_S) t≤3 s .. 230°C
Reverse Voltage (V_R) .... 32 V
Power Dissipation (P_{TOT}) T_A=25°C ...... 150 mW

Characteristics (T_A=25°C)

Parameter		Sym.	Value	Unit	Condition
Spectral Sensitivity		S	60 (≥42)	µA	V_R=5 V, E_E=1.0 mW/cm^2
Wavelength, Maximum Photosensitivity	SFH205F	λ_{Smax}	950	nm	
	SFH205FA		900		
Photosensitivity, Spectral Range	SFH205F	λ	800 to 1100		S=10% of S_{MAX}
	SFH205FA		740 to 1100		
Radiant Sensitive Area		A	7.00	mm^2	
Radiant Sensitive Area Dimensions	L x W	2.65 x 2.65	mm	
Distance, Chip Surface to Case Surface		H	2.3 to 2.5		
Half Angle		φ	±60	Deg.	
Dark Current		I_R	2 (≤30)	nA	V_R=10 V
Spectral Sensitivity	SFH205F	S_λ	0.59	A/W	λ=950 nm
	SFH205FA		0.63		λ=870 nm
Quantum Yield	SFH205F	η	0.77	electrons photon	λ=950 nm
	SFH205FA		0.9		λ=870 nm
Open Circuit Voltage	SFH205F	V_O	330 (≥250)	mV	E_E=0.5 mW/cm^2,
	SFH205FA		350 (≥280)		E_E=1 mW/cm^2
Short Circuit Current		I_{SC}	56	µA	E_E=1 mW/cm^2
Rise and Fall Time, Photocurrent		t_R, t_F	20	ns	R_L=50 Ω, V_R=5 V, λ=850 nm, I_P=800 µA
Forward Voltage		V_F	1.3	V	I_F=100 mA, E_E=0
Capacitance		C_O	72	pF	V_R=0 V, f=1 MHz, E=0
Temperature Coefficient V_O		TC_V	−2.6	mV/K	
Temperature Coefficient I_{SC}	SFH205F	TC_I	0.18	%/K	λ=950 nm
	SFH205FA		0.03		λ=870 nm
Noise Equivalent Power	SFH205F	NEP	4.3 x 10^{-14}	W/√Hz	V_R=10 V
	SFH205FA				
Detection Limit	SFH205F	D*	6.2 x 10^{12}	cm•√Hz/W	V_R=10 V
	SFH205FA				

SFH 205F Data Sheet reprinted with permission from Siemens Components, Inc. Optoelectronics Division.

Figure 1. Directional characteristic $S_{REL}=f(\varphi)$

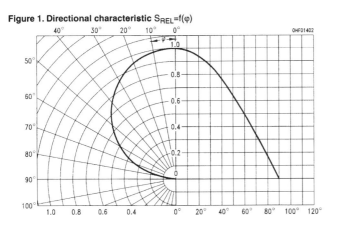

Figure 6. Dark current $I_R=f(V_R)$, E=0

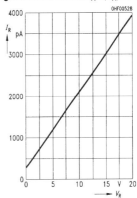

Figure 2. Relative spectral sensitivity $S_{REL}=f(\lambda)$ **SFH205**

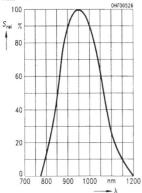

Figure 4. Photocurrent $I_P=f(E_E)$ $V_R=5\,V$
Photocurrent $I_P=f(E_E)$ $V_R=5\,V$

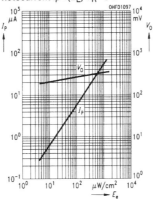

Figure 7. Capacitance C=f(V_R),f=1 MHz, E=0

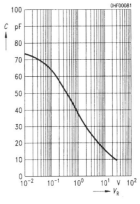

Figure 3. Relative spectral sensitivity $S_{REL}=f(\lambda)$ **SFH205FA**

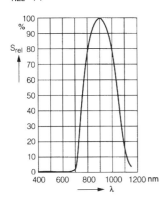

Figure 5. Power dissipation $P_{TOT}=f(T_A)$

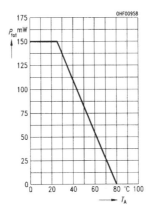

Figure 8. Dark current $I_R=f(T_A)$, $V_R=10\,V$, E=0

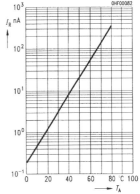

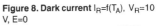
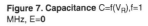

13

Optical Receivers

Photodiodes are manufactured to be used in a large number of applications. They are available in numerous configurations to meet the particular application. The specific applications and device configurations are too numerous to mention here, so we will begin our study with a general purpose silicon pin photodiode used as a receiver element. Silicon pin photodiodes are available in a wide range of active areas to meet the required light collection specification of a receiver design. Applications such as position sensing require multiple pin photodiode elements in more than one package. Photodiodes are also available in arrays of 50 or more elements for applications such as spectroscopy. Whatever the application may be, certain interfacing techniques must be used in almost every case. One of the objectives of this chapter is to present these basic interfacing techniques in such a way that the reader may apply them to his or her own situation.

Of primary importance when using photodiodes is the accurate conversion of the photocurrent to an electrical signal. In most cases, this signal will be linear over several decades of light intensity. As we will see, there is the usual trade-off between speed and sensitivity. The various noise sources studied in the last chapter determine the upper limit to this sensitivity.

13.1 GENERAL PURPOSE DETECTORS

The receiver section of a typical electro-optical system contains the detector element and its associated electronic circuitry. In the last chapter, we emphasized the various detector types and the conversion of optical energy into an electrical current. The detector is usually considered the "front end" or first electrical element in the receiver circuit. The optical receiver converts the influx of photons that fall upon the detector element into a usable electrical signal to be amplified and then electrically processed further. Since this optical input signal has usually been modulated in some way, it must then be demodulated into a usable signal.

A typical application of a high speed receiver can be found in an optical communication system using amplitude modulation. In this example, we switch an LED on and off at specified time intervals corresponding to a high and a low or a 1 and a 0. The high or 1 interval results from a step increase in optical power from the low or 0 level. The receiver portion of this system must be sensitive to the intensity level and wavelength of the emitted optical radiation from the LED. It must also be capable of responding to the switching speed of the emitted signal. Switching the LED on allows the emitted photons from the LED to fall upon the receiver's detector element assuming correct alignment. This optical energy in the form of photons produces electron-hole pairs within the semiconductor material of the detector. The resultant electron-hole pairs then set up a photocurrent within the detector. This photocurrent flows through a load resistor to produce a potential difference or voltage that can then be amplified by the receiver circuit electronics. In the off state, no photons from the LED strike the receiver, and thus the converted voltage becomes zero or some other predetermined bias level. In this manner, the voltage variations from the receiver will follow the action of the LED. For example, a simple message in Morse Code could be transmitted from an infrared LED to a receiver circuit through space or an optical fiber link. This coded message would not be visible to the human eye. At the receiver, the voltage output that matches the variations of the infrared LED could be used to turn on or off a simple transistor in the same fashion. This output could then be connected to a visible LED. In this way, one would be able to see a visible reproduction of the original infrared pulses.

As mentioned before, one objective of this chapter is to show how to correctly integrate detector elements such as photodiodes into an electronic circuit, thus providing a desired electrical output. A simple circuit that performs the basic function of photocurrent to voltage conversion in a high speed digital receiver will be the first practical example considered. We will rely on our previous discussions of photodiode characteristics, and then use some simple circuit theory. The reader may wish to refer to Appendix B for a basic discussion on operational amplifiers at this time.

The first receiver circuit to be considered uses a photodiode in the photoconductive mode. As you remember, in this case the photodiode must be reverse-biased by a specified voltage. This mode results in photocarriers that travel at a faster speed in the photodiode's intrinsic region than when not reverse-biased. Before we consider the photoconductive mode of operation, we need to discuss some basic circuit parameters. Figure 13.1

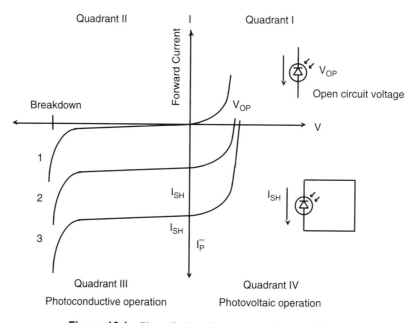

Figure 13.1   Photodiode voltage-current characteristics.

shows an electrical schematic in the lower right of the figure. In this simple circuit, a conductive wire connects the photodiode's anode and cathode together. The operating point for this circuit can be found along the vertical axis. When light falls upon the active surface of the device, a current will flow as indicated. This current is known as the photodiode's short circuit current, or I_{SH}. This short circuit current displays extreme linearity, and is proportional to the incident light intensity. As the light intensity increases, the point will move further down on the vertical axis. Another parameter, the open circuit voltage V_{op}, results by disconnecting the wire in the short circuit example, and then placing a voltmeter or equivalent device across the photodiode. The electrical schematic for this example is shown in the upper right of the figure. Figure 13.1 shows the operating point for this configuration to be along the horizontal axis. When taking voltage measurements on an open circuited photodiode under various light intensities, you will find that V_{op} has a logarithmic relationship as shown in the voltage-current characteristic curve in Figure 13.1. We must have a good understanding of the curve displayed in Figure 13.1 before considering the process of integrating a photodiode into an electrical circuit. The information presented on this curve allows us to understand how the optical input to the photodiode affects its electrical output. Once we understand how the photodiode responds electrically to a specified range of light input, it can then be treated as a typical electrical component. Each photodiode has its own particular voltage-current transfer curve since device configurations vary greatly. This curve gives important electrical information about the current flow within the device for both forward and reverse-bias voltage conditions.

We will now use the curves presented in Figure 13.1 to explain the operations of a photodiode in both the photoconductive and photovoltaic modes. Curve 1, in the figure, shows the voltage-current characteristics for the conditions of no light incident upon the device and different bias voltage conditions. As we know, when reverse-biasing the device, a dark current I_d flows within the photodiode. This can be measured at the specified voltage at which the device will be operating. The photodiode operates as a typical diode under these conditions. Reverse-biasing the photodiode places its operating point on the curve in quadrant III. You can see the small amount of leakage or dark current that results in this mode; the curve is not parallel to the horizontal axis. This current remains relatively constant until the reverse-bias voltage reaches the breakdown level. If we apply voltage in the forward direction, the operating point now moves to the first quadrant. In the forward biased condition, current flows freely from the anode to the cathode with a forward voltage level of about 1.0 volt. The forward current specification depends upon photodiode parameters such as active area size and the semiconductor material.

When light falls upon the photodiode's sensitive area, electron-hole pairs are generated within the semiconductor structure. The number of electron-hole pairs generated depends upon the flux of optical radiation falling upon the active area. Of course, this assumes that the incident optical radiation matches the photodiode's wavelength sensitivity. The voltage-current characteristics for this situation can be found by reviewing curve 2 in Figure 13.1. As you can see, curve 2 keeps the same shape but becomes shifted down according to the amount of incident radiation falling upon the device. The current generated in the photoconductive mode is composed mostly of photocurrent, I_p, with a small amount of dark current, I_d. This places the operating point for the photodiode in Quadrant III as specified by curve 2. In this quadrant, the current changes only by a very small amount with large changes in reverse-bias voltage. The photodiode can thus be considered a constant current source when operated in this mode. If we remove the reverse-bias voltage from the photodiode, its operating point now is in the fourth quadrant as specified by curve 2. This quadrant gives data for a photodiode operating in the photovoltaic mode. The generated photocurrent flows through the internal shunt resistance, R_{sh}, causing a voltage to appear across the photodiode. The value of R_{sh} decreases exponentially with increasing optical radiation to the active area. Thus, at high light intensity levels, this photo-generated current displays a logarithmic relationship making it unsuitable for linear operation. The advantages to using a photodiode operating in the photovoltaic mode are increased sensitivity and less noise due to the absence of dark or leakage current. It may be possible to obtain linear operation in the photovoltaic mode if close attention is paid to the maximum amount of incident light striking the active area, the internal shunt resistance characteristic, and the load resistance value. Circuits using both operating modes will be considered in this chapter.

As more light strikes the device's active area, the operating point will still be in the third quadrant for the photoconductive mode. The curve specifying its operation will shift down further due to an increase in photocurrent, I_p. This case can be found by following the curves down until you reach curve 3 in Figure 13.1. The point labeled I_{SH} for short circuit current is a very important operating point. The relationship between short circuit current and incident radiation shows extreme linearity. As the incident light intensity increases, this curve changes position resulting in increased I_{SH}. This photodiode characteristic can be used to make a circuit displaying linearity over a large range of input light intensities.

Photoconductive Mode—This circuit, in its basic form, consists of two electrical components, a silicon pin photodiode and a load resistor. To show its output graphically, we will use a portion of Figure 13.1, specifically the third quadrant. Figure 13.2 shows this electrical schematic and the operating points. Once we select the photodiode for a particular application, we must then select the load resistor, R_L. Applying a reverse-bias voltage to this combination results a faster response time. As you remember, this reverse-bias voltage makes it possible for electrons and holes to travel at saturated velocities within the intrinsic region of the photodiode. When light strikes the active area of the photodiode, a signal current will flow through R_L. As you can see from the graph in Figure 13.2, when the amount of incident radiation increases, the operating point moves closer and closer to the vertical axis. A line drawn through each operating point on the different illumination curves will produce the "load line." The exact operating point depends upon the value of R_L, the reverse-bias voltage, and the amount of illumination. The slope of this load line is inversely proportional to the value of the load resistance. Large values of R_L usually yield a linear response. A voltage will appear across the load resistor as determined by the equation below:

$$V_S = I_p R_L$$

This signal voltage, V_S, must be amplified electronically to be useful in a practical circuit. Thus, the selection of both the photodiode and the load resistor must be carefully considered, as they will determine the amount of current-to-voltage conversion in the circuit. Next, a simple example will be given to show this relationship.

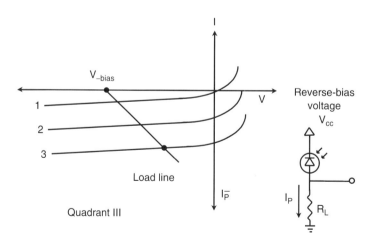

Figure 13.2 Photoconductive mode of operation showing load line.

Example 13.1(a)

The photodiode used in Example 12.1 in Chapter 12 received 5 µWatts of optical power. We found that this resulted in a photocurrent of 2.84 µamps. What signal voltage will appear across a 10,000 ohm load resistor if this photodiode is connected in the photoconductive mode as shown in Figure 13.2? Assume that this signal voltage is due only to the photocurrent generated by the received optical power.

To solve for the signal current, we must use Ohm's law. This solution is shown below:

$$V_S = I_p R_L = (2.84 \times 10^{-6})(10^4) = 2.84 \times 10^{-2} \text{ volt}$$

You can easily see that this voltage will require electronic amplification in a useful circuit.

Example 13.1(b)

We find that the pin photodiode chosen in the last example has a junction capacitance, C_J, of 24 picofarads with an applied reverse-bias voltage of 10 volts. The circuit configuration is the same as shown in Figure 13.2. The value of the load resistor is 10,000 ohms. Find the rise time and the 3 dB bandwidth of this circuit due to the combined $R_L C_J$ product.

From the last chapter, we use the expression for rise time to get the result shown below:

$$f_c = \frac{1}{2\pi RC} = \frac{.35}{\tau_r}$$

Rearranging terms we get:

$$\tau_r = 2.2 R_L C_J = (2.2)(24 \times 10^{-12})(10^4) = 530 \text{ nsec.}$$

Using this result, we can now find the 3 dB bandwidth due to the combined $R_L C_J$ product:

$$B = 0.35/\tau_r = 0.35/.53 \times 10^{-6} = 660 \text{ KHz}$$

Figure 13.3 gives a graph for the relationship between response time and the $R_L C_J$ product for this device. The line labeled "V = 10V" gives data for this case (photoconductive mode). When checking this graph, we find a good agreement with the calculated response time. If we find this response time to be too long, we can change the value of R_L or C_J or both. Reducing the value of R_L will result in a shorter response time at the expense of a smaller voltage signal. Circuit noise may increase slightly due to the increase in thermal noise when the resistance decreases. If this can be tolerated, then the proper compromise has been reached. The value of C_J can be decreased by increasing the reverse-bias voltage. This may not be possible in all cases as some photodiodes have a relatively low reverse-bias specification. Another way to decrease C_J is to select another photodiode having a smaller junction capacitance value. This choice usually results in a smaller active area device that provides less light collection. The ultimate solution may be to use a lens to collect more light for the smaller area photodiode. You can see that this process must be considered very carefully so that circuit optimization can be achieved for the particular application. The line labeled "photovoltaic" gives data for when this device is not reverse-biased (photovoltaic mode). Notice that the response times are somewhat more.

RISE/FALL TIMES - NON SATURATED

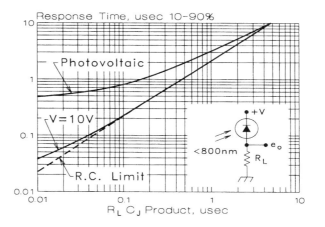

Figure 13.3 Response times for the $R_L C_J$ product. Reproduced with permission from EG &G Vactec.

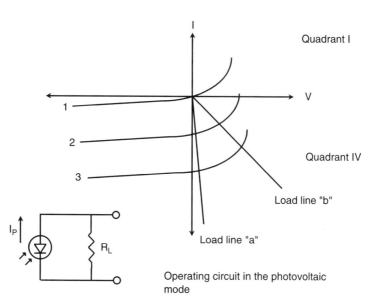

Figure 13.4 Photovoltaic mode of operation showing load lines for two different value load resistors.

Photovoltaic Mode—Another way in which the photodiode can be connected electrically into the circuit is the photovoltaic mode. In this case, the load resistor connects in parallel with the photodiode as shown in the circuit of Figure 13.4. This mode offers the most sensitivity, but at the expense of a somewhat reduced bandwidth. When using a photodiode in this mode, you run the risk of operating in the nonlinear range. Operating points are given in Figure 13.4 for two different resistor values under increasing amounts of illumination. Load line "b" uses a higher value resistor. In this case, the circuit runs the risk of operating in the nonlinear range when given a large amount of illumination. As you remember, the photodiode in this configuration has no reverse-bias voltage. When photons strike the photodiode, the generated photocurrent becomes divided between the internal shunt resistance, R_{SH}, of the pn junction, and the load resistance, R_L. The portion of photocurrent flowing through R_L will produce a voltage across the photodiode in the forward bias direction. If this forward voltage becomes great enough, the operating point for the circuit will display a nonlinear output with input light. As a rule of thumb, this load voltage should not exceed 10 millivolts. Of course, the precise operating voltage depends upon the application and the specifications of the photodiode.

One solution is demonstrated by load line "a" in Figure 13.4. With a much smaller value load resistor, only a small amount of voltage is produced under low illumination. This voltage will increase in a linear fashion with an increase in illumination provided that the operating point remains close to the vertical axis. Another way to insure linear operation over a wide illumination range is to feed the photocurrent to the virtual ground of the amplifier. This case will be discussed in detail later in this chapter. We will next present an example of this circuit using the same photodiode and load resistor as in our previous example.

Example 13.2

We find that after some testing, the photoconductive circuit in the last example does not yield the desired sensitivity. What will be the trade-off, in terms of bandwidth, if we use this same photodiode and load resistor in a photovoltaic mode in an attempt to achieve better sensitivity?

First, we must find the value for C_J for the condition of no reverse-bias voltage. From the manufacturer's data book on this device, we find this value to be 120 picofarads. The resultant rise time can be calculated as shown below:

$$\tau_r = 2.2R_L C_J = (2.2)(120 \times 10^{-12})(10^4) = 2.64 \times 10^{-6} \text{ sec}$$

We now use this result to find the 3 dB bandwidth:

$$B = 0.35/\tau_r = 0.35/2.64 \times 10^{-6} = 130 \text{ KHz}$$

This result shows that the bandwidth, in this case, will decrease by a factor of 5 when compared with the previous example. This trade-off in bandwidth will yield a higher sensitivity due to the absence of leakage current. But there are still other alternatives to consider. Next, we will take a look at other photodiode parameters that must be considered when selecting the best photodiode and load resistor for a particular application.

13.2 SELECTION OF A PHOTODIODE AND LOAD RESISTOR FOR CIRCUIT OPERATION

Photodiode Selection

The photodiode or detector element is usually the first circuit component selected when designing an electro-optical receiver circuit. There are many photodiode parameters to consider during this selection process. The following outline serves only as a guide. This selection process usually varies significantly depending upon the application. The characteristics listed below can be found for the Siemens SFH 205F photodiode by reviewing its data sheet at the end of Chapter 12. Most of these parameters were initially introduced in Chapter 12. For convenience, we list them here.

> *Spectral Sensitivity*—This is usually the most important parameter for a detector component. The photodiode must be sensitive to the incident radiation for photocurrent production. Thus, the peak spectral response of the photodiode must be as close as possible to that of the radiation trying to be detected. Photodiodes are available with a wide variety of spectral response characteristics. For example, blue-enhanced photodiodes are more sensitive to visible light than general purpose photodiodes. Some photodiodes have integrated filters that block out unwanted light in a particular spectral region. Still others, like germanium and indium gallium arsenide photodiodes, have spectral sensitivities at much longer wavelengths than silicon.
>
> *Radiant Sensitive Area*—This parameter usually involves answering the question, how much light will be at the photodiode for current-to-voltage conversion? It must be carefully balanced with the acceptable speed of response. As a rule of thumb, junction capacitance increases with active area. If the active area becomes too small, the result may be reduced light collection. The solution may be a small area photodiode with a large lens for increased light collection. Another solution may be to operate the photodiode in a particular mode, such as the photoconductive mode, for larger bandwidth. While the input signal can usually be amplified greatly, a limit exists due to system noise and amplifier frequency response. Most photodiode specifications also give a list of typical applications for that device. This is a very good place from which to start. Initially, several devices may be selected as possibilities for the actual circuit. This selection may be reduced significantly after performing tests with these devices separately in the circuit.
>
> *Package*—Packaging of the semiconductor is another important consideration. As stated in the beginning of the chapter, there are as many packages as there are circuit applications. Some devices come with an integral lens that assists in the light collection. The acceptance angle characteristic must be evaluated before using a device with an integral lens. There may also be a size limitation to a particular package, preventing it from being used on a particular circuit board. A metal can package may help greatly in reducing noise pick-up due to radiation emitted at radio frequencies. Still other photodiode packages come with multiple sensing elements for applications such as position sensing.

Responsivity—This is a measure of the amount photocurrent generated by the photodiode per unit of input optical power. It is usually expressed in the practical units of amps/watt.

Quantum Efficiency—This parameter is a measurement of the percentage of incident optical power that results in a photocurrent, I_P, within the photodiode. The quantum efficiency will also vary with wavelength, so when selecting a photodiode to sense light at a wavelength other than the specified one, the efficiency at a particular wavelength becomes important. The conversion factor must be known before a circuit can be designed using this device, otherwise there may not be enough photocurrent generated to make it practical.

Capacitance—This parameter must be carefully considered, as it will affect the bandwidth of the device. The junction capacitance of a photodiode results from device construction, and varies considerably from device type. Generally speaking, as the active area increases, so does the junction capacitance. Junction capacitance can be reduced by applying a reverse-bias voltage to the photodiode. This results in a decrease in the rise and fall times from those obtained with the zero-bias voltage condition. At a certain voltage, called the full depletion voltage, the junction capacitance becomes constant for increasing values of applied reverse-bias voltage. Increasing the bias voltage beyond this point may result in excess leakage current, and thus more noise due to this current source. In the extreme case, the device can be destroyed by exceeding the maximum reverse voltage rating.

Dark Current—This is the current that flows in the photodiode in the absence of light with the application of a reverse-bias voltage. The amount of dark current will increase with voltage. A graph of dark current vs. reverse-bias voltage can be found on the Siemens SFH 205F data sheet. Generally, the smaller the active area of the photodiode, the smaller the dark current will be for the same reverse-bias voltage. When using the photodiode in a true photovoltaic mode, no dark current will be present. This allows for greater device sensitivity due to reduced noise.

Selection of a Load Resistor for Circuit Operation

This is the next important component that must be selected. When using a photodiode in the photoconductive mode, the load resistor R_L will be connected in series. The flow of photocurrent in the photodiode will produce a signal voltage across this resistor. If the value of R_L is too large, the RC time constant will be increased, thus increasing the response time. The signal can also become current limited by using a high value resistor under conditions of high illumination. If the value of R_L is too small, the signal voltage may be limited in low light applications. Small value resistors also contribute more thermal noise to the circuit than higher value ones. This increased noise may be tolerated when the signal is strong enough. When using the photovoltaic mode, a resistor having too high of a value may result in a nonlinear response. Thus, close attention must be paid to the current-voltage transfer characteristics for this part of the circuit. This can be done by considering the minimum and maximum light levels that the photodiode will experience, and then applying this information to the current-voltage transfer curve.

13.3 PHOTOCONDUCTIVE OPERATION USING AN OPERATIONAL AMPLIFIER

In the circuit described in this section, the amplifier used will be an FET type input op-amp. This type of op-amp is commonly used with photodiodes. The FET input op-amp offers very low input bias current and good transimpedance characteristics. There are many other types of op-amps that can be used. The factors influencing the selection of an amplifier include input bias current, noise, speed, and quiescent operating current.

When operating a photodiode in the photoconductive mode, more noise may be present than in the photovoltaic mode of operation. This results from the presence of a dark leakage current when applying a reverse-bias voltage. Improved bandwidth becomes the reason for using this mode of operation, if the increased noise can be tolerated. The circuit schematic for a typical high speed light sensor is given in Figure 13.5. For simplicity, we will not consider specific amplifier characteristics such as offset voltage, bias current, and temperature effects. These parameters must eventually be considered in the final design though.

In the circuit in Figure 13.5, a signal voltage develops across R_L when photocurrent flows in the photodiode. The op-amp's noninverting input monitors this signal voltage. The inverting input to the op-amp has a resistor network consisting of R_1 and R_2. These two resistors set the voltage gain for the circuit. This gain can be calculated by the following equation:

$$V_o = V_s[1 + R_2/R_1]$$

In this equation, V_o is the output voltage from the op-amp, V_s is the input signal voltage, and R_1 and R_2 are resistors. The circuit gain can be increased by increasing the ratio of R_2 to R_1. As the value of the feedback resistor increases, there will be a corresponding increase in thermal noise voltage. This relationship is expressed in the equation below:

$$V_n = (4KTBR)^{1/2}$$

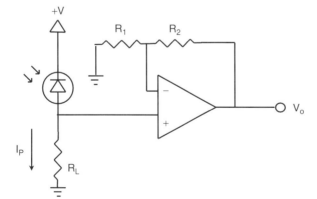

Figure 13.5 Photodiode used in the photoconductive mode with an op-amp.

In the last equation, K is Boltzmann's constant, T is the absolute temperature, B is the bandwidth, and R is the value of the feedback resistor. The transimpedance gain of the signal increases with resistance in accordance with ohm's law:

$$V_o = I_s R$$

From these two relationships, it can be seen that the signal-to-noise ratio will improve by $(R)^{1/2}$. There is a limit to how much gain you can apply in this circuit. As the gain increases, the frequency response decreases. This means that rise and fall times will increase. The maximum value of the feedback resistor must be determined by considering the bode plot for the device. This plot allows the designer to see the trade-off between amplification and bandwidth. Information on this subject can be found in Appendix B and in most electronics books. In practice, component leakage currents and noise sources will probably become the limiting factors to the feedback resistor value, even when satisfying the bandwidth requirements.

If the signal current is not sufficient, a larger area photodiode can be considered, but at the expense of a larger capacitance. This increased capacitance will cause a longer response time, thus reducing the bandwidth. As another option, a lens can be added to increase the amount of light reaching the photodiode. Thus, all of the above parameters must be carefully considered to optimize the circuit for a particular application. We will next consider a simple circuit example using an op-amp.

Example 13.3

A circuit described by the schematic diagram in Figure 13.5 is used to amplify an input voltage signal of 1 millivolt. This input signal voltage appears across R_L. If R_1 is 10 KΩ and R_2 is 1 MΩ, find the amplified output voltage.

Using the above equation for voltage gain, we solve to get the output voltage:

$$V_o = (1 \times 10^{-3})[(1) + (1 \times 10^6 / 1 \times 10^4)] = 0.101 \text{ volt}$$

If this amount of amplification does not meet the output requirement, increasing the value of R_2 may be an option. As discussed above, one consequence to increasing this resistance value will be added thermal noise. To find out just how much extra thermal noise there will be, we consider the following example.

Example 13.4

If the value of R_2 in Example 13.3 above is increased to 100MΩ, what will be the subsequent increase in the thermal noise voltage? How does this affect the signal-to-noise ratio?

To answer this question, we must calculate the thermal noise voltage in each case by using the relationship established above for the thermal noise of a resistor. For simplicity, use 1 Hz as the bandwidth in each case.

For the case when R_2 is 1MΩ, we get the following noise voltage:

$$V_n = [(4)(1.38 \times 10^{-23})(300)(1)(10^6)]^{1/2} = 12.9 \times 10^{-8} \text{ volt}$$

For the case when R_2 is 100 MΩ, we get the following noise voltage:

$$V_n = [(4)(1.38 \times 10^{-23})(300)(1)(10^8)]^{1/2} = 12.9 \times 10^{-7} \text{ volt}$$

This result shows a ten-fold increase in noise level due to thermal noise voltage when increasing the resistance by a factor of 100. Thus, the signal-to-noise level actually improves by a factor of $(R)^{1/2}$. This does not take into account the other affects that may occur as this resistance value increases. Factors such as component leakage currents, external noise sources, and the reduction in bandwidth may make this resistor change impractical.

13.4 PHOTOVOLTAIC OPERATION USING AN OPERATIONAL AMPLIFIER

As with the circuit previously discussed, an FET input operational amplifier will be used to amplify the signal voltage for the circuits discussed in this section. The photovoltaic mode of operation provides for the highest sensitivity when the speed of response is of secondary importance. When you look at the specifications for most photodiodes, you will see that they were designed to operate with an applied reverse-bias voltage. Nevertheless, these same photodiodes will also work quite effectively without any reverse-bias voltage.

The first circuit considered in this section uses the photovoltaic mode of operation by placing the photodiode and load resistor configuration as shown in Figure 13.6. In this circuit, photocurrent produced in the photodiode flows through R_L resulting in a voltage across the load resistor and the photodiode. This signal voltage then appears at the noninverting input of the op-amp. As we know, the sensitivity of the photodiode varies with its voltage. Great care must be exercised to insure that this voltage does not increase to a level where the circuit operates in a nonlinear mode. This can be accomplished by observing the minimum and maximum light levels to be experienced by the photodiode. Resistors R_1 and R_2 determine the voltage gain for the circuit. This voltage gain can then be calculated by using the same equation introduced in our last circuit example.

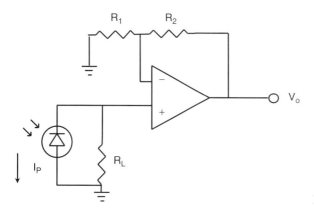

Figure 13.6 Photodiode used in the photovoltaic mode with an op-amp.

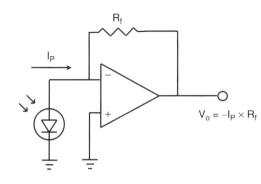

Figure 13.7 Photocurrent is fed to the virtual ground of the op-amp providing a linear response with input light.

The circuit in Figure 13.6 has the drawback of nonlinear operation for the reasons previously discussed. As an alternative, the simple circuit shown in Figure 13.7 can be used to assure that nonlinear operation does not occur when using a photodiode in the photovoltaic mode. In this circuit, the photodiode connects to the virtual ground of the op-amp. This effectively provides a short circuit to the load of the photodiode, thus removing voltage produced across it. The junction of the photodiode does not become forward-biased. A linear response with input light will now occur since its operating point will be along the vertical axis as shown in Figure 13.1. In this case, the photocurrent I_P is monitored by the op-amp. The output voltage can be determined by this photocurrent and the value of the feedback resistor, R_F. Ignoring any offset voltage, the relationship is also shown in Figure 13.7.

Another problem with the circuit of Figure 13.6 is its lack of noise immunity. If stray noise is introduced into the noninverting input, it will be amplified with the signal. There is an improved circuit using the photovoltaic mode that has good noise immunity by taking advantage of the differential input capability of the op-amp to reduce certain noise inputs. This circuit configuration also results in a more efficient amplifier. Since the signal generated from a photodiode is a current, it can be used to drive both input terminals of the op-amp as shown in Figure 13.8. Here the signal from the photodiode drives

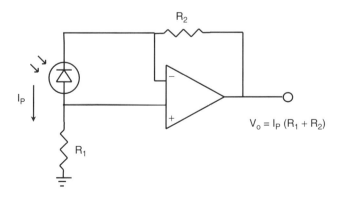

Figure 13.8 Photodiode taking advantage of the differential input capability of the op-amp. $V_o = I_p(R_1 + R_2)$

both amplifier inputs, effectively doubling the available signal voltage when $R_1 = R_2$. The voltage gain is given by the equation in Figure 13.8.

As mentioned before, this circuit offers good noise immunity due to the op-amp's common-mode rejection. By connecting the photodiode as shown in this configuration, both inputs will receive an equal amplitude in noise current, thus canceling the noise voltage effects. Techniques to reduce noise caused by interference sources will be discussed later in this chapter.

Example 13.5

Find the output voltage when using the circuit described the schematic diagram in Figure 13.8. The following parameters apply: $I_P = 0.1$ μamp, and $R_1 = R_2 = 10$ MΩ.

Using the above equation, we get the following result:

$$V_o = (1 \times 10^{-7})(2 \times 10^6) = 0.2 \text{ volt}$$

13.5 CONTROLLING NOISE

In this area of electronic design, there is no substitute for experience. Nevertheless, we will discuss some well-known circuit design techniques that reduce noise. When confronted with the requirement to greatly boost the amplification of an input signal, one may opt to use a very large feedback resistor. After all, the equations show this to be possible. If we take this option, the surprise will be at least two-fold. First, we already know that as the feedback resistor value increases, bandwidth will decrease. The second, not so obvious, result of using a very large feedback resistor shows up as a loss in stability. This instability results from the combined effect of the feedback resistor and the junction capacitance of the photodiode. We must understand the physical principles involved before we can solve this instability problem. Some of the most common variables involved are listed below.

1. R_F and C_J used in the circuit
2. The circuit board layout
3. The type of photodiode used and bias voltage

When interfacing a device such as a photodiode to an op-amp, the input characteristics of both devices must be considered. A rigorous mathematical treatment of this subject will not be attempted here. Instead, we will use the results and then apply them to common photodiode circuit designs. We will find that trade-offs in bandwidth, stability, and noise must be made to achieve the appropriate compromise.

Increasing R_F and C_J has some serious consequences in terms of stability. The combination of the photodiode's junction capacitance and the feedback resistance produces a feedback pole. If left unchecked, the circuit may even go into oscillation. This feedback pole causes gain peaking to occur in the amplified signal as shown in Figure 13.9. This

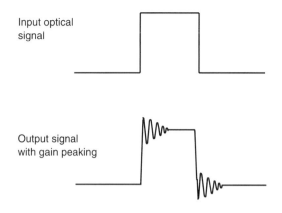

Input optical
signal

Output signal
with gain peaking

Figure 13.9 The affect of gain peaking upon an input pulse.

manifests itself as a sharp increase in gain in the high frequency region of the signal. A
typical input square wave will show sharp voltage spikes at the step increase and decrease
portion of each pulse. A ringing or dampened sine wave will appear after these voltage
spikes.

Gain peaking can be reduced or eliminated by adding a small compensation capacitor, C_F, in parallel with the feedback resistor as shown in Figure 13.10. This capacitor provides phase compensation to counteract the effect of the feedback pole. It must be carefully selected, as it will tend to decrease the circuit bandwidth. To achieve the desired
effect, the value of this capacitor is usually very small, typically a few picofarads or so.
The capacitance value can be empirically selected to a level achieving the best compromise between stability and bandwidth. As an alternative, select a photodiode with the
smallest possible junction capacitance while meeting the active area requirement. If this
option does not seem like a good compromise, apply more reverse-bias voltage to the
photodiode to reduce the junction capacitance.

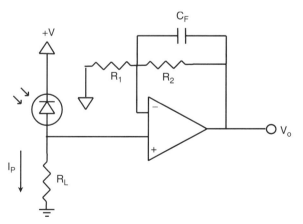

Figure 13.10 A small feedback capacitor, C_F, provides the phase compensation to reduce or eliminate gain
peaking.

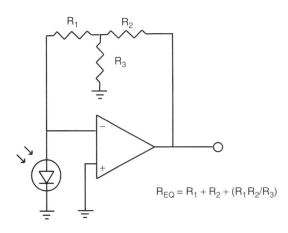

$$R_{EQ} = R_1 + R_2 + (R_1 R_2 / R_3)$$

Figure 13.11   Feedback "T" network.

For applications such as maximum sensitivity at low illumination, the high value feedback resistor can be replaced with a feedback bypass network having a much lower impedance. This option may look promising if a relatively large gain is needed. Figure 13.11 shows a simple circuit schematic using a feedback bypass or "T" network. This network takes the place of the larger feedback resistor in Figure 13.7. With this configuration, lower value resistors can be used instead of a high value one. This will reduce stray capacitance in the feedback loop.

The pattern and location of the electrical components on the circuit board must be carefully considered. A component placed incorrectly could cause unwanted circuit noise. For example, you don't want a high speed switching component next to the photodiode. Noise will couple into the circuit from this source. Circuit board layout and component placement will be discussed later in this chapter.

13.6 BANDWIDTH

Thus far, we have discussed several circuit examples involving photodiodes interfaced to other electronic components. This interfacing required attention to many circuit parameters to result in successful operation. Another important parameter used in circuit design is known as bandwidth. The circuit bandwidth can be defined as the range of frequencies over which it responds within specified limits. For example, a receiver circuit used for digital communications must respond to the input optical pulses within a specified time limit. If not, no meaningful output signal will be produced. This frequency response of the device is related to the input signal's rise time. As the bandwidth of the device increases, the total output noise will also increase by the square root of this increase. This occurs because we are sampling a broader noise spectrum. The bandwidth requirement applies not only to communication circuits but to any circuit where response time is critical. For example, in certain sensor circuits, the receiver has only a certain amount of time to respond to an input optical pulse that has been greatly amplified. Thus, the bandwidth of the receiver determines critical operating parameters such as modulation rate and pulse width.

There are many factors affecting the bandwidth of a typical photodiode circuit. Four of these factors that we will consider in this section are listed below.

1. Frequency response of the photodiode
2. Parasitic capacitance
3. Bandwidth of the op-amp or amplifier used
4. Phase compensation used to achieve stability

To consider the frequency response of the photodiode, we use an equivalent circuit of this device with a load resistor and op-amp as shown in Figure 13.12. The equivalent circuit replaces these devices with their fundamental elements. A review of the equivalent circuit of a photodiode can be found in Chapter 12. As previously discussed, the photodiode has a junction capacitance, C_J, that limits the frequency response, and thus its bandwidth. In most cases, as the active area of the photodiode increases, so does this capacitance. As the figure shows, a signal voltage develops across the photodiode and its capacitance, C_J, with the production of photocurrent. This condition will result in the photocurrent becoming shunted at higher frequencies. A bandwidth limit will then be established. Thus, large values of C_J and R_L tend to restrict bandwidth. The capacitance of the amplifier, C_{int}, will also shunt the signal in the same way as with C_J, thus both must be taken into account. A discussion of why this shunting of photocurrent occurs is beyond the scope of this book.

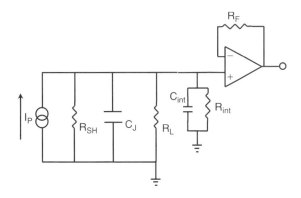

R_{SH} = Shunt resistance of photodiode
C_J = Junction capacitance
R_L = Load resistance
C_{int} = Amplifier input capacitance
R_{int} = Amplifier input resistance

Figure 13.12 Equivalent circuit for photodiode and amplifier.

Parasitic capacitance usually results when using a high value feedback resistance to increase gain. The high frequency response can become limited with high values of R_F. In the high frequency limit, this capacitance will shunt photocurrent away from the feedback resistor. This action reduces the voltage gain of the circuit. To reduce parasitic or stray capacitance, good construction practices such as mounting the feedback resistor on teflon-insulated standoffs must be followed. But, sometimes this stray capacitance can be useful. This small amount may be just enough to provide the phase compensation required for stabile circuit operation by counteracting the feedback pole.

The bandwidth of the op-amp itself can limit the response in lower gain applications. Here, the restrictions imposed by high resistance value feedback resistors and stray capacitance are not present. In practice, a very small signal voltage appears across the photodiode in the circuit of Figure 13.12. At low frequencies, the effect of this voltage is negligible. But at higher frequencies, this voltage imposes a response limit. To determine the bandwidth limit for a particular op-amp, its bode plot must be considered. This plot gives the designer information relative to the op-amp's frequency response. A bode plot can usually be constructed from the information provided on the manufacturer's data sheet for the device.

For many high gain applications, phase compensation will be required by placing a small capacitor, C_F, across the feedback resistor. The use of this capacitor provides stable circuit operation. Unfortunately, this capacitance also limits the frequency response of the circuit by shunting the output signal from the photodiode as previously discussed. This capacitor must be selected to provide circuit stability without limiting the bandwidth for a particular application.

In conclusion, interfacing a photodiode to an amplifier circuit involves paying attention to many circuit parameters due to the photodiode's dynamic nature. If the photodiode behaved as a linear resistive element, this interfacing would be much simpler. The current-to-voltage converter may at first seem simple to deal with, but offers a challenge in many circuit applications. For example, nonlinear operation may result if voltage in the forward-bias direction appears across the photodiode. This can be eliminated by following some of the suggested circuit design techniques previously discussed. Another area for concern involves the junction capacitance of the photodiode. This capacitance can react with the feedback resistor at high frequencies, thus causing unstable circuit operation. Increasing the value of this feedback resistor in an attempt to provide more gain can also cause instability. To prevent this occurrence, a compensation capacitor must be added across the feedback resistor in most cases. This small capacitor, while helpful in compensating for the unstable condition caused by the input capacitance of the photodiode, results in a reduction in bandwidth. Finally, in an attempt to increase this bandwidth, we can select the reverse-bias mode of operation. This mode can result in a substantial reduction in photodiode junction capacitance at the expense of increased leakage current. As we know, this leakage current will show up as increased noise in the output circuit. This noise will limit the sensitivity of the current-to-voltage converter. After learning about the available trade-offs, optimum circuit operation can usually be achieved with the correct application of these parameters.

13.7 ELECTROMAGNETIC INTERFERENCE

We are all familiar with the affects of interference, such as when a condition of high voltage between two conductors causes a spark to jump a gap while a radio is on. When this occurs, the electromagnetic energy released becomes coupled into the electronic appliance. We then hear a momentary noise or static instead of the clean signal from the radio station. External noise sources must be considered in a practical circuit design. A well designed circuit can easily be made inoperative with the application of external noise. In Section 12.3, we considered the noise sources present in the photodiode, the load resistor, and the input optical signal. We also investigated ways to reduce this noise, and offered circuit alternatives that have less noise due to the application of certain electrical components. The noise sources that we will consider in this chapter originate from time-varying electric and magnetic fields. The energy from these fields can be coupled into the sensitive part of a circuit by mutual capacitance or inductance. Once coupled into the circuit, this energy shows up as a background noise. We need to understand how these noise sources cause this background noise so that we can prevent this undesirable effect. To accomplish this, we will consider each noise source separately with ways to minimize its effects on a typical receiver circuit.

In the last section, we briefly discussed a circuit that offered more noise immunity by taking advantage of the differential nature of the op-amp. When high impedance components such as current-to-voltage converters are used, external noise sources can easily become coupled into the circuit. These external sources are electrostatic, magnetic, and radio frequency. To reduce the effects of these noise sources on a typical receiver circuit, consideration must be given to component layout, circuit design (as demonstrated in our last circuit example), grounding, and shielding. Basically, we want to separate the source of the noise from the sensitive parts of the circuit.

The first noise source that we will consider is present in every building that has electrical power. The electric field from this relatively low frequency source can be coupled to a circuit through the mutual capacitance that exists between source and the circuit. In this case, the noise source and the circuit board act like the plates of a capacitor. The dielectric material is the air between them. When a voltage difference occurs between the two objects, a noise current will flow from the object of higher voltage to the object of lower voltage. This process is known as electrostatic coupling. For the case above, the circuit receives the noise current that is then amplified by the op-amp. Figure 13.13 shows how this coupling can occur when using an op-amp in the configuration discussed in Figure 13.6. For simplicity, we will assume a zero photo signal here so that only noise coupling effects are illustrated. The electric field noise source, E_n, couples noise current, i_n, into the op-amp's feedback loop through the mutual capacitance, C_M. This noise current flows through the feedback resistor, R_2, to produce a noise signal. An electrostatic shield placed around the sensitive portion of the circuit can be used to collect this noise current, and then shunt it to ground. The shield, in this case, must be made of highly conductive material so that little or no voltage drop occurs across the shield itself. It should be connected to circuit ground or signal common so that further coupling can be avoided.

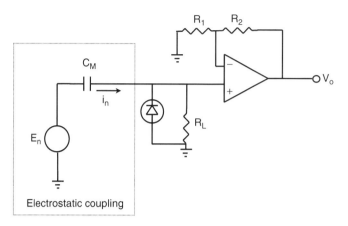

Figure 13.13   Electrostatic coupling of noise onto a circuit.

When using photodiodes with very high resistance op-amps, such as FET types, there will be a high sensitivity to noise coupling. The circuit configuration used can help to reduce or eliminate noise coupling. By using a differential input configuration, as shown in Figure 13.14, the noise can be removed due to the op-amp's common mode rejection or CMR. At the relatively low frequencies associated with power line energy, the op-amp does a good job at eliminating this noise through CMR. Thus, consider a differential input rather than a single-ended input to reduce noise from this source. It should be noted that the op-amp's CMR cannot do the same job as a metal shield for at least two

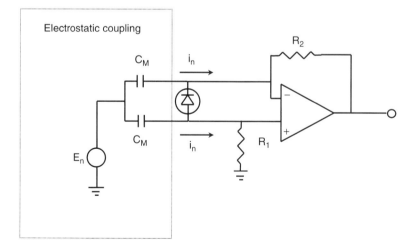

Figure 13.14   Common mode rejection of op-amp can remove external noise.

reasons. First, the noise source usually has a broad spectrum. And secondly, the distribution of the noise is not always the same at every point on the circuit board. This configuration is most helpful in reducing noise that passes through holes in a shield or from the coupling of secondary sources.

In Figure 13.14, electrostatic coupling occurs at both op-amp inputs. We considered this circuit configuration in Figure 13.8. The noise currents, i_n, flow from the noise source, E_n, to the current-to-voltage converter through the two mutual capacitances, C_M. For the purpose of illustrating this coupling effect, a zero photo signal is assumed at the photodiode. The noise currents, i_n, that develop are both equal since the two inputs are the same distance from the noise source, thus making the two capacitances, C_M, also the same. This result occurs due to amplifier feedback that places the op-amp's inputs at the same voltage. Thus, the noise signal is rejected due to the op-amp's differential configuration.

The second noise source to consider is magnetic-noise coupling. Unlike the electrostatic noise, this noise becomes coupled into a circuit by mutual induction. Instead of our previous capacitor model used to describe electrostatic noise coupling to the circuit, the model of a transformer must be used. This model illustrates how mutual induction occurs. As we know, a transformer has two coils of wire called primary and secondary windings. The primary and secondary windings are wound around a common iron core to allow the magnetic flux from the primary winding to become linked to the secondary winding. This can occur because the magnetic fields are coupled. The magnetic interference source acts like the primary winding of the transformer, producing flux. The circuit loops or paths act like the secondary winding of a transformer. When a magnetic field is coupled through mutual inductance, a voltage rather than a current signal will be produced when using an op-amp circuit. Metal shields can be used to reduce this coupling. At lower frequencies, such as from power transformers, only ferromagnetic materials provide good shielding. At higher frequencies, such as RF, a copper shield works better. In either case, the shield must be properly grounded. As with electrostatic noise, a good way to minimize magnetic noise is to keep the noise source away from the sensitive portion of the circuit. Components such as the photodiode, the load resistor, and any feedback resistors must be placed as close as possible to the op-amp to reduce mutual inductance. This practice reduces the circuit loops within the field, thus reducing the amount of coupled noise. The length of the photodiode leads must be made as short as possible. Surface mount components will help greatly in reducing this noise due to their small size.

The third noise source to consider is radio frequency interference. At these relatively high frequencies, the op-amp has little or no gain. Thus, the common mode rejection ability of the op-amp provides little help here. To reduce this noise, we must use shielding and filtering techniques. An RF shield made from conductive metal can be used. The copper ground plane layer in the circuit board can also act as a shield. Many designs extend the ground plane under the sensitive area of the circuit for further protection. Consideration must also be paid to the placement of certain circuit components on the circuit board. Digital circuitry can become a noise source due to the high switching speeds involved. Further separation of digital components from sensitive circuit areas must be considered.

13.8 REDUCTION OF EMI AT THE CIRCUIT BOARD LEVEL

Electromagnetic interference or EMI is complex, but if we understand the physical principles behind its generation and effects, the problem areas can easily be solved in most
cases. Some important physical principles involved with EMI were introduced in the last
section. In this section, we will consider some applications of these physical principles.
Several guidelines will be presented that have been proven effective in reducing EMI.
After a discussion of these guidelines, we will apply them to a typical receiver circuit.

Before you begin the design for the component layout on a circuit board, some questions need to be answered. These questions deal with the application involved, and are not
necessarily limited to what is outlined here. Some typical questions that a circuit board designer should ask are as follows. Is EMI present in the environment where this circuit will
be used? Typical sources include power line noise, transformers, radio transmitters, etc.
What are the consequences to a circuit failure due to EMI? Are there any specific EMI requirements? We will next consider some proven guidelines that are affective in reducing the
effects of EMI. The discussion here does not cover the complete list of guidelines.

1. Know which circuits emit EMI, and which are susceptible to it.

    EMI can couple from other parts of the circuit. Sources of EMI on a typical
    circuit board include devices such as oscillators and other devices that produce
    repetitive signals. These signals produce EMI due to the time-varying voltages in-
    volved with their operation. Power buses and input/output (I/O) lines also emit
    EMI. Thus, fast clock rates and short rise/fall times produce EMI energy at high fre-
    quencies. Sensitive circuits include high impedance op-amps, transistors, field ef-
    fect transistors (FETs) and their input devices.

    Sensitive circuits must not be placed near oscillators, oscillator lines, bus
    lines, or I/O lines. The high frequency energy could easily couple from these areas
    onto the sensitive high impedance circuits by the processes described in the last sec-
    tion. Thus, identify the problem areas and keep emission sources as far away as pos-
    sible from sensitive circuit portions.

2. Use devices that emit less EMI, if possible.

    We have seen that signals with increased rise/fall times have more potential
    for causing EMI problems. Sometimes the circuit design will allow the use of a
    slower device for an oscillator or clock. For example, high speed CMOS devices are
    sometimes substituted for the lower speed devices because they consume less cur-
    rent. Some circuit designers think that since high speed CMOS devices consume
    less current, they must also emit less EMI. This assumption has been found to be
    false, since high speed CMOS consumes current in narrow spikes or pulses. These
    narrow spikes, by their nature, have a much faster rise/fall time. This can create
    more EMI problems. If the circuit design can accept the slower devices, this source
    of EMI can be easily reduced.

3. Power decoupling is a necessity.

    When using digital circuitry, pulses of current are produced at the switching
    rate. These current pulses can result in EMI emission. The usual practice involves

the addition of decoupling capacitors at the devices where the EMI originates. These capacitors must be placed as close as possible to the IC. They must also have the correct capacitance value to control the EMI. For example, placing the capacitors across the input power line while keeping the leads as short as possible is common practice. Typical values for these capacitors range from .01 to .1 µf.

4. Input/output lines must be positioned correctly.

These lines transmit and receive electrical energy to and from the outside world. High frequency decoupling may also be required on these lines. Keep in mind that these lines can emit EMI and become susceptible to its emission from other sources. Thus, the location of these lines may be critical to the overall circuit operation. It is best to separate these lines from the rest of the circuit if at all possible.

Design Example: A Receiver Circuit Using a Photodiode and Op-Amp

The example presented here is a typical receiver circuit similar to the wireless remote control for a television. We will present a description of the circuit board layout with a list of concerns. This is only one example, and some of the concerns involved with this example may not apply in other cases. Conversely, other applications may involve issues and concerns not covered here. The suggestions given here for circuit board layout are by no means the only ones that can be used.

After the circuit schematic has been worked out, the circuit board layout must be considered next. This is a very important process in the receiver design. The collection of components in this schematic diagram must somehow be placed in one common area allowing for the production of a desired electrical output from an anticipated optical input. This layout will depend upon the type of devices used, whether they are conventional or surface mount. As part of this design, the actual paths or etch runs connecting these devices must be carefully considered. A well-designed circuit may not operate at all if EMI couples to sensitive portions of the circuit due to poorly placed circuit paths.

The photodiode location becomes the first item to consider. It must be in the best physical location on the circuit board to receive the most possible incident optical radiation from the signal source. The use of a lens may be required to increase its light gathering ability. This is usually a good choice over increasing the active area size. As you remember, the frequency response will suffer as you increase the active area of the photodiode. A lens can provide an increase in received optical energy without affecting the frequency response of the circuit. When using a lens, the throughput of the optical radiation must be optimized by considering the optical parameters associated with the lens itself. This process will be considered later in this chapter.

After the location for the photodiode has been determined, the circuit path to it must be considered. The photodiode lead length must be as short as possible, preferably less than 1 centimeter. The circuit path or etch must also be as short as possible, and as far away as possible from any other circuit path. Figure 13.15 shows a PC board layout for an IrDA data receiver. It uses a 16 pin surface mount IC (LT 1319) as U1 with a pin photodiode, D1. Pin 2 of this IC serves as the input for the photodiode. The component side of the

PC Board Layout for IRDA-SIR/FIR and Sharp or TV Remote Data Receiver

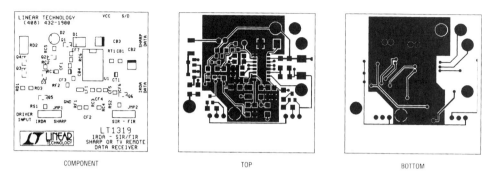

COMPONENT   TOP   BOTTOM

Figure 13.15 PC board layout for an IrDA receiver circuit. Reprinted with permission of Linear Technology Corporation.

layout shows the location of the surface mounted components. The other two views show top and bottom circuit etch paths.

The circuit layout shown here helps to minimize parasitic signal coupling to the input stage, pin 2, and the other sensitive portions of the circuit. The ground plane provides this function by surrounding pins 1 to 9 and pin 16. It connects to a low impedance point, pin 1 of the IC. A close inspection of the figure labeled "TOP" shows that the circuit pads for the photodiode are placed in close proximity to the amplifier's signal inputs. All low level signal components are placed as close as possible to the IC.

The resistor(s) used in the feedback loop should be placed as close to the amplifier IC as possible. All low level signal etch runs and components should also be as short as possible to reduce noise input due to coupling. This means that digital ICs should always be placed as far away as possible from analog ICs. You can see from the component layout that the driver input (lower left), Q5, and Q6 are as far as possible from photodiode, D1.

The circuit board should have a ground plane area extending under the photodiode and load resistor. This practice may not be possible in all cases. When using surface mount ICs and resistors, a ground plane beneath the board may act as an RF shield. The sensitive portion of the circuit should always be shielded to reduce any EMI that may be present. Shields can be placed around the amplifier IC and feedback resistors. If possible, shield the entire circuit except for the photodiode, since it must receive the incoming optical signal. The shield material will depend upon the type of EMI that needs to be attenuated. You can see from the top and bottom circuit etch views that plenty of ground plane is used in this circuit design. The ground plane also extends under the IC circuitry.

13.9 INTEGRATED AMPLIFIER AND PHOTODIODE

In our previous receiver circuits, we saw that the placement of the various circuit components such as the amplifier, feedback resistor, and the photodiode itself, required careful consideration. The correct placement of these components will help to reduce noise pick-up and optimize frequency response. In this section, we will consider integrated devices that have all three of the above components on one chip. These hybrid amplifiers can virtually eliminate many of the problems that confront the circuit designer when using discrete components. Specifically, hybrid amplifiers display a dramatic reduction in noise pick-up, leakage current errors, and gain peaking. Electronically, this results in an improved signal-to-noise ratio. All of this is possible due to the compact placement of key amplifier components that greatly reduces lead lengths. As we learned from the previous sections, lead lengths must be kept to an absolute minimum when using high impedance devices. Also, due to recent advances in the area of device integration, many of these hybrids have become very cost effective. The choice of a hybrid amplifier will assure the proper match of photodiode and amplifier, thus allowing the circuit designer to concentrate on other portions of circuit.

A hybrid amplifier that has all of the advantages previously stated is the OPT 101 manufactured by Burr-Brown Corp. The data sheet for the OPT 101 can be found at the end of this chapter. This device operates on a single power supply with a range of +2.7 to + 36 Volts. The schematic diagram on the front page of the data sheet shows the basic circuit configuration for this device. Information on spectral responsivity is also given for different package versions. In the simplest application using the 8-pin DIP package, only a jumper wire from pins 4 to 5 and power connections to pins 1 and 8 are required. The internal 1 Megohm resistor performs the feedback function required for amplification. In this configuration, the output voltage will be the product of the photodiode current times the feedback resistor, R_F, plus an offset voltage of about 7.5 mV due to single supply operation. Since the photodiode operates in the photoconductive mode, the output current will be very linear with input optical power. According to the data sheet, the frequency response for this basic circuit is 14 KHz.

This device has great versatility. The amplification can easily be increased by placing a resistor in series with the internal 1 Megohm resistor. This resistor must be placed across pins 4 and 5 for either package. The amplification can be reduced by placing a resistor across pins 2 and 5 for the DIP package. This effectively forms a parallel combination of the internal 1 Megohm resistor and R_{EXT}. Figure 13.16 gives some typical values for R_{EXT} for both SIP and DIP packages. Note that for some configurations, an external capacitor, C_{EXT}, must be used to eliminate or reduce gain peaking. It is interesting to notice that as R_{EXT} increases, the bandwidth decreases.

Consideration must be paid to how the input light illuminates the active area of the device since the amplifiers in the IC have a certain sensitivity to light. Try to keep the light focused only upon the sensitive photodiode area. This practice will improve settling times compared to when the light illuminates the full area of the integrated circuit. The photodiode area appears very dark compared to the surrounding op-amp circuitry. Hybrid amplifiers are available in numerous configurations and packages. If one does not exist for a particular application, it is usually possible to have a manufacturer custom design a device. Variations in these devices include sensor chip size, amplifier type, feedback resistor value, package type, and optical filtering. Miniaturization of the photodiode/amplifier circuit has allowed sensor chips with active areas of up to 5 mm^2 in a TO5 package. Figure 13.17 shows such a hybrid amplifier with 10 leads. These devices find important applications in low light conditions. The reduction of the lead lengths and

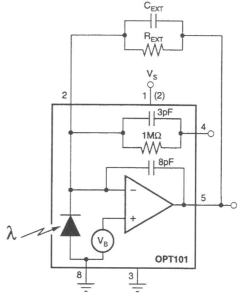

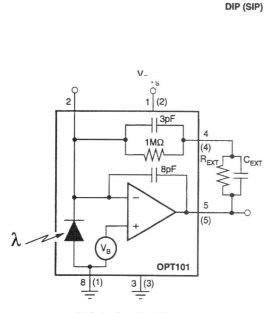

(a)-Series R_{EXT} (for SIP package).

R_{EXT} (MΩ)	C_{EXT} (pF)	DC Gain (x10⁶V/A)	Bandwidth (kHz)
1	50	2	8
2	25	3	6
5	10	6	2.5
10	5	11	1.3
50	—	51	0.33

(b)-External Feedback (for DIP package).

R_{EXT} (MΩ)	C_{EXT} (pF)	DC Gain (x10⁶V/A)	Bandwidth (kHz)
0.05[1]	56	0.05	58
0.1[1]	33	0.1	44
1	—	1	23
2	—	2	9.4
5	—	5	3.6
10	—	10	1.8
50	—	50	0.34

Note: (1) May require 1kΩ in series with pin 5 when driving large capacitances.

Figure 13.16 DC Gain and Bandwidth parameters for various values of R_{EXT} and C_{EXT}. Copyright or © 1994 Burr-Brown Corporation. Reprinted, in whole or in part, with the permission of Burr-Brown Corporation.

stray capacitances allows for lower noise and faster response as compared to using discrete components.

This device can be configured into many different modes of operation, as shown in Figure 13.18. With the addition of short jumper wires and the minimum of external components, application options such as the photovoltaic and photoconductive modes can be easily configured.

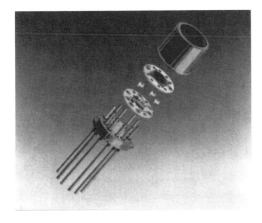

Figure 13.17 View showing the internal features of an OSI series photodiode/op-amp hybrid manufactured by Centronic Inc. Reprinted with permission of Centronic Inc., Newbury Park, CA. 1-800-700-2088

13.10 OPTICAL CONSIDERATIONS

Thus far in this chapter, we have considered mostly the electrical parameters involved in the design of an optical receiver. In this section, we will discuss some optical parameters related to the flux-gathering capability of the receiver system. As we said before, a lens

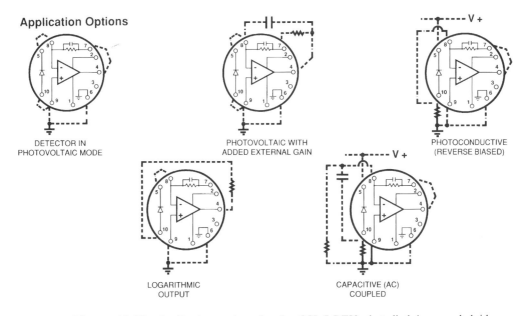

Figure 13.18 Application options for the OSI 5-DEV photodiode/op-amp hybrid. Reprinted with permission of Centronic Inc., Newbury Park, CA. 1-800-700-2088

may be used to provide more light to the detector portion of the receiver circuit without the added electrical noise. The lens and its housing can also be used to block out or divert unwanted light from entering into the detector. Figure 13.19 shows a simple optical system designed to increase light collection and reduce unwanted reflections from the interior walls of the lens housing. In this design, the lens serves as the aperture, thus determining the diameter of the light cone. The flux of light that contains the optical signal can be found within this cone. The size of the lens will also determine the amount of irradiance at the detector or photodiode.

Sometimes unwanted optical energy may enter through the objective lens to be reflected from the internal walls of the lens housing. These reflections can be reduced or eliminated by adding internal baffles as shown in Figure 13.19. They must be arranged and sized such that the light flux cone is not compromised. The placement of baffles as shown prevents stray light from entering at angles other than at the cone of acceptance. In the design example shown, the metal shield placed around the circuit to reduce EMI can also serve as a baffle. The metal shield in this area may require a flat black finish to reduce reflections from the side facing the lens. The geometry involved here will depend upon the aperture lens size, its focal length, and the active area of the photodiode. Next, we will consider how to optimize this part of the optical design. The idea involved here is to use a lens having the smallest possible diameter to achieve optimum light input to the photodiode or detector element. Since the cost of the lens itself can be a significant portion of the final cost of the completed system, this part of the design will have an important impact.

When using a lens to direct light from an emitter to a detector element, its throughput must be considered. The lens will limit the detector's field of view to a cone having an acceptance angle. In the case of using a lens with a transmitter, it will form an emitted cone of light. We will consider the case for selecting a lens to collect light for a photodiode. To optimize receiver performance, we must know the throughput of the lens to the photodiode's active area. We usually have physical constraints, so the smallest lens with the largest amount of throughput must be selected. The cost of the lens will also dictate our final decision.

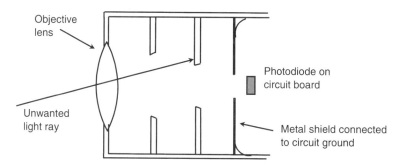

Objective lens

Unwanted light ray

Photodiode on circuit board

Metal shield connected to circuit ground

Figure 13.19 Unwanted stray light can be stopped from entering the receiver element by using internal baffles.

The basic parameters required to determine the throughput of a lens are given in Figure 13.20. We will use a plano-convex objective lens of radius r_1 and focal length f. The photodiode in this case has an active surface area with radius r_2. The throughput of the lens can be calculated as shown below by using these parameters.

Using the simple lens system shown in Figure 13.20, we can find the variables to solve the throughput equation:

$$x = r_2/f \qquad y = f/r_1 \qquad z = 1 + (1 + x^2)y^2$$

$$\text{Throughput (TP)} = F\pi A_{lens}$$

$$\text{where } F = 0.5[z - (z^2 - 4x^2y^2)^{1/2}]$$
$$A = \text{Area of objective lens}$$

The derivation of the mathematical expression for throughput can be found in the book by Clair L. Wyatt, *Electro-Optical System Design*. The serious reader is encouraged to consult this book for more detailed information. Here, we use only the result from this book to determine the throughput. We will next consider a typical design example.

Example 13.6

A plano-convex lens with a focal length of 20 mm and diameter of 20 mm is used to limit the field of view of a photodiode with an active area diameter of 4 mm. If the separation distance between the two elements is 20 mm, find the throughput.

First, we will convert the dimensions to centimeters to solve for x, y, and z. Then, we can find F to use in the throughput equation:

$$r_1 = 1.0 \text{ cm}, r_2 = 0.2 \text{ cm}, f = 2.0 \text{ cm}.$$

$$x = 0.2/2.0 = 0.1 \quad y = 1.0/2.0 = 0.5 \quad z = 1 + [1 + (0.1)^2](0.5)^2 = 1.25$$

Now, we can solve for F:

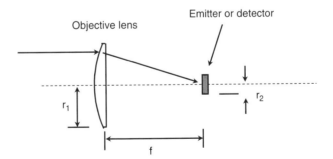

r_1 = radius of objective lens
r_2 = radius of receiver or emitter (assumed circular)
f = focal length of objective lens

Figure 13.20   A simple lens system used to define throughput.

$$F = 0.5 \,[1.25 - (1.56 - 4 \times 0.0025)^{1/2}] = 2.5 \times 10^{-3} \text{ (unitless)}$$

The final step is to solve for the throughput:

$$\text{Throughput} = F\pi A_{\text{lens}} = (2.5 \times 10^{-3})(3.14)(3.14)(1)^2$$
$$\text{Throughput} = 2.47 \times 10^{-2} \text{ cm}^2\cdot\text{sr}$$

This exercise can be done for the different lenses considered in an optical design to determine the best compromise between small diameter and large throughput.

SUMMARY

In this chapter, we discussed ways to integrate photodiodes into useful circuits. In most cases, a resistor must be used to convert the photocurrent produced within the photodiode's structure into a useful signal voltage. This load resistor must be carefully chosen and placed in the circuit. For example, if the resistance value is too large, the bandwidth may suffer due to the increased RC time constant. Very large value resistors can also cause leakage current problems when placed improperly on a circuit board. If the resistance value is too small, the signal voltage will not be adequate. We also saw the distinct advantages to using an integrated amplifier and photodiode as it makes circuit design easier and reduces noise pick-up.

We have now discussed the major components of a general electro-optical system. In these discussions, you can see why it was important to use both electrical and optical parameters. It is hoped that a good understanding of the basic physics behind the operation of the chosen optoelectronic devices and their use in electronic circuits has been achieved. In the final chapter, we will consider four different applications of electro-optical systems. The first application considered will be the audio CD player. We have already studied several of the critical components found in a typical audio CD player. These components include the diode laser, lenses, polarizing beam splitter, quarter-wave plate, diffraction grating, and various photodiodes. We will see how these components and more are all integrated together to produce the electro-optical portion of the CD player. The other examples will also use much of the material covered in the last 13 chapters.

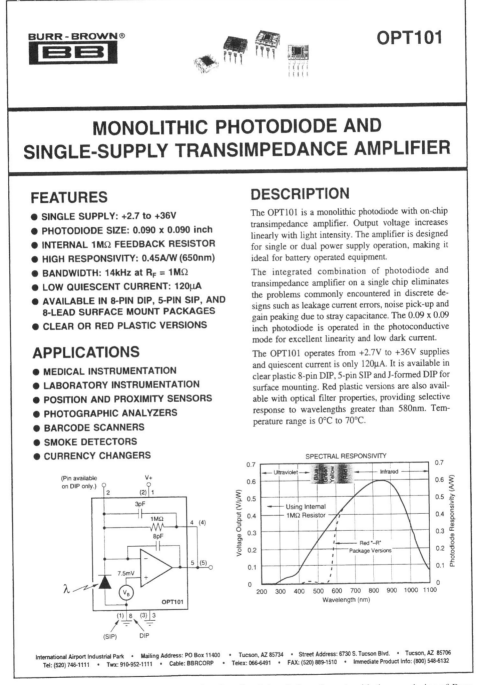

BURR - BROWN®

OPT101

MONOLITHIC PHOTODIODE AND SINGLE-SUPPLY TRANSIMPEDANCE AMPLIFIER

FEATURES

- SINGLE SUPPLY: +2.7 to +36V
- PHOTODIODE SIZE: 0.090 x 0.090 inch
- INTERNAL 1MΩ FEEDBACK RESISTOR
- HIGH RESPONSIVITY: 0.45A/W (650nm)
- BANDWIDTH: 14kHz at R_F = 1MΩ
- LOW QUIESCENT CURRENT: 120μA
- AVAILABLE IN 8-PIN DIP, 5-PIN SIP, AND 8-LEAD SURFACE MOUNT PACKAGES
- CLEAR OR RED PLASTIC VERSIONS

APPLICATIONS

- MEDICAL INSTRUMENTATION
- LABORATORY INSTRUMENTATION
- POSITION AND PROXIMITY SENSORS
- PHOTOGRAPHIC ANALYZERS
- BARCODE SCANNERS
- SMOKE DETECTORS
- CURRENCY CHANGERS

DESCRIPTION

The OPT101 is a monolithic photodiode with on-chip transimpedance amplifier. Output voltage increases linearly with light intensity. The amplifier is designed for single or dual power supply operation, making it ideal for battery operated equipment.

The integrated combination of photodiode and transimpedance amplifier on a single chip eliminates the problems commonly encountered in discrete designs such as leakage current errors, noise pick-up and gain peaking due to stray capacitance. The 0.09 x 0.09 inch photodiode is operated in the photoconductive mode for excellent linearity and low dark current.

The OPT101 operates from +2.7V to +36V supplies and quiescent current is only 120μA. It is available in clear plastic 8-pin DIP, 5-pin SIP and J-formed DIP for surface mounting. Red plastic versions are also available with optical filter properties, providing selective response to wavelengths greater than 580nm. Temperature range is 0°C to 70°C.

International Airport Industrial Park   •   Mailing Address: PO Box 11400   •   Tucson, AZ 85734   •   Street Address: 6730 S. Tucson Blvd.   •   Tucson, AZ 85706
Tel: (520) 746-1111   •   Twx: 910-952-1111   •   Cable: BBRCORP   •   Telex: 066-6491   •   FAX: (520) 889-1510   •   Immediate Product Info: (800) 548-6132

SPECIFICATIONS

T_A = +25°C, V_S = +2.7V to +36V, λ = 650nm, internal 1MΩ feedback resistor, and R_L = 10kΩ unless otherwise noted.

PARAMETER	CONDITIONS	OPT101P, W			UNITS
		MIN	TYP	MAX	
RESPONSIVITY					
Photodiode Current	650nm		0.45		A/W
Voltage Output	650nm		0.45		V/μW
vs Temperature			100		ppm/°C
Unit to Unit Variation	650nm		±5		%
Nonlinearity[1]	FS Output = 24V		±0.01		% of FS
Photodiode Area	(0.090 x 0.090in)		0.008		in²
	(2.29 x 2.29mm)		5.2		mm²
DARK ERRORS, RTO[2]					
Offset Voltage, Output		+5	+7.5	+10	mV
vs Temperature			±2.5		μV/°C
vs Power Supply	V_S = +2.7V to +36V		10	100	μV/V
Voltage Noise, Dark, f_B = 0.1Hz to 20kHz	V_S = +15V, V_{PIN3} = −15V		300		μVrms
TRANSIMPEDANCE GAIN					
Resistor			1		MΩ
Tolerance, P			±0.5	±2	%
W			±0.5		%
vs Temperature			±50		ppm/°C
FREQUENCY RESPONSE					
Bandwidth	V_{OUT} = 10Vp-p		14		kHz
Rise Fall Time, 10% to 90%	V_{OUT} = 10V Step		28		μs
Settling Time, 0.05%	V_{OUT} = 10V Step		160		μs
0.1%			80		μs
1%			70		μs
Overload Recovery	100%, Return to Linear Operation		50		μs
OUTPUT					
Voltage Output, High		(V_S) − 1.3	(V_S) − 1.15		V
Capacitive Load, Stable Operation			10		nF
Short-Circuit Current	V_S = 36V		15		mA
POWER SUPPLY					
Operating Voltage Range		+2.7		+36	V
Quiescent Current	Dark, V_{PIN3} = 0V		120	240	μA
	R_L = ∞, V_{OUT} = 10V		220		μA
TEMPERATURE RANGE					
Specification		0		+70	°C
Operating		0		+70	°C
Storage		−25		+85	°C
Thermal Resistance, θ_{JA}			100		°C/W

NOTES: (1) Deviation in percent of full scale from best-fit straight line. (2) Referred to Output. Includes all error sources.

PHOTODIODE SPECIFICATIONS

T_A = +25°C, V_S = +2.7V to +36V unless otherwise noted.

PARAMETER	CONDITIONS	Photodiode of OPT101P			UNITS
		MIN	TYP	MAX	
Photodiode Area	(0.090 x 0.090in)		0.008		in²
	(2.29 x 2.29mm)		5.2		mm²
Current Responsivity	650nm		0.45		A/W
	650nm		865		μA/W/cm²
Dark Current	V_{DIODE} = 7.5mV		2.5		pA
vs Temperature			doubles every 7°C		
Capacitance			1200		pF

14

Electro-Optical Systems

14.1 THE COMPACT DISC PLAYER

The compact disc or CD was introduced to the consumer market in 1982, and has since become the standard recording medium in the music industry. CD technology offers noiseless playback, fast random access, and an extremely durable recording medium. Since 1982, more than 6 billion optical discs have been sold.

The optical disc contains recorded information in microscopic pits along spiral tracks. These tracks form concentric rings on the plastic disc. The pitted side of the disc contains a thin coat of reflective aluminum. The pits themselves are 0.5 μm wide by 0.9 to 3.3 μm long, and are located on a flat-topped ridge. These ridges form tracks separated by grooves 1.6 μm apart. A clear plastic covering of thickness 1.2 mm protects the surface containing the pits. A 4.75 inch disc can store more than 70 minutes of audio.

The mechanism that reads the optical disc, known as the actuator, remains stationary while the disc spins on a spindle. An objective lens in this actuator assists in the functions of focusing and tracking. A diode laser beam must be accurately focused onto a particular track so its optical energy can reflect from the pits as they move. The reflected beam contains the signal that produces the playback of the music. To accomplish this, the actuator allows motion in two degrees of freedom for the objective lens. It can move up and down on a shaft for focusing control of the laser spot, and horizontally (radially to the disc) for tracking control. Photodiodes perform many important functions in the operation of a CD player.

The remainder of this section will be devoted to explaining the basic operation of the optical assembly found in a typical audio CD player. Due to the complexity involved in its operation, these discussions will be divided into three separate areas of concentration. The first area for consideration will be the optical path of the beam from the diode laser to the optical disc, and to the detector. A typical Fabry-Perot diode laser operating at 780 nm will be used for this purpose. Next, the function of tracking control will be considered. Many schemes exist to keep the diode laser's beam on the correct track during playback. A discussion of all schemes will not be attempted. Instead, we will consider a tracking scheme used by most audio CD players. The third function that will be considered is focusing. The diode laser's small spot must remain in constant focus to produce the high fidelity sound output that we associate with the CD recording medium. To achieve correct focusing, the objective lens must be able to move along the optical axis of the laser beam to keep the spot size constant. This assumes that the tracking control function keeps the spot on the correct track of the optical disc. Again, many different methods exist to accomplish this task. In this section, we will consider one of the most popular ways that this can be accomplished.

Optical Path

We will first consider the specifications for a diode laser before discussing the details involved with the optical path the beam must take. Listed below are typical specifications for a diode laser for this application. The diode laser used in a typical audio CD player must have the following features.

1. Low operating current due to battery operation. (Typically 40 mA or less)
2. The laser chip must be accurately positioned in its housing to allow for correct beam alignment.
3. The diode laser must have a long operating life.
4. The beam must be coherent enough to allow for the formation of a diffraction-limited spot that can be easily focused.
5. The cost and availability must be conducive to mass marketing.

The diode laser used in most audio CD players emits light in the wavelength region from 780 to 840 nm. A shorter wavelength emission, such as found with the DVD (digital versatile disc) format, allows for a smaller spot size. DVD players use a diode laser with an emission wavelength of 635 to 650 nanometers. We know from our discussions on the diode laser that its cross sectional beam shape is elliptical. To reduce or eliminate this elliptical shape, specialized optics must be used. Upon leaving the diode laser structure, this elliptical beam goes through a collimating lens and circularizing optics such as an anamorphic prism pair. The anamorphic prism pair provides magnification along only one meridian. Its ellipticity can be corrected by positioning this optical component correctly in the beam. Thus, the beam's cross section becomes nearly circular when this prism expands the beam along its shorter dimension. A circular beam shape will allow for a tighter focus at the optical disc. Typically, a 3 mW diode laser runs continuous wave (CW) since the modulation or signal input occurs when the beam reflects from the pits on the optical disc. Figure 14.1 shows the optical path of the beam emitted from a diode laser. The actual components may not look the same, as this is only a schematic illustration.

As the beam leaves the circularizing optics, it encounters a polarizing beam splitter. In this particular design, the polarizing beam splitter prevents reflected light at the optical disc from returning back into the diode laser. If this occurs, the laser can easily become unstable due to unwanted light striking the monitor photodiode. As you remember, the diode laser emits a polarized beam of light. We will use this fact to direct the beam to the optical disc, and then back to the detector element without going back into the diode laser structure. Notice that the polarization of the emitted beam is initially along the y-axis before it enters the polarizing beam splitter. This direction of polarization corresponds to the plane of the semiconductor junction. As the beam exits the polarizing beam splitter, it encounters a quarter-wave plate. After going through the quarter-wave plate, the beam's polarization state changes to circularly polarized. Next, the beam passes though a lens to form the small spot of laser light required to "read" the pits. When the reflected circularly polarized beam passes back through the quarter-wave plate on its return path, it becomes linearly polarized along the x-axis. The beam has now become shifted by 90° from its original state. This shift in polarization prevents the beam from going through the polariz-

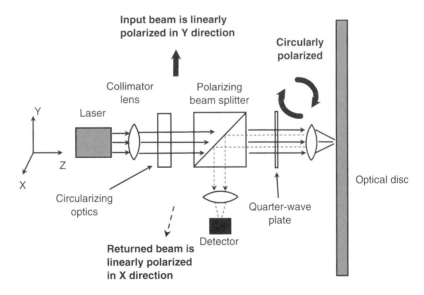

Figure 14.1   The optical path of the diode laser in a CD player. Notice how the beam changes its polarization state after passing through the quarter-wave plate twice.

ing beam splitter on its return trip. Instead, the beam reflects from the polarizing beam splitter, and then into the detector where the variations in laser light caused by the pits become processed into an audio signal. It is very important that the polarizing beam splitter prevents the beam from entering the diode laser. As discussed in previous chapters, if laser light enters the diode laser due to reflection, its output will become unstable.

Now, we will discuss the formation of the small spot on the optical disc from the emitted laser beam. The beam must be focused to a small spot so that it can "read" the individual microscopic pits on the optical disc. In Chapter 6, we discussed just how small this spot can become in the theoretical limit. The minimum spot size, D, is limited by diffraction. The mathematical expression below allows us to calculate the spot size:

$$D = 1.22 \; \lambda/NA$$

In the above equation, NA is the numerical aperture of the objective lens in Figure 14.1. Since the wavelength used is 780 nm, a typical lens with a numerical aperture of 0.55 will produce a spot size of about 1.7 μm (see Example 6.4 of Chapter 6). Typical focal lengths will be from 3 to 5 millimeters. This spot size determines the maximum storage capacity of the optical disc. As the spot size becomes smaller, more pits can be placed on the optical disc since the size requirement for the pit dimensions also decreases.

The material used for the objective lens is usually plastic instead of glass for lighter weight. This lighter weight results in less inertia, thus allowing the actuator to access data more quickly. A glass lens would tend to slow this motion down somewhat.

Tracking

The old record player used a plastic spinning disc with a needle to pick up audio information on a track. The optical disc is similar in that it also has to spin to pick up data on tracks or lands. This similarity soon vanishes when we look at how the CD player accomplishes tracking control of the disc. As this optical disc spins, reflections from the continuous wave diode laser are produced by the pits. The focused laser beam must stay on one particular track to produce digital data at the receiver without going into adjacent tracks. A few tracking schemes exist to keep the laser spot on the correct track during playback. One of these schemes, known as the three beam tracking method, uses a diffraction grating placed between the circularizing optics and the polarizing beam splitter. As you remember from Chapter 6, a diffraction grating will disperse the beam into different orders corresponding to maxima (see Example 6.5 of Chapter 6). The number of rulings per inch will determine the angular separation of these maxima or orders. This grating splits the laser beam into one zero-order and two first-order beams as shown in Figure 14.2. The two first-order beams are aligned so that they strike the track at slightly different angles as shown. The zero-order beam, which is brighter, reads the digitized data as described previously. This leaves the two first-order beams to reflect from the pits. The intensities from these two beams are independently monitored by fast response photodiodes as shown in the diagram. The movement of either beam results in an unbalanced condition. The difference produced due to this unbalanced signal determines how much correction in the + or − direction is required. This tracking error signal (TES) gets fed into servo motors that keep this offset within tolerance.

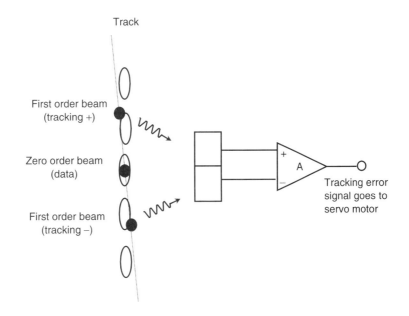

Figure 14.2   One method used for tracking control in a CD player.

Focusing

Now that we have control over the tracking of the optical disc, how does the laser beam maintain a precise focus on a particular track so that the pits can be read in correct sequence? As we mentioned before, the spot size must be maintained by the objective lens so that it can remain on only one track. It must keep the tolerance within about 0.5 μm while the disc spins. This is not an easy feat since the disc itself may not be completely flat. At the same time, the tracking must be kept to an even tighter tolerance of 0.1 μm or less. An objective lens with an NA of 0.5 has a depth of focus of approximately ± 1 μm at 780 nm. This fact imposes another stringent requirement on the focusing of the laser beam.

The modulated returned light can be used to achieve correct focusing of the laser spot on the moving disc. After reflecting from the optical disc, and then passing through the polarizing beam splitter, the light goes through an astigmatic lens and then to a quadrant detector. This astigmatic lens is used to determine the focusing of the laser spot. Depending upon how out of focus the spot is, the beam at the quadrant detector will change shape accordingly. Figure 14.3 shows how this can be accomplished. The quadrant detector has four photodiodes that are used to determine the shape of the spot incident upon its active surface. For example, a correctly focused beam will form a circular spot on the detector's active surface after passing through the astigmatic lens. In this case, the output from the four photodiodes should be equal. When the outputs are placed into an averaging circuit, a zero error signal results. This signal can be fed to a servo motor that corrects the focus by moving the objective lens up or down relative to the optical disc. When the beam is out of focus, the light passing through the astigmatic lens will form an oval shape as

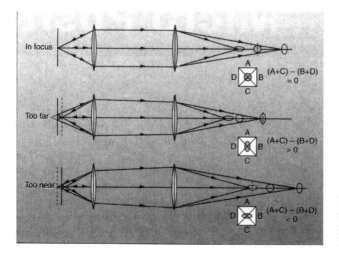

Figure 14.3 Focusing is accomplished by using a quadrant detector to define the shape of the laser spot. Reprinted with permission. ©PennWell Publishing Company.

shown in Figure 14.3. This will cause an imbalance in the four photodiode outputs. If the focusing is too far, the averaging circuit will produce an error signal greater than zero. This error signal to the servo motor will tell it the amount of correction needed to focus spot back to a circular shape at the quadrant detector. This procedure, known as signal nulling, is used in many other applications to maintain constant beam alignment or focusing. The current output from each of the four photodiodes in the quadrant detector must be converted to a useable voltage by using load resistors and amplifiers.

14.2 IrDA TRANSCEIVER

The electro-optical system covered in this section finds applications in data transmission. This transceiver uses the IrDA protocol initially discussed in Chapter 11. IrDA protocol provides a standard for infrared data transmission between devices such as laptop computers, laser printers, and calculators. Since its acceptance in 1993 as a standard signaling protocol, many companies have developed ICs and integrated optoelectronic components for the circuit designer to use in IrDA compliant devices. One such optoelectronic component, developed by Hewlett-Packard, offers many of the advantages found in most integrated optoelectronic devices. We will discuss how to integrate the HSDL-1000 into a system that can be used for an IrDA data link. This data link has a maximum range of 1 meter for operating speeds up to 115.2 Kb/sec. Many of the concepts introduced in the last few chapters will be applied in the discussion of this application.

Device Description

Before going into the details of a typical circuit using the HSDL-1000, we will first describe both the electrical and optical specifications of this important device. The HSDL-

1000 is a fully integrated transceiver module designed to be compliant with the IrDA standard. This integrated optoelectronic component provides the functions of transmitting and receiving IrDA data. The interested reader may wish to turn to Chapter 11 for a review of the IrDA parameters for compliant devices.

A schematic diagram and IC pin out description are given in Figure 14.4. The HSDL-1000 contains the transmitter driver, receiver, and associated receiver circuitry, all in an 8 pin package. The transmitter section of the IC consists of a high efficiency IR LED with high speed drive circuitry. Meeting the optical specifications for IrDA transmission has been made easier for the designer with the use of a built-in lens. This lens provides for infrared transmission within an optical half angle of ± 15 to ± 30 degrees, the limits for IrDA compliance. The output from the transmitter consists of relatively narrow, high speed pulses of infrared light consistent with the IrDA specification. To provide the required radiant intensity, the current through the transmitter must be approximately 250 mA. At this current level, the resultant radiant intensity is guaranteed to be a minimum of 44 mW/sr up to a maximum of 250 mW/sr. The peak wavelength emission will be at 875 nanometers with a spectral half width of 35 nanometers.

The transmitter section has electrical specifications that must be taken into consideration. To prevent premature IR LED failure, the average current should be 100 mA or

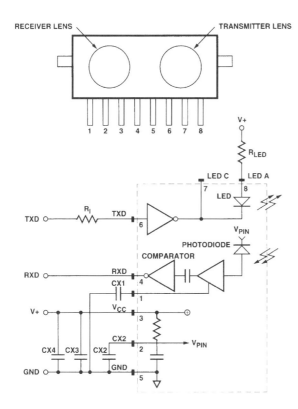

Figure 14.4 HSDL-1000 schematic diagram and pin out description. Reprinted with permission from Hewlett-Packard Company.

less. A peak current of 500 mA can be applied for pulses of 90 µsec or less in width at a 20 percent duty cycle. A maximum of 1.0 ampere peak current is allowed for pulses of 2 µsec or less in width at a duty cycle of no more than 10 percent. The current through the IR LED is adjusted using an external resistor, R_{LED}.

The receiver section contains the pin photodiode, lens, and other circuitry. The receiver circuit is designed to automatically adjust the incoming infrared light level to eliminate pulse width distortion. Without this feature, strong input pulses from a very close transmitter would saturate the receiver causing transmission errors to occur. The integrated lens provides the pin photodiode with a viewing angle of 30 degrees, and has a daylight filter that rejects light below a wavelength of 850 nm. You can see from Figure 14.4 that the pin photodiode operates in the photoconductive mode. As we learned previously, this mode provides for a fast response at the expense of leakage current. The reverse bias voltage is applied at the point labeled V_{PIN}.

Circuit Description

We will now discuss the circuit schematic detailed in Figure 14.4. This circuit uses only a few external components, thus making the circuit designer's job easier while providing for less susceptibility to EMI. With the successful design of this transceiver section, an interfacing to an IR Encoder/Decoder and UART completes the major portion of the design for the IrDA compliant device. We will briefly discuss the use of these interfacing ICs. What follows next is a discussion of the individual components used in the schematic diagram.

Starting with pin 1, the capacitor CX1 should have a capacitance value of 0.22 µf ± 10%. The function of this circuit section has to do with ambient light immunity. When increasing CX1 to a larger capacitance value, the received light pulses may actually merge together at the receiver's output, resulting in unstable operation. This daylight cancellation takes place in the first stage of the receiver's amplifier circuitry. If the capacitance value decreases substantially from the recommended amount, the result will be a loss of receiver sensitivity.

Pin 2 uses a 0.4 µf capacitor, CX2, connected to ground. This capacitor serves as a bypass capacitor for the pin photodiode, since it also connects to V_{PIN}. The value of this capacitor can be increased if the circuit board allows.

Pin 3 uses two capacitors connected from this pin to ground (pin 5). The capacitor CX3 functions as a bypass capacitor for V_{CC} or the input voltage to the IC. This capacitor should be a 0.1 µf ceramic type, and positioned as close as possible to the HSDL-1000. Capacitor CX4 serves to increase noise immunity. It has a suggested value of 4.7 µf. Both of these capacitors can be increased, if necessary, to provide for more filtering. V_{CC} must be derived from a clean source, not a noisy switching power supply.

Pin 4 (RXD) provides the receiver's digital signal output to an interface IC such as an Encoder/Decoder.

Pin 5 is ground.

Pin 6 is the transmitter data input required to modulate the IR LED. Resistor R1 has a nominal value of 300 ohms. The Encoder/Decoder IC provides the input modulation data.

Pin 7 connects to the cathode of the IR LED.

Pin 8 connects to the anode of the IR LED. The load resistor, R_{LED}, has a value from 4 to 10 ohms. The exact value used in the circuit will depend upon V_{CC}. The range over which V_{CC} can vary must be taken into account when calculating this value. A minimum current value to the IR LED of 250 mA must be considered when doing this calculation.

PC Board Layout

Now that we have selected the various electrical components to use in the circuit, the important job of designing the placement and electrical connections for the components remains. A well-designed electrical schematic can be rendered inoperative if the layout of the components is not carefully considered. The standard practice for PC board layout of surface mount devices should be followed. We will review some of the most basic considerations in this section that minimize EMI and noise coupling when using optoelectronic devices. The list of suggested practices given here is by no means complete. The serious reader may want to consult an electronic text that specializes in this area.

As we discussed in Chapter 13, there can be many sources of EMI on the circuit board itself. These sources include switching power supplies, I/O ports, clock generators, and other digital circuits. The sources external to the PC board must also be considered. For example, will this device be used in an environment where high level radio emissions exist? If so, EMI shielding may be required. Once we have identified the component locations, the configuration of the ground plane must then be determined. The correct design of the ground plane will contribute to error free operation. The goal for EMI immunity should be for receiver operation with a bit error rate (BER) of 10^{-9} or less.

To implement the proper type of ground plane, a multilayer PC board design is recommended. The HSDL-1000 module must have ground plane metal beneath it for at least 3 centimeters in any direction. This requirement must be implemented due to the analog electronics integrated within the IC. As we know, analog electronics can be very susceptible to EMI. This ground plane area made for the HSDL-1000 must then be connected to the circuit board using a very low impedance path. The external components shown in the schematic of Figure 14.4 must be placed as close as possible to the module to reduce noise pick-up. Of course, these are very general recommendations, and the precise board design will depend upon the particular application. Finally, keep all noise sources, such as those previously mentioned, at least 5 centimeters from the module. This includes I/O ports as well. These sources must be placed external to the special ground plane created for the module.

Optomechanical Design

The position of the module in relation to the product case will depend upon the particular application. For example, greater immunity to ambient light sources, such as fluorescent lamps and sunlight, can be attained by recessing the module within the case of the device. An IR transparent window can then be placed in front of the module. Figure 14.5 shows a typical arrangement for the situation described above. To optimize the transmitter's radiant intensity and the receiver's irradiance, a simple relationship exists between the positioning of the module and the window aperture size. Figure 14.5 shows the Y and X dimensions required for this relationship. The Y distance from the tip of the receiver element to the plane of the front panel has a corresponding distance X. This X distance is the half width of the aperture as shown, and can be calculated using the mathematical relationship below:

$$X = 2.87Y + 7.0$$

For example, when the Y distance is adjusted for 6 mm, the X distance must be about 24 mm.

The aperture window should permit an optimum amount of IR transmission. The window should not be made too large, as EMI could couple from external sources. In some cases, a compromise situation will exist between the window size and EMI susceptibility. If the exterior case acts as a shield, then any opening will compromise EMI susceptibility.

The transceiver discussed can then be placed in typical devices such as laptop computers and printers to provide wireless communications. The point and shoot approach of IrDA is shown in Figure 14.6.

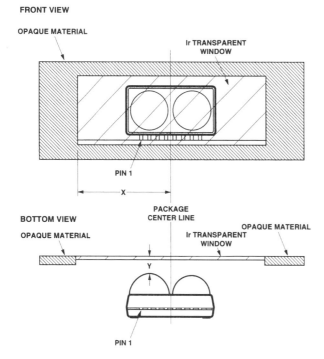

Figure 14.5 Position of module with respect to product case. Reprinted with permission from Hewlett-Packard Company.

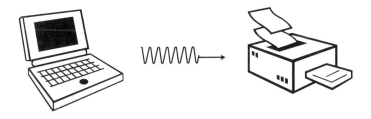

Figure 14.6   Typical application for an IrDA data link.

14.3 A SENSING APPLICATION

The previous applications used optoelectronic components in communications or data transmission applications. Another important application for these devices can be found in sensing. The use of optoelectronic components can be found in biomedical, petrochemical, and household appliances. This next application uses an integrated optoelectronic device that converts light into a digital pulse train. The output from this sensor allows it to be interfaced directly to digital-based circuitry.

Previous applications of photodiode-based receivers required a load resistor to convert the photocurrent generated by the optical signal into a useful signal voltage. Instead of comparing the amplitude of this signal to some reference value, it can be converted to a pulsed frequency instead. This process, known as analog to digital (A to D) conversion, forms the basis for the operation of our next device. The process of A to D conversion usually requires many electrical components with careful board layout and EMI shielding. A better solution to this circuit problem can be found by using an integrated optoelectronic device from Texas Instruments, the TSL 230. It is available in a clear, 8-pin DIP package.

The TSL 230 has many of the unique features found in integrated optoelectronic devices. Light-to-current conversion is accomplished with up to 100 photodiodes connected in parallel. This configuration provides for maximum sensitivity. The circuit designer can select two other sensitivities that are reduced by using 1 or 10 of these photodiodes. As we know, the photocurrent generated is proportional to the light intensity. The TSL 230 converts this photocurrent into a pulse train having a frequency proportional to the incident light intensity. This conversion eliminates the A to D process used in previous circuit designs, thus making it compatible with digital logic circuitry.

Circuit Operation

The TSL 230 consists of two basic sections as shown in the simplified block diagram of Figure 14.7. These sections are the photodiode and current-to-frequency converter. A matrix of 100 photodiode elements in a 10×10 arrangement accomplishes the light sensing function. Pins S0 and S1 are used to control the device sensitivity. Table 1a of Figure 14.7 shows that 1, 10, or 100 photodiode elements can be used for this purpose. These two pins can be controlled by an analog multiplexer. This type of sensitivity control can be thought of

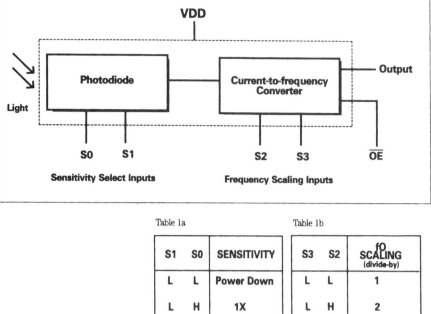

S1	S0	SENSITIVITY
L	L	Power Down
L	H	1X
H	L	10X
H	H	100X

S3	S2	fO SCALING (divide-by)
L	L	1
L	H	2
H	L	10
H	H	100

Table 1a   Table 1b

Figure 14.7 Simplified internal block diagram of the TSL230. Reprinted by permission of Texas Instruments.

as electronic aperturing. It also allows the IC to respond to a wide dynamic range of input light intensities. The amount of photocurrent produced will be proportional to the number of photodiodes activated and the light intensity. Photocurrent then becomes input to the current-to-frequency converter. The output from the converter section is a train of pulses with a frequency proportional to the photocurrent. Frequency scaling can be employed by using inputs S2 and S3 according to Table 1b of Figure 14.7. This provides a converter frequency divided by 1, 2, 10, or 100. The resultant output square wave has a 50% duty cycle. Since the output is a serial stream of bits, it can be easily interfaced to a microcontroller.

Light Measurements

The measurement of light intensity can be accomplished with relatively high resolution when the light level does not vary appreciably. In this case, the pulses produced from the current-to-frequency converter will be counted for a given time period. Resolution will increase with increased measurement time.

For a light source or output that varies rapidly with time, the measurement of intensity will involve a loss in resolution due to the shorter time measuring interval. In this case, the frequency scaling inputs, S2 and S3, can be used.

The TSL 230 can be found in applications like process control, chemical analysis, and water turbidity sensing. Figure 14.8 shows how this optoelectronic device is used in the GE Profile dishwasher to determine the clarity of the water. By monitoring the turbidity of the water, water usage can be controlled. For example, a cycle can be lengthened, shortened, or skipped altogether depending upon the degree of turbidity and rate of cleaning of the water. A fuzzy-logic system makes these decisions based upon the sensor output. GE reports up to 30% energy savings with this system.

14.4   OPTICAL PARAMETERS IN A FIBER OPTIC SYSTEM

The use of fiber optic systems has become common place over the last few years. Increasing demands on bandwidth, transmission speed, and channel capacity have given the fiber optic system a clear advantage over copper-based transmission systems. In this section,

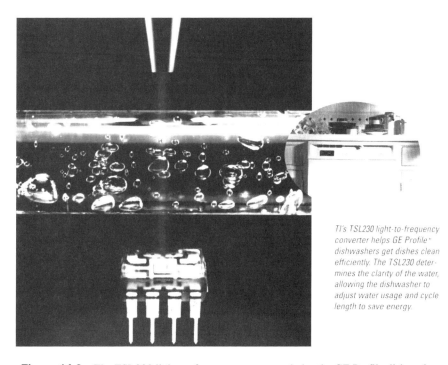

TI's TSL230 light-to-frequency converter helps GE Profile™ dishwashers get dishes clean efficiently. The TSL230 determines the clarity of the water, allowing the dishwasher to adjust water usage and cycle length to save energy.

Figure 14.8   The TSL230 light-to-frequency converter helps the GE Profile dishwasher to get dishes clean efficiently. Reprinted by permission of Texas Instruments.

we will discuss some of the limitations of fiber optic systems, and some of the important optical parameters that need to be addressed for reliable operation.

Optical Losses

In a fiber optic system, the total optical losses must be known so that a reliable signal transmission can occur. These losses are due to the fiber itself, connectors, splices, bends, etc. In Chapter 4 we discussed the mechanisms responsible for these losses. The amount of light reaching the receiver depends upon all of the losses mentioned above. A study of these losses is known as the link power budget. The individual losses can be calculated, and then added together to determine an acceptable operating power margin for the system. This margin should take into account component aging and temperature changes.

The total attenuation of the input optical signal can be measured with an optical time domain reflectometer (OTDR). This important tool can also characterize the fiber's uniformity, splicing losses, and sights where breaks have occurred. It uses a nondistructive measuring technique, and requires access to only one end of the fiber link.

The ODTR operates in a similar manner to optical radar. Light pulses of at least 10 milliwatts are launched from a laser into one end of the fiber. When a pulse reaches a connector, splice, or the fiber's end, a certain amount of optical energy will be reflected or back-scattered back. An analysis of this returned optical energy is used to determine the properties of the fiber. Pulse frequency and width must be selected depending upon the parameters to be tested. For example, the pulse frequency or repetition rate must be chosen such that pulses reflecting at the fiber's end do not overlap any transmitted pulses. Figure 14.9 shows a schematic representation of the OTDR. Notice that the returning pulses are taken out of the optical path by the use of a beam splitter.

The receiver portion of the OTDR must be extremely fast to deal with signals transmitted at the speed of light. To accomplish this, an APD is often used as the light-to-current converter. Its output can be amplified by conventional electronics. Two basic phenomena are involved with the returning pulses. Fresnel reflection occurs when light strikes a region of two different refractive indices. This can occur at a splice, a connector, or a break in the fiber. Rayleigh back-scattering happens when light hits imperfections within the fiber. The returned optical energy is used to display a waveform as shown in Figure 14.10. Here, the plot of returned light intensity as a function of time identifies places in the fiber's length where excess optical losses occur. A faulty splice, connector, or even a break can be located within a few meters.

We will next take a close look at Figure 14.10 to see just what information can be obtained from the returned optical energy. From this plot, we can determine several features of an optical data link. The slope of this line indicates the loss of the fiber. A list of some features is given below.

1. The "dead zone" at the beginning of the fiber results from the relatively large optical signal due to Fresnel reflection at this point. The receiver becomes saturated due to this large input energy from the diode laser.

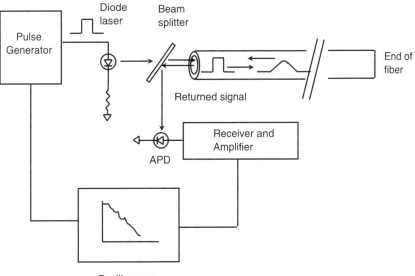

Figure 14.9   The principle behind the operation of the Optical Time Domain Reflectometer (OTDR).

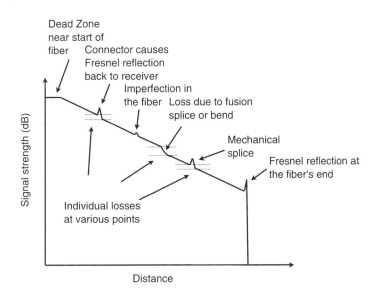

Figure 14.10   A typical waveform from an OTDR.

2. Fresnel reflections occur at the splices. These show up as peaks in the sloped curve. As you remember, this results when light travels between an interface of two dielectrics having different indices of refraction. Light is also attenuated here due to this situation. This shows up as a displacement of the curve.

3. Fusion splices show up as optical loss due to the mechanism of Rayleigh scattering. Light can also couple out of the fiber at a bend without causing a back reflection.

4. When the pulse finally reaches the end of the fiber, it experiences Fresnel reflection. The remainder of the optical energy exits at this interface.

System Bandwidth

In Chapter 4, we spoke about the mechanism of dispersion in optical glass fiber being responsible for the spreading out of an input pulse. Specifically, a square digital optical pulse will become rounded and wider due to dispersion as it travels within the optical fiber. This rounded pulse has a finite rise time, thus providing a frequency response limit. If two separate light pulses travel far enough through the fiber, they will eventually overlap, resulting in intersymbol interference. The light energy or ray spending most of its time near the center of the fiber will experience fewer reflections than one traveling near its edges. This fact will cause different arrival times for rays depending upon the order of the mode. Thus, dispersion limits the distance that a digital pulse can travel within the fiber. Typical dispersion values for multimode fiber are on the order of nanoseconds per kilometer. This dispersion affects the digital transmission rate, and thus the bandwidth of the system.

Four basic elements of a typical fiber optic system contribute to the total system rise time. The first two elements, the transmitter and receiver rise times, were discussed in the last few chapters. To study the last two contributors, we must consider the optical realm. Specifically, rise times due to material dispersion, τ_{mat}, and modal dispersion, τ_{mod}, will be considered next. We considered the mechanisms of modal and material dispersions in Chapter 4. The rise time calculations for these mechanisms in a typical multimode fiber will allow us to determine the system bandwidth. For an NRZ code, the total rise time should not be in excess of 70% of the bit period. We will next consider the rise times due to the transmitter and receiver.

In a system with a relatively short optical transmission distance, the limitations due to bandwidth will come from the transmitter and the receiver. For the transmitter, its rise time results primarily from the light source and the drive circuitry. The receiver's rise time results primarily from the photodetector element and its circuitry. As you remember, we calculated the 3 dB electrical bandwidth, B, from the rise time, τ_r, using the following mathematical expression:

$$B = 0.35/\tau_r$$

For multimode fiber, the rise time will depend primarily upon the amount of material and modal dispersion present in the fiber. As we know, dispersion depends upon the fiber length, the spectral width of the optical source, and wavelength. In our discussions, we will use the example of an 8 kilometer multimode optical fiber transmitting digital pulses from an IR LED with peak spectral emission at 900 nm. Once we have calculated the two dispersion rise times, we will consider the transmitter and receiver rise times to give us a system rise time.

The rise time due to material dispersion can be calculated by using the linear function given below:

$$\tau_{mat} = D_{mat}\sigma_\lambda L$$

In the above expression, D_{mat} is the material dispersion factor given in units of nanoseconds/nm·km. The spectral width, σ_λ, for a typical IR LED is 40 nm. Figure 4.19 in Chapter 4 displays this data. For the IR LED in our example, we get a material dispersion factor or 0.06 nanoseconds/nm·km. The last factor is the length of the optical link. If we substitute these values into the material dispersion equation, we get the desired value for the rise time due to this mechanism:

$$\tau_{mat} = (0.06)(40)(8) = 19.2 \text{ nanoseconds}$$

We must now consider the case for modal dispersion. The empirical expression given below is usually used to determine the rise time due to this mechanism:

$$\tau_{mod} = 0.44L^{(0.7)}/B_O$$

In this expression, L is the length of the optical data link. A typical multimode fiber will have an optical bandwidth, B_O, of 400 MHz·Km. Using our 8 kilometer data link with the IR LED, we get the following value for the rise time due to modal dispersion:

$$\tau_{mod} = (0.44)(8^{0.7})/400 = 4.7 \text{ nanoseconds}$$

The rise times for the transmitter and receiver are usually well-known to the system designer. To determine the total system rise time, we must take the square root of the sum of the squares for each individual response time. We shall make the following assumptions for the transmitter and receiver. These rise times are given below:

$$\tau_{transmitter} = 10 \text{ nanoseconds}$$
$$\tau_{receiver} = 14 \text{ nanoseconds}$$

For our 8 kilometer optical data link, the total rise time can now be calculated as shown below:

$$\tau_{sys} = [(\tau_{tx})^2 + (\tau_{rec})^2 + (\tau_{mat})^2 + (\tau_{mod})^2]^{1/2}$$
$$\tau_{sys} = [(10)^2 + (14)^2 + (19.2)^2 + (4.7)^2]^{1/2}$$
$$\tau_{sys} = 26 \text{ nanoseconds}$$

A quick calculation shows that this value for system rise time allows for an NRZ data transmission rate of 25 Mbits/second while staying within the 70% degradation time requirement.

SUMMARY

In this chapter, we studied four electro-optical systems by applying what was learned in the previous chapters. This approach involved considering both the electrical and optical design issues for each system. I hope you have found this approach to be a useful tool in understanding the operation of electro-optical systems.

In the short time since the early 1960s to the present there has been a tremendous growth in the area of electro-optics. This change in technology has affected our lives in many ways. For example, electro-mechanical devices such as cash registers have been replaced with barcode scanners. The analog telephone transmission network of the 1960s has been replaced with a digital network. In this network, fiber optic cable makes up much of the terrestrial long-haul links. The record player has been replaced with the audio CD player to provide us with better sound quality due to its digital format. But what does the future hold for electro-optics? Will this growth continue? Research is currently being done to find ways to improve upon the emission process at shorter visible wavelengths. As you know, using a shorter wavelength increases the storage capacity in optical disc media. Optical media is quickly becoming the media of choice due to its superior storage capacity and the use of a no-contact read head. But, one thing is almost certain: the use of fiber optics will increase dramatically in the near future. The need for increased bandwidth will eventually result in an all-fiber network due to the limitations of copper-based technologies. Increased bandwidth will be needed for digital services such as personal computers with Internet access, and high definition television (HDTV). An all-fiber network can easily handle this increase in bandwidth with room to spare. We can see that the importance of the electron in the twenty-first century will become overshadowed by that of the photon.

Appendix A

Data Sheets From Various Manufacturers

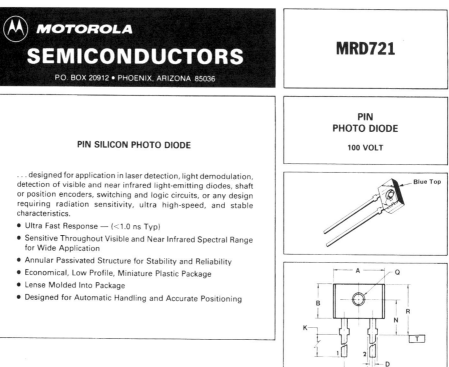

M) MOTOROLA

SEMICONDUCTORS

P.O. BOX 20912 • PHOENIX, ARIZONA 85036

MRD721

**PIN
PHOTO DIODE**

100 VOLT

PIN SILICON PHOTO DIODE

. . . designed for application in laser detection, light demodulation, detection of visible and near infrared light-emitting diodes, shaft or position encoders, switching and logic circuits, or any design requiring radiation sensitivity, ultra high-speed, and stable characteristics.

- Ultra Fast Response — (<1.0 ns Typ)
- Sensitive Throughout Visible and Near Infrared Spectral Range for Wide Application
- Annular Passivated Structure for Stability and Reliability
- Economical, Low Profile, Miniature Plastic Package
- Lense Molded Into Package
- Designed for Automatic Handling and Accurate Positioning

MAXIMUM RATINGS (T_A = 25°C unless otherwise noted)

Rating	Symbol	Value	Unit
Reverse Voltage	V_R	100	Volts
Total Power Dissipation @ T_A = 25°C Derate above 25°C	P_D	100 1.33	mW mW/°C
Operating and Storage Junction Temperature Range	T_J, T_{stg}	−40 to +100	°C

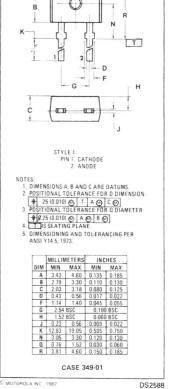

STYLE 1:
PIN 1. CATHODE
2. ANODE

NOTES:
1. DIMENSIONS A, B AND C ARE DATUMS.
2. POSITIONAL TOLERANCE FOR D DIMENSION:
3. POSITIONAL TOLERANCE FOR Q DIAMETER:
4. ⊥ IS SEATING PLANE.
5. DIMENSIONING AND TOLERANCING PER ANSI Y14.5, 1973.

	MILLIMETERS		INCHES	
DIM	MIN	MAX	MIN	MAX
A	3.43	4.60	0.135	0.185
B	2.79	3.30	0.110	0.130
C	2.03	3.18	0.080	0.125
D	0.43	0.56	0.017	0.022
F	1.14	1.40	0.045	0.055
G	2.54 BSC		0.100 BSC	
H	1.52 BSC		0.060 BSC	
J	0.23	0.56	0.009	0.022
K	12.83	19.05	0.505	0.750
N	3.05	3.30	0.120	0.130
Q	0.76	1.52	0.030	0.060
R	3.81	4.60	0.150	0.185

CASE 349-01

DS2588

FIGURE 1 — TYPICAL OPERATING CIRCUIT

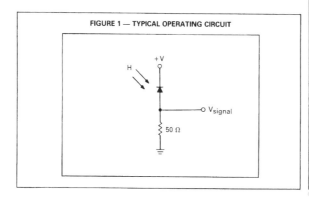

MRD721

STATIC ELECTRICAL CHARACTERISTICS (T_A = 25°C unless otherwise noted)

Characteristic	Fig. No.	Symbol	Min	Typ	Max	Unit
Dark Current (V_R = 20 V, R_L = 1.0 MΩ; Note 2) T_A = 25°C T_A = 100°C	 3 and 4 	I_D		 0.06 14	 10 —	nA
Reverse Breakdown Voltage (I_R = 10 μA)	—	$V_{(BR)R}$	100	200	—	Volts
Forward Voltage (I_F = 50 mA)	—	V_F	—	—	1.1	Volts
Series Resistance (I_F = 50 mA)	—	R_S	—	8.0	—	ohms
Total Capacitance (V_R = 20 V; f = 1.0 MHz)	5	C_T	—	3.0	—	pF

OPTICAL CHARACTERISTICS (T_A = 25°C)

Characteristic	Fig. No.	Symbol	Min	Typ	Max	Unit
Light Current (V_R = 20 V, Note 1)	2	I_L	1.5	4.0	—	μA
Sensitivity (V_R = 20 V, Note 3)	 — —	 $S(\lambda = 0.8\ \mu m)$ $S(\lambda = 0.94\ \mu m)$	 — —	 5.0 1.2	 — —	$\mu A/mW/cm^2$
Response Time (V_R = 20 V, R_L = 50 Ω)	—	$t_{(resp)}$	—	1.0	—	ns
Wavelength of Peak Spectral Response	6	λ_S	—	0.8	—	μm

NOTES: 1. Radiation Flux Density (H) equal to 5.0 mW/cm^2 emitted from a tungsten source at a color temperature of 2870 K.
2. Measured under dark conditions. (H ≈ 0)
3. Radiation Flux Density (H) equal to 0.5 mW/cm^2.

Ⓜ MOTOROLA *Semiconductor Products Inc.*

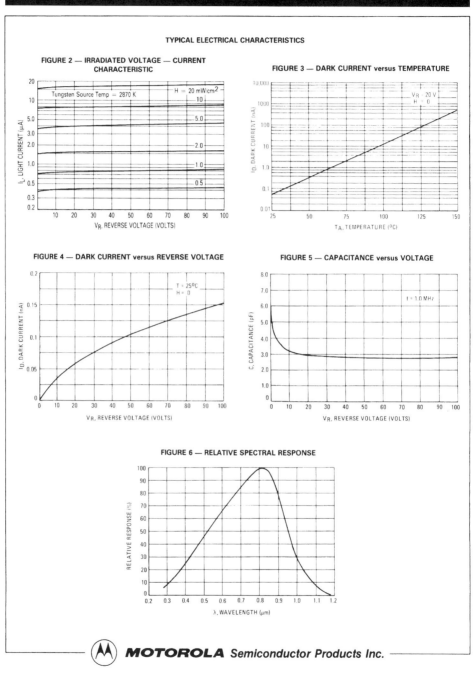

TYPICAL ELECTRICAL CHARACTERISTICS

FIGURE 2 — IRRADIATED VOLTAGE — CURRENT CHARACTERISTIC

FIGURE 3 — DARK CURRENT versus TEMPERATURE

FIGURE 4 — DARK CURRENT versus REVERSE VOLTAGE

FIGURE 5 — CAPACITANCE versus VOLTAGE

FIGURE 6 — RELATIVE SPECTRAL RESPONSE

MOTOROLA *Semiconductor Products Inc.*

VTP Process Photodiodes | VTP8350, 50S, 51

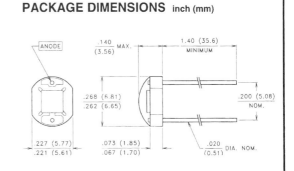

PACKAGE DIMENSIONS inch (mm)

ANODE

.140 (3.56) MAX.

1.40 (35.6) MINIMUM

.268 (6.81)
.262 (6.65)

.200 (5.08) NOM.

.227 (5.77)
.221 (5.61)

.073 (1.85)
.067 (1.70)

.020 (0.51) DIA. NOM.

PRODUCT DESCRIPTION

Planar silicon photodiode mounted on a two lead ceramic substrate and coated with a thick layer of clear epoxy. These diodes exhibit low dark current under reverse bias and fast speed of response.

CASE 11   CERAMIC
CHIP ACTIVE AREA: .012 in^2 (7.45 mm^2)

ABSOLUTE MAXIMUM RATINGS

Storage Temperature: -20°C to 75°C
Operating Temperature: -20°C to 75°C

ELECTRO-OPTICAL CHARACTERISTICS @ 25°C

SYMBOL	CHARACTERISTIC	TEST CONDITIONS	VTP8350			VTP8350S			VTP8351			UNITS
			Min.	Typ.	Max.	Min.	Typ.	Max.	Min.	Typ.	Max.	
I_{SC}	Short Circuit Current	H = 100 fc, 2850 K	65	80			80		65	80		μA
TC I_{SC}	I_{SC} Temp. Coefficient	2850 K		.20			.20			.20		% / °C
V_{OC}	Open Circuit Voltage	H = 100 fc, 2850 K		350			350			350		mV
TC V_{OC}	V_{OC} Temp. Coefficient	2850 K		-2.0			-2.0			-2.0		mV / °C
I_D	Dark Current	H = 0, V_R = 50 V			30 [1]			—			18	nA
R_{SH}	Shunt Resistance	H = 0, V = 10 mV		100		50	100			100		MΩ
C_J	Junction Capacitance	H = 0, V = 3 V			50			120 [2]			24 [3]	pF
R_a	Responsivity	940 nm		.06		.04	.06			.06		A /(W/cm^2)
S_R	Sensitivity	@ Peak		.55			.55			.55		A/W
$λ_{range}$	Spectral Application Range		400		1150	400		1150	400		1150	nm
$λ_p$	Spectral Response - Peak			925			925			925		nm
V_{BR}	Breakdown Voltage		33	140		50	140		50	140		V
$θ_{1/2}$	Ang. Resp. - 50% Resp. Pt.			±60			±60			±60		Degrees
NEP	Noise Equivalent Power			1.8×10^{-13} (Typ.)			1.8×10^{-13} (Typ.)			1.4×10^{-13} (Typ.)		W/√Hz
D^*	Specific Detectivity			1.5×10^{12} (Typ.)			1.5×10^{12} (Typ.)			2.0×10^{12} (Typ.)		cm√Hz/W

[1] V_R = 10 V   [2] V = 0 V   [3] V = 15 V

0991

VTP Process Photodiodes

RELATIVE DARK CURRENT VS TEMPERATURE

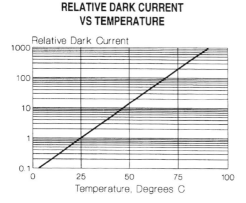

RELATIVE JUNCTION CAPACITANCE VS BIAS VOLTAGE
(REFERRED TO ZERO BIAS)

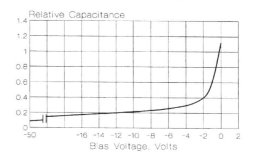

TEMPERATURE COEFFICIENT OF LIGHT CURRENT VS WAVELENGTH

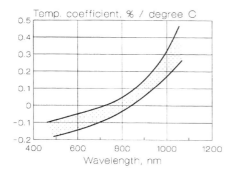

RELATIVE SHORT CIRCUIT CURRENT VS ILLUMINATION

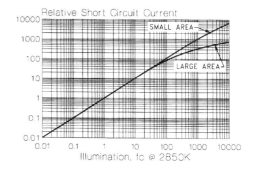

.050" NPN Phototransistors
TO-46 Flat Window Package

VTT1015, 16, 17

PACKAGE DIMENSIONS inch (mm)

.206 (5.23)
.195 (4.95)

.015 MAX.
(0.38)

.155 NOM.
(3.94)

.188 (4.78)
.178 (4.52)

.020 DIA. NOM.
(0.51)

.50 (12.7)
MINIMUM

.025 NOM.
(0.64)

.145 (3.68)
.139 (3.53)

45°

.043 (1.09)
.037 (0.94)

.100 DIA. NOM.
(2.54)

1 EMITTER
2 BASE
3 COLLECTOR, CASE

.215 (5.46)
.205 (5.21)

CASE 1   TO-46 HERMETIC (FLAT WINDOW)
CHIP TYPE: 50T

PRODUCT DESCRIPTION

A large area high sensitivity NPN silicon phototransistor in a flat lensed, hermetically sealed, TO-46 package. The hermetic package offers superior protection from hostile environments. The base connection is brought out allowing conventional transistor biasing. These devices are spectrally and mechanically matched to the VTE10xx series of IREDs.

ABSOLUTE MAXIMUM RATINGS
(@ 25°C unless otherwise noted)

Maximum Temperatures
   Storage Temperature: -40°C to 110°C
   Operating Temperature: -40°C to 110°C

Continuous Power Dissipation: 250 mW
   Derate above 30°C: 3.12 mW/ °C

Maximum Current: 200 mA

Lead Soldering Temperature: 260°C
   (1.6 mm from case, 5 sec. max.)

ELECTRO-OPTICAL CHARACTERISTICS @ 25°C

Part Number	Light Current I_C			Dark Current I_{CEO}		Collector Breakdown $V_{BR(CEO)}$	Emitter Breakdown $V_{BR(ECO)}$	Saturation Voltage $V_{CE(SAT)}$	Rise / Fall Time t_R / t_F	Angular Response $\theta_{1/2}$
	mA		H fc (mW/cm²) V_{CE} = 5.0 V	H=0		I_C =100 µA H=0	I_E =100 µA H=0	I_C =1.0 mA H=400 fc	I_C =1.0 mA R_L=100 Ω	
	Min.	Max.		(nA) Max.	V_{CE} (Volts)	Volts, Min.	Volts, Min.	Volts, Max.	µsec, Typ.	Typ.
VTT1015	0.4	—	100 (5)	25	20	40	6.0	0.40	5.0	±35°
VTT1016	1.0	—	100 (5)	25	20	30	6.0	0.40	5.0	±35°
VTT1017	2.5	—	100 (5)	25	10	20	4.0	0.40	8.0	±35°

.050" NPN Phototransistor Chip   VTT-C50

DESCRIPTION

EG&G Vactec fabricates its silicon photosensor chips using state-of-the-art planar diffusion technology. All chips are nitride passivated to ensure long term stability. Collector contact can be made through the backside of the chip. With some devices an additional collector contact is available on the top surface. Base and emitter contacts are available on the top surface of the chip.

A nickel metallization system, suitable for conductive epoxy die attach, is employed on the backside of the chip. Aluminum metallization is used for the bond pads on the top surface of the die.

Chips can be specially probed for current gain, breakdown voltage, dark current, etc., to satisfy a specific application. Please contact Vactec with your requirements.

CHIP DIMENSIONS inch (mm)

CHIP 50T
.050 (1.27) x .050 (1.27) x .017 (0.43) Thick
.00152 in² (0.981 mm²) Exposed Sensitive Area
Collector Contact Is Also Back Side Of Chip

ABSOLUTE MAXIMUM RATINGS

Maximum Temperatures
  Storage Temperature: -65°C to 150°C
  Operating Temperature: -65°C to 125°C

Nominal Maximum Continuous
Power Dissipation @ 25°C: 50 mW *

Maximum Collector Current: 50 mA

  * Exact maximum power dissipation capabilities are determined by customer packaging and are not guaranteed by Vactec.

ELECTRO-OPTICAL CHARACTERISTICS @ 25°C (See also 50T curves, pg. 26)

Symbol	Characteristic	Test Condition	Min.	Typ.	Max.	Units
H_{FE} (Beta)	dc Current Gain	I_B = 6.0 μA, V_{CE} = 5.0 V	200	350		
I_D	Dark Current	V_{CE} = 10 V, I_B = 0			100	nA
$V_{BR(CEO)}$	Collector Breakdown Voltage	I_C = 100 μA	30			Volts
$V_{BR(ECO)}$	Emitter Breakdown Voltage	I_E = 100 μA	6.0			Volts
$V_{CE(SAT)}$	Collector-Emitter Saturation Voltage	I_C = 1.0 mA, I_B = 50 μA			0.4	Volts
t_R, t_F	Rise / Fall Time	I_C = 1.0 mA, R_L = 100 Ω		5		μsec
$S_{P (CBO)}$	Collector-Base Photometric Sensitivity	V_{CB} = 5.0 V, 2850 K		70		nA / fc
$S_{R (CBO)}$	Collector-Base Radiometric Sensitivity	V_{CB} = 5.0 V, 940 nm		4.0		nA /(μW/cm²)
C_J	Collector-Base Capacitance	V_{CB} = 5.0 V, 1 MHz		23		pF

0394

| PHOTOCONDUCTIVE CELL | VT300 Series |

PACKAGE DIMENSIONS inch (mm)

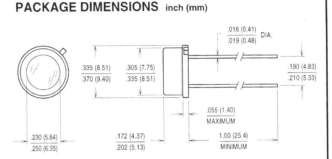

ABSOLUTE MAXIMUM RATINGS

PARAMETER	SYMBOL	RATING	UNITS
CONTINUOUS POWER DISSIPATION	P_D	125	mW
DERATE ABOVE 25° C	$\Delta P_D / \Delta T$	2.5	mW/°C
TEMPERATURE RANGE OPERATING AND STORAGE	T_A	− 40 to +75	°C

ELECTRO-OPTICAL CHARACTERISTICS @ 25° C (16 HRS. LIGHT ADAPT, MIN.)

Part Number	Resistance (Ohms)						Material Type	Sensitivity (γ, typ.) $\frac{\text{LOG (R10/R100)}}{\text{LOG (100/10)}}$	Maximum Voltage (V, pk)	Response Time @ 1 fc (ms, typ.)	
	10 lux 2850 K			2 fc 2850 K	DARK					Rise (1-1/e)	Fall (1/e)
	Min.	Typ.	Max.	Typ.	Min.	sec.					
VT3ØN1	6.0 k	12 k	18 k	6.0 k	200 k	5	Ø	0.75	100	78	8
VT3ØN2	12 k	24 k	36 k	12 k	500 k	5	Ø	0.80	200	78	8
VT3ØN3	24 k	48 k	72 k	24 k	1.0 M	5	Ø	0.80	200	78	8
VT3ØN4	50 k	100 k	150 k	50 k	2.0 M	5	Ø	0.80	300	78	8
VT33N1	20 k	40 k	60 k	20 k	500 k	5	3	0.90	100	35	5
VT33N2	40 k	80 k	120 k	40 k	1.0 M	5	3	0.90	200	35	5
VT33N3	80 k	160 k	240 k	80 k	2.0 M	5	3	0.90	200	35	5

0195

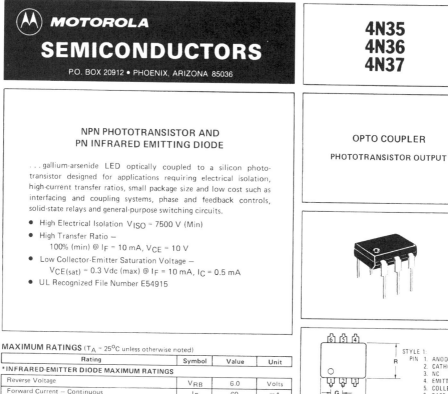

MOTOROLA SEMICONDUCTORS
P.O. BOX 20912 • PHOENIX, ARIZONA 85036

4N35
4N36
4N37

NPN PHOTOTRANSISTOR AND PN INFRARED EMITTING DIODE

. . . gallium-arsenide LED optically coupled to a silicon photo-transistor designed for applications requiring electrical isolation, high-current transfer ratios, small package size and low cost such as interfacing and coupling systems, phase and feedback controls, solid-state relays and general-purpose switching circuits.

- High Electrical Isolation V_{ISO} = 7500 V (Min)
- High Transfer Ratio —
  100% (min) @ I_F = 10 mA, V_{CE} = 10 V
- Low Collector-Emitter Saturation Voltage —
  $V_{CE(sat)}$ = 0.3 Vdc (max) @ I_F = 10 mA, I_C = 0.5 mA
- UL Recognized File Number E54915

OPTO COUPLER

PHOTOTRANSISTOR OUTPUT

MAXIMUM RATINGS (T_A = 25°C unless otherwise noted)

Rating	Symbol	Value	Unit
***INFRARED-EMITTER DIODE MAXIMUM RATINGS**			
Reverse Voltage	V_{RB}	6.0	Volts
Forward Current — Continuous	I_F	60	mA
Forward Current — Peak Pulse Width = 1.0 μs, 2.0% Duty Cycle	I_F	3.0	Amp
Total Power Dissipation @ T_A = 25°C Negligible Power in Transistor Derate above 25°C	P_D	100 1.3	mW mW/°C
Total Power Dissipation @ T_C = 25°C Derate above 25°C	P_D	100 1.3	mW mW/°C
***PHOTOTRANSISTOR MAXIMUM RATINGS**			
Collector-Emitter Voltage	V_{CEO}	30	Volts
Emitter-Base Voltage	V_{EBO}	7.0	Volts
Collector-Base Voltage	V_{CBO}	70	Volts
Output Current — Continuous	I_C	100	mA
Total Power Dissipation @ T_A = 25°C Negligible Power in Diode Derate above 25°C	P_D	300 4.0	mW mW/°C
Total Power Dissipation @ T_C = 25°C Derate above 25°C	P_D	500 6.7	mW mW/°C
TOTAL DEVICE RATINGS			
*Total Power Dissipation @ T_A = 25°C Derate above 25°C	P_D	300 3.3	mW mW/°C
Input to Output Isolation Voltage, Surge 60 Hz Peak ac, 5 seconds JEDEC Registered 4N35 = 3500 V Data @ 8 ms 4N36 = 2500 V 4N37 = 1500 V	V_{ISO}	7500	Volts V_{pk}
*Junction Temperature Range	T_J	−55 to +100	°C
*Storage Temperature Range	T_{stg}	−55 to +150	°C
*Soldering Temperature (10 s)	—	260	°C

STYLE 1:
PIN 1. ANODE
2. CATHODE
3. NC
4. EMITTER
5. COLLECTOR
6. BASE

SEATING PLANE

NOTES:
1. LEADS WITHIN 0.25 mm (0.010) DIAMETER OF TRUE POSITION AT SEATING PLANE AT MAXIMUM MATERIAL CONDITION.
2. DIMENSION "L" TO CENTER OF LEADS WHEN FORMED PARALLEL.

DIM	MILLIMETERS		INCHES	
	MIN	MAX	MIN	MAX
A	8.13	8.89	0.320	0.350
B	1.27	2.03	0.050	0.080
C	2.92	5.08	0.115	0.200
D	0.41	0.51	0.016	0.020
F	1.02	1.78	0.040	0.070
G	2.54	BSC	0.100	BSC
H	1.02	2.16	0.040	0.085
J	0.20	0.30	0.008	0.012
K	2.54	3.81	0.100	0.150
L	7.62	BSC	0.300	BSC
M	0⁰	15⁰	0⁰	15⁰
N	0.38	2.54	0.015	0.100
P	0.81	0.97	0.032	0.038
R	6.10	6.60	0.240	0.260

CASE 730-01

ELECTRICAL CHARACTERISTICS

Characteristic	Symbol	Min	Typ	Max	Unit
LED CHARACTERISTICS (T_A = 25°C unless otherwise noted)					
*Reverse Leakage Current (V_R = 6.0 V)	I_R	—	0.05	10	μA
*Forward Voltage	V_F				Volts
(I_F = 10 mA)		0.8	1.2	1.5	
(I_F = 10 mA, T_A = –55°C)		0.9	—	1.7	
(I_F = 10 mA, T_A = 100°C)		0.7	—	1.4	
Capacitance (V_R = 0 V, f = 1.0 MHz)	C	—	150	—	pF
*PHOTOTRANSISTOR CHARACTERISTICS (T_A = 25°C and I_F = 0 unless otherwise noted)					
Collector-Emitter Dark Current	I_{CEO}				
(V_{CE} = 10 V, Base Open)		—	3.5	50	nA
(V_{CE} = 30 V, Base Open, T_A = 100°C)		—	—	500	μA
Collector-Base Dark Current (V_{CB} = 10 V, Emitter Open)	I_{CBO}	—	—	20	nA
Collector-Base Breakdown Voltage (I_C = 100 μA, I_E = 0)	BV_{CBO}	70	—	—	Volts
Collector-Emitter Breakdown Voltage (I_C = 1.0 mA, I_B = 0)	BV_{CEO}	30	—	—	Volts
Emitter-Base Breakdown Voltage (I_E = 100 μA, I_B = 0)	BV_{EBO}	7.0	—	—	Volts
*COUPLED CHARACTERISTICS (T_A = 25°C unless otherwise noted)					
Current Transfer Ratio	I_C/I_F				—
(V_{CE} = 10 V, I_F = 10 mA)		1.0	—	—	
(V_{CE} = 10 V, I_F = 10 mA, T_A = –55°C)		0.4	—	—	
(V_{CE} = 10 V, I_F = 10 mA, T_A = 100°C)		0.4	—	—	
Input to Output Isolation Current (2) (3)	I_{IO}				μA
(V_{io} = 3550 V_{pk}) 4N35		—	—	100	
(V_{io} = 2500 V_{pk}) 4N36		—	—	100	
(V_{io} = 1500 V_{pk}) 4N37		—	—	100	
Isolation Resistance (2) (V = 500 V)	R_{IO}	10^{11}	—	—	Ohms
Collector-Emitter Saturation Voltage (I_C = 0.5 mA, I_F = 10 mA)	$V_{CE(sat)}$	—	—	0.3	Volts
Isolation Capacitance (2) (V = 0, f = 1.0 MHz)	—	—	1.3	2.5	pF
*SWITCHING CHARACTERISTICS (Figure 1)					
Turn-On Time (V_{CC} = 10 V, I_C = 2.0 mA, R_L = 100 Ω)	t_{on}	—	—	10	μs
Turn-Off Time (V_{CC} = 10 V, I_C = 2.0 mA, R_L = 100 Ω)	t_{off}	—	—	10	μs

*Indicates JEDEC Registered Data.

NOTES: 1. Pulse Test: Pulse Width = 300 μs, Duty Cycle ⩽ 2.0%.
      2. For this test LED pins 1 and 2 are common and phototransistor pins 4, 5, and 6 are common.
      3. Pulse Width ⩽ 8.0 ms.

(M) MOTOROLA *Semiconductor Products Inc.*

4N35 ● 4N36 ● 4N37

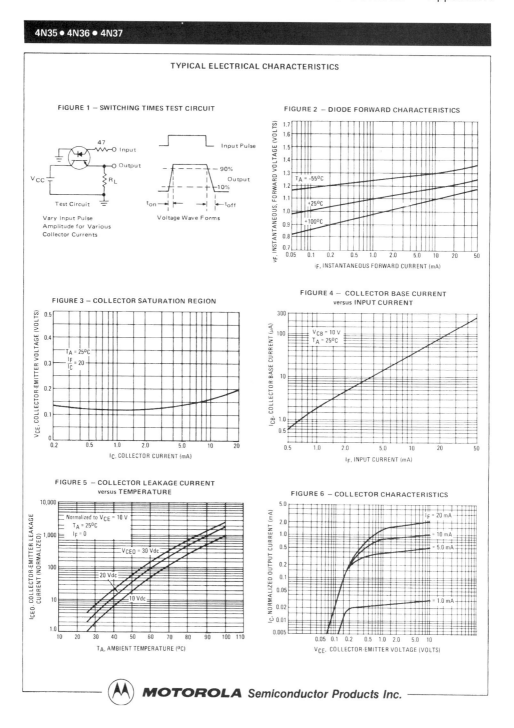

TYPICAL ELECTRICAL CHARACTERISTICS

FIGURE 1 — SWITCHING TIMES TEST CIRCUIT

FIGURE 2 — DIODE FORWARD CHARACTERISTICS

FIGURE 3 — COLLECTOR SATURATION REGION

FIGURE 4 — COLLECTOR BASE CURRENT versus INPUT CURRENT

FIGURE 5 — COLLECTOR LEAKAGE CURRENT versus TEMPERATURE

FIGURE 6 — COLLECTOR CHARACTERISTICS

MOTOROLA *Semiconductor Products Inc.*

4N35 ● 4N36 ● 4N37

TYPICAL APPLICATIONS

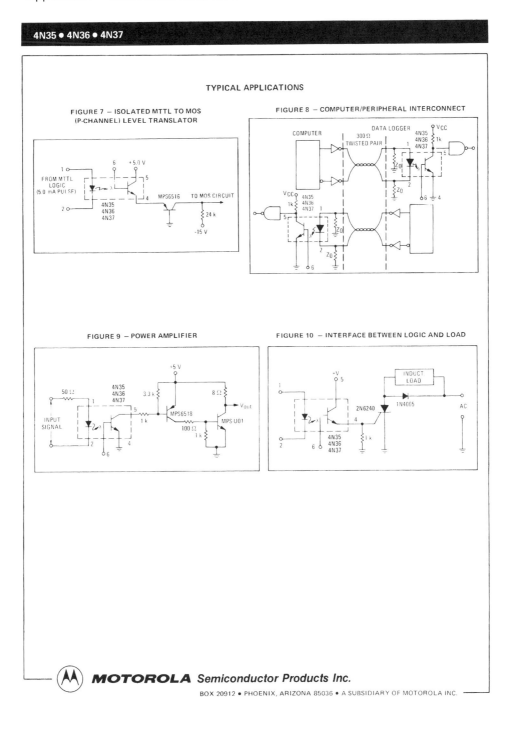

FIGURE 7 – ISOLATED MTTL TO MOS
(P-CHANNEL) LEVEL TRANSLATOR

FIGURE 8 – COMPUTER/PERIPHERAL INTERCONNECT

FIGURE 9 – POWER AMPLIFIER

FIGURE 10 – INTERFACE BETWEEN LOGIC AND LOAD

MOTOROLA *Semiconductor Products Inc.*
BOX 20912 ● PHOENIX, ARIZONA 85036 ● A SUBSIDIARY OF MOTOROLA INC.

Appendix B

Op-Amp Basics

B.1 THE IDEAL OP-AMP

Our discussion of the op-amp in this section will be limited to the ideal case. We will not be concerned with the construction of this component. A block diagram will be used to represent the op-amp with the associated circuitry external to it. The various electrical parameters of the op-amp will also be defined. We will consider only the inverting and non-inverting configurations using simple resistors.

Figure B–1 shows a block diagram used to represent the op-amp. You will notice that it has two inputs and one output. These inputs are known as the (−) or inverting and the (+) or non-inverting. Ideally, the input resistance is effectively infinite, the output is zero ohms, and the voltage gain is infinite. The output will be such that a zero voltage difference appears between the two input terminals. In practice though, the input resistance is on the order of 10^6 to 10^9 ohms, the output has a resistance of a few ohms, and the gain is on the order of 10^6. The op-amp responds only to a voltage difference between the two input terminals. If a positive voltage pulse enters the inverting terminal, the output will be a negative-going pulse. When this same pulse enters the non-inverting input, a positive-going pulse results. This differential voltage can be amplified by the gain characteristic of the amplifier.

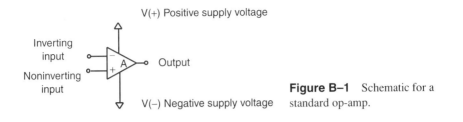

Figure B–1   Schematic for a standard op-amp.

The Inverting Amplifier

We will use this simple configuration to introduce some of the electrical parameters. When configured according to Figure B–2, a differential voltage will occur between the two input terminals allowing the op-amp to amplify this voltage. The input (+) terminal connects to circuit ground. The signal voltage is applied to the (–) input terminal through resistor R_1 with feedback returned from the output terminal through resistor R_2. In the ideal case, the input draws no current since the input resistance is effectively infinite. Also, both inputs are at ground potential since the noninverting input is connected to ground. This occurs for the reason stated above: the output will be such that a zero voltage difference will appear between the two input terminals. The (–) input is a virtual ground at the same potential as the (+) input terminal. Since the input impedance is zero, the same current flows through both resistors ($I_{in} = I_f$). We can see that the voltage across R_1 is V_{in}. Using the same logic, the voltage across R_2 is V_{out}. This situation can be expressed mathematically using Ohm's law as shown below:

$$V_{out}/R_2 = -V_{in}/R_1$$

If we rearrange the terms above, we get the following expression for voltage gain in terms of the two external resistors:

$$Gain = V_{out}/V_{in} = -R_2/R_1$$

The output voltage that appears across R_2 is negative due to the sign convention of the amplifier. To change the gain of the amplifier, we can change the ratio of R_2 to R_1.

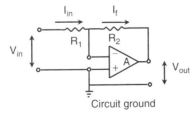

Figure B–2   Inverting amplifier configuration.

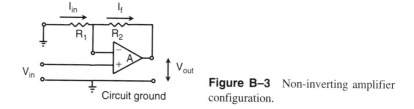

Figure B–3   Non-inverting amplifier configuration.

The Non-Inverting Amplifier

The non-inverting amplifier configuration uses the same two resistors connected as shown in Figure B–3. In this case, the signal input, V_{in}, connects directly to the (+) input of the op-amp. Resistor R_1 connects from circuit ground to the (−) input terminal of the op-amp. Resistor R_2 provides feedback by connecting to the (−) input and to the op-amp's output. The gain, in this case, can be expressed in terms of the two external resistors as shown below:

$$\text{Gain} = R_1 + R_2/R_1$$

As with our last example, the gain can be varied by selecting the values of R_1 and R_2. When the feedback resistor equals zero ohms (a short), this yields a unity gain amplifier. The unity gain amplifier finds important applications as a buffer circuit. A buffer circuit becomes useful when trying to isolate two portions of a circuit of different impedances.

We summarize below the conditions that hold true for the ideal op-amp.

1. The voltage difference between the two inputs is zero.
2. No current flows into either input terminal (infinite input impedance).
3. The voltage gain is infinite.
4. $V_{out} = 0$ when both inputs are at the same voltage (in other words, there is a zero "offset voltage").
5. The output can change instantaneously (infinitely fast slew rate).

B.2   THE NON-IDEAL OP-AMP

We used the ideal case to introduce the basic concepts associated with this device. In practice, the ideal op-amp does not exist. A typical op-amp has characteristics approaching those of the ideal case, but with small differences. These small differences account for the various limitations and errors that will be discussed next.

The first limitation deals with voltage gain. Typical op-amps have what is called a dc or direct current open loop gain (A_{vol}) of about 100 dB (10^5). Notice that this number may be very large, but it is not infinite as in the ideal case. This gain can be defined as the ratio of the change in output voltage to the change in input terminal voltage. Since the open loop gain is finite, the voltage gain when using the configuration in Figure B–2 or

B–3 will begin dropping at a frequency determined by the ratio of R_2/R_1. The open loop gain becomes attenuated when using a feedback resistor producing a characteristic called closed loop gain. Figure B–4 shows open loop gain and closed loop gain on the same graph. The closed loop gain voltage is determined by the resistance ratio. This gain amplifies the input signal and is less than the open loop gain of the amplifier since it results from attenuation. To find where its frequency response starts to roll off, we must first make a bode plot for the amplifier. Figure B–4 shows this bode plot for the op-amp under consideration as a graph of open loop gain vs. frequency. The information given in the data sheet for this device is needed to make this graph. The frequency rolls off at a rate of 20 dB/decade beginning at about 10 Hz. A typical op-amp has this type of roll off. In the figure, this curve is labeled "open loop gain." To find the closed loop gain, we take the value for the gain as determined by the two resistors for the circuit of Figure B–2 or B–3. Next, we draw a horizontal line at the level of the voltage gain value obtained by this calculation (closed loop gain) to intersect the open loop gain curve. We now notice another limitation to the op-amp. The amplifier has a finite gain and bandwidth. As we increase the gain by using a larger ratio of R_2 to R_1, the frequency response of the op-amp will decrease. This roll off also results when using a simple RC low pass filter to suppress higher frequencies.

When adding a feedback resistor to an op-amp to achieve voltage gain, we change the gain/frequency relationship. For example, when $R_2 = 100$ KΩ and $R_1 = 1$ KΩ, we get

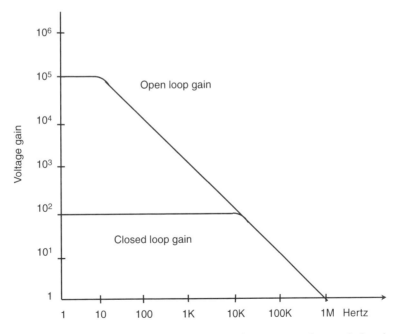

Figure B–4   Bode plot showing relationship between open loop and closed loop gain.

an output gain of 100 by using the ratio of the two resistors. This closed loop gain can be shown graphically in Figure B–4. We see that for this value of closed loop gain, the output does not begin to roll off until approximately 10 KHz. By contrast, the open loop gain begins to roll off at approximately 10 Hz. The closed loop gain curve provides for a relatively precise circuit up to about 10 KHz. You can see from this curve that as the amplification increases, the frequency response decreases. As we know, the frequency response determines the bandwidth of the device. This typical trade-off situation must be considered in almost every circuit design when using op-amps to amplify input voltage signals.

Appendix C

Derivation of the Stefan-Boltzmann Law

In Chapter 8, we introduced Planck's formula for blackbody radiation. This formula will give the intensity of the radiation at one particular wavelength. For convenience, we give that formula below:

$$I_\lambda = \frac{2\pi hc}{\lambda^5} \frac{1}{\left(e^{hc/\lambda kT} - 1\right)}$$

Although being a very useful relationship, it may be important to know the total radiation due to all wavelengths. To determine this mathematically, we must integrate the Planck function above with respect to λ. The following exercise details how to perform this integration. First, we change the variable of integration from λ to the dimensionless variable $x = hc/\lambda kT$. After rearranging the terms we get:

$$I = \int I_\lambda d\lambda = \frac{2\pi k^4 T^4}{h^3 c^2} \int \frac{x^3 dx}{e^x - 1}$$

$$\text{where } x = \frac{hc}{\lambda kT}$$

The value for the definite integral is $\pi^4/15$. Solving for I we get the desired result. This new relationship is known as the Stefan-Boltzmann law which is shown below:

$$I = \sigma T^4$$

In this equation, σ is the Stefan-Boltzmann constant given by the relationship below:

$$\sigma = \frac{2\pi^5 k^4}{15c^2 h^3}$$

The numerical value for this constant in MKS units is 5.67×10^{-8} watts/meter^2/K^4. According to the above relationship, a blackbody radiator at 1000 Kelvins will emit energy at a rate of 56.7 Kilowatts per square meter of surface area.

Appendix D

Physical Constants

Table D–1

Physical Constant	Symbol	Numerical Value
Speed of light in free space	c	2.9979×10^{8} m/s
Permittivity in free space	ε_0	8.8542×10^{-12} F/m
Permeability in free space	μ_0	$4\pi \times 10^{-7}$ H/m
Planck's constant	h	6.6262×10^{-34} Js
Boltzmann constant	k_B	1.3807×10^{-23} J/K
Electron charge	q or e	$1.6021917 \times 10^{-19}$ C
Electron mass	m_0	9.1095×10^{-31} kg
Electron charge to mass ratio	e/m	1.7588028×10^{11} C/kg
Bohr radius	a_0	5.29×10^{-11} m
Energy of one electron volt	eV	$1.6021917 \times 10^{-19}$ J
Stefan-Boltzmann constant	σ	5.66961×10^{-8} W/m$^2 \cdot$K^4

Selected Bibliography

Chuang, Shun Lien. *Physics of Optoelectronic Devices.* New York: John Wiley & Sons, Inc., 1995.

Dereniak, Eustace L., and Boreman, Glenn D. *Infrared Detectors and Systems.* New York: John Wiley & Sons, Inc., 1996.

Graeme, Jerald. *Photodiode Amplifiers–Op Amp Solutions.* New York: McGraw-Hill, 1996.

Hawke, John. *Lasers: Theory and Practice.* Englewood Cliffs, NJ: Prentice-Hall International Series in Optoelectronics, 1995.

Hecht, Jeff. *Understanding Lasers–An Entry-Level Guide.* 2nd ed. Piscataway, NJ: IEEE Press, 1994.

Horowitz, Paul, and Hill,Winfield. *Art of Electronics.* 2nd ed. New York: Cambridge University Press, 1989.

Meyer-Arendt, Jurgen R. *Introduction to Classical and Modern Optics.* 4th ed. Englewood Cliffs, NJ: Prentice-Hall, Inc., 1995.

Miller, John L., and Friedman, Ed. *Photonics Rules of Thumb*: *Optics, Electro-Optics, Fiber Optics, and Lasers.* New York: McGraw-Hill, 1996.

Riaziat, Leonard. *Introduction to High Speed Electronics and Optoelectronics.* New York: John Wiley & Sons, Inc., 1995.

Singh, Jasprit. *Optoelectronics: An Introduction to Materials and Devices.* New York: McGraw-Hill, 1996.

Smith, Warren J. *Modern Optical Engineering–The Design of Optical Systems.* 2nd ed. New York: McGraw-Hill, 1990.

Wood, David. *Optoelectronic Semiconductor Devices.* Englewood Cliffs, NJ: Prentice-Hall International Series in Optoelectronics, 1994.

Wyatt, Clair L. *Electro-Optical System Design For Information Processing.* New York: McGraw-Hill, 1991

Index